MECHANICS OF
SHEET METAL FORMING

Material Behavior and Deformation Analysis

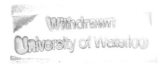

MECHANICS OF
SHEET METAL FORMING
Material Behavior and Deformation Analysis

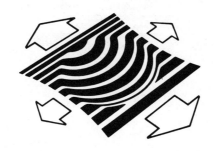

Edited by
DONALD P. KOISTINEN and NENG-MING WANG

General Motors Research Laboratories

PLENUM PRESS • NEW YORK–LONDON • 1978

Library of Congress Cataloging in Publication Data

Symposium on Mechanics of Sheet Metal Forming, Warren, Mich., 1977
  Mechanics of sheet metal forming.

  (General Motors symposia series)
  Includes index.
  1. Sheet-metal work — Congresses. 2. Deformation (Mechanics) — Congresses.
I. Koistinen, Donald P. II. Wang, Neng-Ming, 1933-   III. Title.
TS 250.S95   1977   671   78-21857
ISBN 0-306-40068-5

Proceedings of a Symposium on Mechanics of Sheet Metal Forming
held at the General Motors Research Laboratories, Warren,
Michigan, October 17–18, 1977

© 1978 Plenum Press, New York
A Division of Plenum Publishing Corporation
227 West 17th Street, New York, N.Y. 10011

PREFACE

This volume records the proceedings of an international symposium on "ME-CHANICS OF SHEET METAL FORMING: Material Behavior and Deformation Analysis." It was sponsored and held at the General Motors Research Laboratories on October 17–18, 1977. This symposium was the twenty-first in an annual series.

The objective of this symposium was to discuss the research frontiers in experimental and theoretical methods of sheet metal forming analysis and, also, to determine directions of future research to advance technology that would be useful in metal stamping plants. Metal deformation analyses which provide guidelines for metal flanging are already in use. Moreover, recent advances in computer techniques for solving plastic flow equations and in measurements of material parameters are leading to dynamic models of many stamping operations. These models would accurately predict the stresses and strains in the sheet as a function of punch travel. They would provide the engineer with the knowledge he needs to improve die designs.

The symposium papers were organized into five sessions: the state of the art, constitutive relations of sheet metal, role of friction, sheet metal formability, and deformation analysis of stamping operations. We believe this volume not only summarizes the various viewpoints at the time of the symposium, but also provides an outlook for materials and mechanics research in the future.

The editors gratefully acknowledge the cooperation of the authors and session chairmen. Also, we thank Professor Bernard Budiansky of Harvard University and Dr. Stuart Keeler of National Steel Corporation for serving on the Organizing Committee. We appreciate the support of Dr. Nils L. Muench, Technical Director, Dr. Frank E. Jamerson, Head, Physics Department, and Mr. Allen V. Butterworth, Head, Mathematics Department. We also thank Ms. Kay Sutherland for secretarial assistance, Mr. David Havelock for assistance with manuscripts, and Mr. Tom Beaman for physical arrangements.

Donald P. Koistinen
Neng-Ming Wang

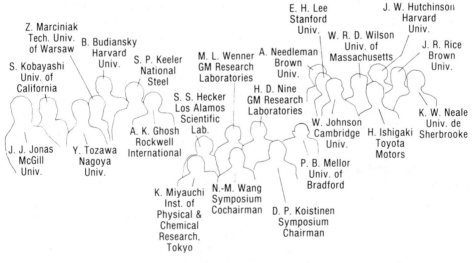

Z. Marciniak
Tech. Univ.
of Warsaw

S. Kobayashi
Univ. of
California

B. Budiansky
Harvard
Univ.

S. P. Keeler
National
Steel

M. L. Wenner
GM Research
Laboratories

A. Needleman
Brown
Univ.

E. H. Lee
Stanford
Univ.

W. R. D. Wilson
Univ. of
Massachusetts

J. W. Hutchinson
Harvard
Univ.

J. R. Rice
Brown
Univ.

S. S. Hecker
Los Alamos
Scientific
Lab.

H. D. Nine
GM Research
Laboratories

K. W. Neale
Univ. de
Sherbrooke

J. J. Jonas
McGill
Univ.

Y. Tozawa
Nagoya
Univ.

A. K. Ghosh
Rockwell
International

W. Johnson
Cambridge
Univ.

H. Ishigaki
Toyota
Motors

P. B. Mellor
Univ. of
Bradford

K. Miyauchi
Inst. of
Physical &
Chemical
Research,
Tokyo

N.-M. Wang
Symposium
Cochairman

D. P. Koistinen
Symposium
Chairman

Organizers, Session Chairmen and Authors
of the
1977 GMR Symposium

"Mechanics of Sheet Metal Forming"

CONTENTS

SESSION I

STATE OF THE ART

Session Chairman
W. JOHNSON

University of Cambridge
United Kingdom

SHEET METAL STAMPING TECHNOLOGY—

NEED FOR FUNDAMENTAL UNDERSTANDING

S. P. KEELER

National Steel Corporation, Ecorse, Michigan

ABSTRACT

Historically, the evolution of a sheet metal stamping from concept through part design to die design and ultimate tryout has been a slow, cautious process based on the trial-and-error experiences and the skills of the artisan. While major technological advances today allow the artisan to produce extremely complex parts, only limited progress has been made in the last twenty-five years on a scientific understanding of the basic formability process. Many new pressures, however, will severely limit the future ability of the artisan to function effectively. A formability analysis system, therefore, must be developed from a number of heterogeneous scientific disciplines. To achieve a performance level superior to the artisan, the system of the future should have eight characteristics. The system must: 1) be analyzed as an interactive system, 2) be modeled with known and unknown variables, 3) incorporate properties of real materials, 4) not be biased by historical rules of thumb, 5) provide predictive capability, 6) improve the interaction between design and manufacturing functions, 7) be responsive to new in-service requirements, and 8) be attuned to end-product economics.

INTRODUCTION

The 23rd Autumn Lecture to the Institute of Metals in England [1] was delivered twenty-five years ago last month by a man whose work in metal deformation was the foundation for many of our current formability studies. The lecturer was Professor H. W. Swift of the University of Sheffield. Professor Swift's goal in that 23rd lecture was—according to the synopsis—

"To survey the activities of the several classes of pure and applied scientists engaged in the general field of metal plasticity, and to discuss the present state of knowledge and the extent to which research is at present able to make its contribution to the various technological processes involving plastic deformation."

References p. 17.

The parallelism between Professor Swift's goal a quarter of a century ago and the intent of this symposium is remarkable. In fact, Professor Swift's keen observations in the introduction of that lecture echoed many of the problem areas identified by the organizers of this symposium. A study of that introduction brings several questions to mind. For example, during the past twenty-five years, have we of the scientific community really contributed much to solving the basic problems of the press shop? Have we been able to significantly transform sheet metal formability from an art to an applied science? Have we mistakenly taken credit for advances which rightfully should be classed as technological improvements? Where are the major contributions to the fundamental understanding of the real world, black box systems which transform flat sheets into automobiles, refrigerators, containers, and the like?

In order to formulate answers to these questions, this opening address will begin on a somewhat unusual tack. A major section of Professor Swift's introductory remarks from twenty-five years ago will be quoted. Two purposes will be served. First, a historical flavor of the problem will be developed as an interesting frame of reference. Second, Professor Swift has already said very adequately in spicy language many of the points which should be developed. In fact, if Professor Swift were giving this opening address, he himself probably would have been sorely tempted to repeat the introduction of his 1952 lecture.

SWIFT'S ALLEGORY

Here then, for careful consideration, are excerpts from Swift's lecture entitled "ON THE FOOT-HILLS OF THE PLASTIC RANGE" [1].

". . . much expense is involved in processes of trial-and-error and in the use of materials capable of withstanding plastic maltreatment . . . there is therefore still much scope for development in technical plasticity, and this involves on the one hand the study of plastic deformation in its physical and mechanical aspects, and on the other the application of the results of this study to industrial problems. And between these there is the need for an important link; scientific results become useful only when they are interpreted in such a way that they can be understood and applied by those concerned with industrial production.

"In this broad field of development there is work for many hands; for the pure scientist and the applied scientist, for the mathematician, the metallurgist, and the . . . engineer. Indeed, since he is directly concerned with the finished product, the engineer is probably most conscious of the need for further knowledge over a wide front.

"He realizes that almost every manufacturing operation from the ingot to the finished product is essentially related to the plastic properties of the material . . . he realizes, moreover, that the plastic properties of materials are becoming increasingly significant in modern engineering design . . . he is beginning to take advantage of the economies made possible by considerations of plastic strain in the limit design of structures. And he is just beginning to realize that in his innocence he has in fact for many years been basing his designs on the plastic properties of ductile metals, for the 'ultimate tensile strength,' to which he has traditionally applied his factor of safety, is itself nothing but a function of the plastic stress/strain curve.

"For his own purposes the engineer needs to be able to assess the possibilities of various processes of formation and fabrication in relation to materials which are economically available. He needs to devise these processes so as to produce results with the minimum demand on these materials and the minimum expense; he needs to design machines and equipment to carry out these processes; and he needs to be able to specify and therefore to test material suitable for these processes. And at all stages he is concerned as far as possible to eliminate the expense and delay involved in trial-and-error methods.

"In this field, if in no other, and however reluctantly, the metallurgist has to accept a role complementary to the engineer. It is his concern to provide material with properties suitable for the engineer to fabricate and process, and to advise him on the specification and treatment of this material. For this purpose he needs to know—and preferably to understand—how suitable plastic properties can be obtained, and he is continually seeking to develop new materials to circumvent the engineer's ingenuity in maltreatment. He also needs to have at his disposal means for the testing of materials, which shall be relevant to their application and acceptable to the engineer.

"If they were left to their own devices, the metallurgist and the engineer would be compelled in the main to have recourse to empirical or trial-and-error methods, methods which are tedious and uneconomical, and uncongenial to the scientific mind. They are therefore gratified to realize that applied mathematicians and such physicists as are prepared to regard the atom as a unit rather than a universe, have been increasingly finding in the field of plasticity material worthy of fundamental inquiry.

"The metal physicist is concerned to understand the nature and mechanism of plastic deformation in relation to atomic and crystal structure, and so to be able to correlate physical and mechanical properties with the structure and treatment of metals. In this way he should be in a position to point directions in which desirable plastic properties are likely to be found, and to guide the metallurgist in the development of materials possessing these properties. . . .

"The mathematician's interest in plasticity can cover a considerable range. He can apply his techniques in the field of metal physics to explain plastic deformation at the atomic level and seek correlation between single-crystal and polycrystalline properties. Or, closer to the field of the engineer, he can study the distribution of stresses and strains in a material of specified properties under a given type of deformation, and he can seek to prescribe the limiting range of this deformation in relation to the intrinsic strength of the material. Engineers in their less generous moments suspect that some mathematicians are also attracted to the subject of plasticity because of the opportunity it provides for analytical virtuosity.

"And so we have four distinct parties each concerned from its own point of approach, by its own route, with its own resources and its own objective, to explore the secrets and exploit the resources of the Plastic Range. The engineer is all too conscious that his own equipment is quite unequal to the task of scaling the real heights, so he must limit his pedestrian efforts to the foot-hills of the Range. Here he can watch and admire the skill of those whose special scientific equipment enables them to climb to greater heights and perhaps even to aspire to the peaks of the Range. Nor need he be entirely an onlooker, for he can make himself useful as base-porter for the pioneers, supplying them with material needs, and at the same time he can do some exploring on his own account at lower levels.

"From his vantage ground on the foot-hills, he is, of course, in a good position, albeit at a respectful distance, to follow the manoeuvres and progress of those who are struggling on the glaciers and ledges above the snowline. As he sees it, the Plastic Range is divided into two main massifs, between which at present there appears to be 'a great gulf fixed.' On one side is the massif which the mathematicians like Hill in this country and Prager in America are exploring. At its higher levels this seems to be wrapped in perpetual cloud, but it is surmounted by a noble peak which may perhaps be called the *Mathahorn*. The other massif, on which the metal physicists are busy, culminates in another peak, which members of this institute will recognize as the *Metahorn*.

"It is a little presumptuous for the engineer to attempt to define the position of the more traditional metallurgist in this terrain, but I hope I shall not be accused of abusing a privileged occasion if I suggest that the main body of this institute is established on the foot-hills below the *Metahorn* at much the same level as the engineer at the foot of the *Mathahorn*. Our general perspectives are therefore much the same, but since the engineer is more directly on the lines of communication with the mathematicians, he feels that his views on their progress are more likely to be acceptable to metallurgists than his occasional sidelong glimpses of incidents on the glide planes and edge dislocations of the *Metahorn*".

Professor Swift's pointed observations are meaningful and enjoyable. Regretfully, however, little has changed in twenty-five years. The *Mathahorn* and the *Metahorn* still stand with their higher levels wrapped in perpetual clouds. The great gulf remains fixed between the two massifs.

If one were to extend Professor Swift's allegory directly to this symposium, one must first define the location of the automotive press shops. The press shops probably exist in the deep valleys far below the Plastic Range. Here, ideally, they are the recipients of the river of science, which originates in the watersheds of the two massifs and flows down through the foot-hills where it is gathered, filtered, and redistributed to the proper tributaries feeding the valley. For some reason, however, only a minute trickle of science has made its way from the Plastic Range to the press shops deep in the valley. One wonders whether scientific understanding applicable to the press shop is not being generated in the Plastic Range, or whether the ground is so barren that any understanding generated is immediately reabsorbed high up the range, or whether the output is being diverted to more lofty regions.

THE ARTISAN SYSTEM

Without the infusion of scientific knowledge, the press shops continue to depend on the art and black magic which have sustained them for multiple decades: the art—which secretly creates a good automotive roof panel instead of a torn one; the black magic—which separates a tight door panel from one that contains an imperceptible buckle seen only by trained eyes with the aid of highlight oil in the special *green room*. In an age where we can set a precise, predictable course for a ten-year journey of space voyagers to the outer planets, or control electronic integrated circuit deposition thicknesses to atomic dimensions, the press shop appears to have eluded the scientists. The press shop remains the realm of the artisan. Quoting again from Swift [1]—

> "But this question must not be held to imply any doubt as to the technological skill of those engaged in the metalforming industries in this country. There can be no doubt on this score in the mind of anyone who has visited an automobile press shop and witnesses the transformation of a flat sheet into a motorcar wing in a single stroke. Neither the man who casually presses the button nor his mate who idly applies an oily rag at the right spot on the sheet has anything to learn from Charlie Chaplin in the matter of masterly nonchalance."

It can not be overemphasized that the press shop is successful today only because of the technological skill developed by artisans. For example, impossible sheet metal stampings are made daily in press shops around the world. A recent example is the cruise control power unit cover (Fig. 1) which looks like a casting or at least two welded components. Yet this part is made from a flat sheet of steel by a nineteen-stage press operation and a quick threading operation without any intermediate anneals. This part is the work of an artisan in a real world press shop today. Can any of us on the Plastic Range accurately characterize, much less duplicate, the skills and techniques used by the artisan in creating this part? In like manner, the skills and techniques used by artisans to create the Stradivarius violins continue to defy the best analytical capabilities of space age science. If we are honest with ourselves, the best of our own current techniques—including finite element analyses, constitutive equations, instability theory, circular grid measurements, forming limit diagrams, strain path corrections, shape analyses, and computer aided programs—can provide no more than a ball park,

Fig. 1.   Cruise control power unit cover formed in nineteen operations from a sheet steel blank and then threaded. (Courtesy of General Motors AC Spark Plug Division.)

postmortem approximation of actual stampings and blueprint stage predictions for only a few, simple, idealized stampings.

In making almost all of the automotive sheet metal stampings (and identical comments would apply to the appliance, container, and other sheet forming industries) the skill of the artisan is directly derived from the trial-and-error process in one of three forms. First, the long trial-and-error work style of the artisan teaches the artisan many successful tricks of the trade which he can usually apply again should an identical set of conditions ever be encountered. One can tap this source of *how-to-do-it* information over a long time frame, such as during extended apprenticeship programs. Most of the time, however, the artisan has no comprehension—or at worst a misunderstanding—of why a specific trick works in a specific case. Unfortunately, the scientific community needs the why answers more than the how answers.

Second, the same long trial-and-error work style also can cause the artisan to develop an instinct, a sixth sense, or a feel for which path he should attempt in a situation which closely approximates but is not identical to previous history. Here the artisan is unable to explain both how and why.

The third trial-and-error situation is that attempted by the artisan when he encounters an entirely new parameter, such as a new metal, a completely foreign part design, or a radically different forming technique. Documentation of the artist's response behavior to such an environment would be interesting, but tedious and hardly useful.

Therefore, while the artisan historically has been eminently successful at his task, his method of operation does not provide a documented base of data, facts, or techniques which the scientific community can interrogate or use as a known reference base. In other words, past history will yield little useful information and the scientific process must start at base zero.

BREAKDOWN OF THE ARTISAN SYSTEM

A logical question one might well ask at this point is, "If the artisan system has achieved such great advances in sheet metal forming technology, why all the excitement, panic, and clamor for scientific involvement?" The answer is quite simple—the current artisan system can not exist effectively under the constantly changing rules imposed by society. Several of these changes are:

1. Lack of young people willing to learn the skill of the artisan through the apprenticeship or any other long-term training system.
2. Earlier retirement opportunities for those skilled artisans who could be the teachers of future generations.
3. Reduced lead times, including die tryout, due to constantly changing targets and requirements.
4. Introduction of newer materials for which no experience bank is available.
5. Greater emphasis on cost effectiveness.
6. Increased complexity of parts.

The last major change, increased part complexity, stems from a number of sources, such as:

- Sheet metal competition for products traditionally produced by other processes, such as casting, forging, powder metallurgy, and welding.
- The combination of multiple part functions into one part.
- Styling characteristics rapidly moving toward the extreme ends of the scale, where the two most desired radii of curvature are infinite and zero.
- Emphasis on weight reduction prompted by discrete inertial weight classes.
- Stringent environmental requirements interacting with part design and function.
- Legislated safety standards imposing additional part functions which affect design and manufacture.
- Regulated mileage goals.
- Demands by consumer groups and legislative bodies for longer warranty life, reduced damage, no corrosion, etc.

As all these changes put greater and greater pressures on the artisan, the artisan finds it more difficult to do the job required of him. Furthermore, these changes are applicable to the artisans responsible for converting part designs into die designs, die designs into initial tooling, initial tooling into usable tooling, and used tooling into refurbished tooling. Thus, no artisans in the pressure shop are immune.

References p. 17.

REPLACEMENT SYSTEMS

For the purpose of this symposium, therefore, let us accept as fact that the trial-and-error techniques of the artisan must be replaced by a system based on scientific understanding of the basic deformation process. Thus, we have arrived at the core questions to which this symposium must address itself.

1. Is there a scientific system today which can replace the artisan at an equal or superior performance level?
2. If such a system is not available today, which system looks the most promising in terms of either the earliest maturity date or the greatest long range potential?
3. What form must this system take?
4. Will the system be derived from a single discipline or will a composite be formed from several disciplines?

Some of the answers to these questions should become evident from the papers and during the formal and informal discussions of this symposium. The ultimate goals of this symposium are to identify consensus answers to these questions and to suggest the optimum system by which the scientific community can solve sheet metal forming problems of the automotive press shop.

BASES FOR AN ANALYSIS SYSTEM

In order to qualify as a candidate system, however, any solution to the press shop formability problems must have certain characteristics. The following is a potential list of these characteristics.

Sheet Metal Forming Operations Must Be Analyzed as an Interactive System

While this requirement could be illustrated by stretching a hemispherical dome or deep drawing a cup, the components of the system are more easily identified by a steady state operation such as roll forming (Fig. 2). This analogy is derived from the sheet rolling analysis presented by Professor Backofen in his 1962 keynote address to the 9th Sagamore Conference on Deformation Processing [2]. The component parts of the system are:

> A—Incoming Metal
> B—Tool-Metal Interface
> C—Deformation Zone
> D—Completed Part

It is no coincidence that the in-depth papers of the symposium bear a direct relation to this analogy.

> Session II   Constitutive Relations = Incoming Metal
> Session III   Role of Friction = Tool-Metal Interface

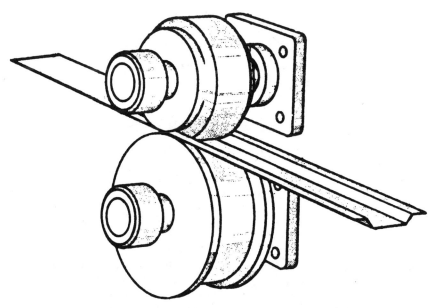

Fig. 2. Generalized deformation processing system [2] represented by a roll forming operation. The four zones are incoming metal, tool-metal interface, deformation zone, and completed part.

Session IV   Instability Processes
                              and   } = Deformation Zone
Session V   Analysis of Deformation

The Completed Part unfortunately remains primarily the inviolable domain of the stylist and designer and is not part of this symposium.

The key point is that the real world forming operation, currently represented by a black box (Fig. 3) is not only interactive, but usually synergistic. One can not isolate a single component and study its behavior independently. Examples of this abound but one should serve to illustrate the point.

Increasing the thickness of sheet steel increases the permissible major strain described by the Forming Limit Diagram [3]. By itself, this would suggest greater formability. However, in the interactive system of complex stampings, draw beads restrict more severely metal flow of thicker sheets. Punch-die clearances decrease. Die radius to sheet thickness and punch radius to sheet thickness ratios decrease. Interface pressures acting on the lubricant increase. Reduced metal flow from the flanges must be compensated by increased major strains over the punch. Is forming severity increased or decreased? Currently, only trial-and-error experiments with actual tooling can summarize all the plus and minus effects. A very sophisticated interactive system is required if there is any hope of accurately modeling formability processes.

References p. 17.

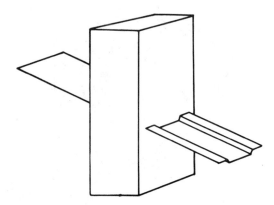

Fig. 3.   Interactive forming system represented by an input-output black box.

The Sheet Metal Forming Analysis System Must Be Modeled After Real World Press Shop Operations with Known and Unknown Variables

The most severe day to day problem of the press shop is unpredictable variables. Metal from multiple source suppliers is made on variable production equipment which results in a statistical distribution of properties. These variations are often overshadowed by intentional supplier manipulation of properties in order to tailor each coil of metal to the designated stamping for the best application as perceived by each supplier. Similar variations can be found in lubricants, which can be applied from calibrated orifices or soaked mops.

Press shop variables can include press speed, hold down pressure, die wear, blanked edge quality, pitman arm adjustment, ambient temperature, air line pressure, humidity, metal storage time, day of week, shift of the day, operating crews, and on and on and on. All of these variables have not yet been identified, much less quantified and measured.

In the real world press shop, the end product is more sensitive to some variables than others. Time and expense can be wasted on narrowing ranges of low sensitivity variables, while high sensitivity variables go unspecified. The analysis system should provide a measure of this sensitivity.

One should not consider the analysis problem with all its variables to be so complex that a solution will never be found, because the foundations for the solution are in place. However, all these variables and many more are known to affect formability and any viable solution must deal with them in one form or another.

Analysis of Sheet Metal Forming Operations Must Incorporate Properties of Real Materials

The extent of this problem is seen in the basic characterization of sheet metal. In many formability research experiments, results are correlated with tensile test data. Tensile specimens have no curvature, almost no strain gradient during uniform elongation, and a uniaxial stress state. The most common deformation conditions in the press shop are sharp curvature, severe strain gradients, and a plane strain or biaxial stress state. Additional metal properties which influence formability are Young's modulus, strain hardening, anisotropy, strain rate, strain rate hardening, surface topography, etc. Simplistic, idealized approximations of the forming operations which do not account for these properties will be inferior to the skill of the artisan and, therefore, unacceptable.

The Analysis Must Not Be Biased by Stereotype Restrictions on Forming or Historical Rules of Thumb.

Cup drawing limits commonly measured in terms of the limiting drawing ratio or LDR, are strongly influenced by the normal anisotropy of the steel. A practical range of $\bar{r}$ for cold-rolled steels would be from one to two. This corresponds to an LDR change from 2.3 to 2.6 [4]. By redesign of the tooling to incorporate expanding segments of the punch or deep drawing with a hydrostatic external pressure, the mode of failure is changed and cup drawing will duplicate quite closely a steady state operation. This change would allow the LDR for steel to change from 2.3 to 3.1 [5], which is several times greater than change for metal improvement. This change is even more spectacular for a 70/30 brass having a high n value and a low $\bar{r}$ value, for which Mellor [6] has shown the experimental LDR to increase from 2.19 to 3.45. Interestingly, these LDR values were predicted by a theoretical analysis in 1959 [5] and experimentally verified in 1962 [6].

Any Analysis System Must Provide Predictive Capability Extending All the Way Back to the Stylist/Designer

To be useful, any formability analysis must allow early decisions to be made before hard tooling is committed. Once hard tooling has been produced, the range of economical alternatives has been greatly decreased. For example, if a switch in metal type or grade has to be made—regardless of the reason—compensation for different metal characteristics is best made in part design and initial die design. Attempting to produce successful parts with a metal different from the one for which the entire process was designed, built, and tuned is extremely difficult, expensive, and time consuming. Rather drastic changes sometimes are required. If a predictive capability were available, the best combination of modifications could be obtained and ordered. Alternatively, an advance analysis might indicate insufficient modifications are possible to allow the proposed change.

The predictive capability must indicate the relative probabilities of success for changes in materials versus changes in tooling. For example, assume an analysis technique which would produce a graph similar to Fig. 4. This graph illustrates the change in part depth as a function of the best and worst workhardening exponent n expected for a given grade of metal. In this case, very little change in stamping depth is obtained as a function of potential n values.

References p. 17.

Assume now that the preliminary part print required a part depth h_d. Three cases might be encountered (Fig. 4). If h_d was in zone A, no problems would be expected in die tryout or production. An h_d in zone B suggests that the range of metals used in producing the n value range may or may not have acceptable breakage limits. However, successful stampings would be expected for minor changes in tooling, lubrication, press adjustments, etc. If, however, h_d fell in zone C or higher, drastic changes in part design or tool design would be required. The sensitivity of h to part and tool parameters could be evaluated to find the correct combination which would allow a successful stamping to be made that would approximate as closely and cheaply as possible the original design.

This predictive capability could be used in the reverse direction for correcting existing problems on the press shop floor. The analysis could indicate the lowest cost, or perhaps the fastest, corrections necessary to eliminate the problem.

The Analysis System Must Be Able to Improve the Interaction Between the Design and Manufacturing Functions

A designer may wish to improve his design by inserting stiffening ribs. This would allow a reduction in sheet thickness, weight, and cost, while improving vehicle miles per gallon. The unknowns for the designer are "can it be made?" and "how much gage reduction can I take?" Thus, limits of manufacturing often provide boundary conditions for design improvement. In addition, the analysis system would allow a reduced lead time, because the verified design and specified tool parameters would shorten die design time, provide for direct machining of patterns and castings, and reduce tryout time.

The Analysis System Must Be Responsive to New In-Service Requirements of the Part

Current criteria for a stamping are that the stamping hold water for unexposed parts and have no observable necks for exposed parts. Future requirements, however, most likely will specify strain control. This means that the strain distribution will not be the resulting summation of all other variables, but that all other variables will be controlled to produce a specified distribution of strain.

The demand for strain control will come from many fronts. In service performance usually is a function of strength and thickness at a critical location. It makes no sense to increase sheet thickness of an entire stamping to correct for localized thinning, when strain control can increase local thickness at a reduced overall thickness, weight, and cost. In addition, strain hardening is the most inexpensive method of strengthening metal. Once knowledgeable designers learn to optimize part design through strength and thickness gradients, strain control must eventually follow.

Strain control is especially important to utilize the benefits of new types of steel, such as the dualphase and age hardening types. The Forming Limit Diagram has been shown to depend on strain paths for non-proportional straining [8].

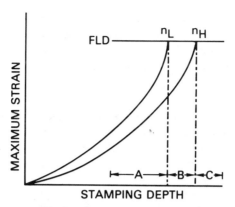

Fig. 4.   Schematic curves [7] showing permissible stamping depths for low and high n workhardening values. Zones A, B, and C depict possible part depths h_d which could be specified by the designer.

Thus, important formability increases may be achieved through careful specification of strain path in multiple stage forming operations.

Any System of the Future—Even Trial-and-Error—Must Be More Attuned to the End Product Economics.

Incoming metal costs may be a minor component of the total economic package.

A glaring example is the cost of offal or the scrap from forming operations. The automotive industry averages 30 percent offal for sheet metal. The U.S. automotive industry consumes about 20 million tonnes of automotive steel. A quick calculation will show that:

20 million tonnes $\times$ 30% = 6 million tonnes $\times$ \$350 rough average cost
= \$2.1 billion of prime steel ends up as scrap.

While some dollar value is retrieved through scrap prices, valuable energy is used in the recycling process. Therefore, a single percent reduction in offal from 30% to 29% saves 200,000 tonnes of prime steels, \$70 million, and 81.1 PJ. That should be tremendous motivation for more optimum control of forming operations. Some panels are even better candidates for offal reduction. A typical quarter panel will have over 50–60 percent offal when trimmed, although some scrap pieces are used for other parts. Reduction in offal could make more efficient utilization of scarce metals and energy.

Economics is reflected in the extremely conservative forming of many sheet metal stampings. A rough rule of thumb in the industry is that 5 percent of the stampings cause 95 percent of the problems. The exact percentages are immaterial. A restatement of the rule of thumb says that a vast majority of the stampings cause no problems at all. This implies that a majority of the stampings

References p. 17.

are not optimized. From an economic standpoint, a formability analysis system may provide its greatest return on investment by more effective material utilization in non-critical stampings. Potential areas of savings might be reduced blank size, offal, and sheet thickness. More severe forming could be incorporated into the design to provide equal structural requirements at a further reduction in sheet thickness.

Discussion of system characteristics could continue, but the complexity of the problem faced by the scientific community should be obvious. The difficulty of the problem is compounded by the time constraints which the automotive industry faces. Assume a total usable system can be ready in five years from today—a very short time period by research standards. To this must be added at least a two-year automotive lead time for design integration, prototype proof testing, component interaction studies, die construction, and integration into actual vehicles. Seven years from today the 1985 vehicles will roll off the assembly lines. We are already too late with a total usable system to assist the automotive industry in meeting a single one of the existing legislated standards which will heavily affect the energy, pollution, economic, and transportation environments in which we will shortly live. Will we be able to help them beyond 1985?

A POTENTIAL SOLUTION

This discussion on the need for fundamental scientific understanding of the formability process would not be complete without some indication of the potential solution. This potential solution then can serve as a reference point against which the pros and cons of the subsequent papers can be tabulated. No adequate solution is available today. In addition, the problem is far too complex for any single discipline to solve in sufficient time. Thus, the most promising solution seems to be a hybrid system—such as mathematical modeling—in which the best of many disciplines can compensate for the weaknesses in all disciplines. For such a hybrid system to achieve operational level within five years, the various heterogeneous disciplines working on the various aspects of the formability problem would have to:

a. Be motivated into a spirit of cooperation and teamwork,
b. Refocus their goals relative to the needs of the hybrid system,
c. Narrow their scope of studies to concentrate on their strongest areas,
d. Establish new lines of communication, and
e. Learn each other's language.

SUMMARY

General Motors Research is to be complimented for sponsoring this symposium, which is a rather unique and bold experiment. In an attempt to solve one of their important problems, they have literally picked experts off the slopes of Swift's two massifs, intermingled them with engineers and metallurgists gleaned from the foot-hills, and transported them down to their valley. It is rare that such a heterogeneous group of professionals be assembled to interact on such a specific problem.

REFERENCES

[1] H. W. Swift, J. Inst. Metals, *81* (1952), 109.
[2] W. A. Backofen, *Fundamentals of Deformation Processing,* Syracuse University Press, Syracuse, New York, 1964, 1.
[3] S. P. Keeler and W. G. Brazier, *Proc. Microalloying 75* (Union Carbide Corporation), New York, 1977, 517.
[4] M. Atkinson, Sheet Met. Ind., *44* (1967), 167.
[5] S. P. Keeler and W. A. Backofen, Trans. ASM, *52* (1960), 166.
[6] M. G. El-Sebaie and P. B. Mellor, 7th Biennial Congress, IDDRG, Hoogovens, Ijmuiden, Holland, (1972), 2.1.
[7] N. M. Wang and M. L. Wenner, This volume.
[8] K. Yoshida and K. Miyauchi, This volume.

DISCUSSION

J. L. Duncan *(McMaster University, Canada)*

I would like to ask a question about the initial analytical system. To illustrate the question, I'll draw the following parallel: Engineers in other fields (i.e. structural design) learn, in a one semester course, "Strength of Materials". Fairly simple things which when applied to structures will, in my view, solve about 95% of the problems. Then they may go on and learn something about the theory of elasticity.

The question I ask is: finding ourselves, as you have indicated, with a need to provide something to the practitioners, shouldn't we, as well, address ourselves to something like "Strength of Materials" where it comes to studying press shop problems? The theory of plasticity is going to provide some of the solution, but I would ask the question, "Have we really addressed ourselves to some of the simple analytical models that could be taught in a one semester course that could help the practitioners in the area of sheet metal forming?"

Keeler

Unfortunately after many courses of "Strength of Materials" we still have the "Galloping Gertie Bridge", and we still have the building that is blowing out its windows in Boston. I think that even in the structural field (which is much simpler than forming the complex parts that you see here) we haven't reached a proper level of competence in the simple structural materials. The job is going to be ten times more difficult when you go into forming of complex parts, and I think we have to go far beyond the simplified "Stength of Materials". We've been trying that for years, and I don't think we've gotten anywhere.

J. W. Hutchinson *(Harvard University)*

Let us consider your own contribution of forming limit diagrams; to what extent has that gotten into the press shop or at least one level above the press shop? Has it gotten there at all?

Keeler

Yes, it is a great postmortem technique. When you are in trouble, that shows you why you got into trouble. But as a predictive capability? No, it is very limited. The die designers in the automotive industry have no hope, really, of predicting how a given die will perform.

Hutchinson

What about in selecting sheet material or characterizing it, do they use r?

Keeler

Specify r? No! The steel companies guarantee to supply steel which will make the part. Since the customers can't tell precisely what is required to make the part, it is an educated guess on both sides.

EXPERIMENTAL STUDIES OF MATERIAL BEHAVIOR AS RELATED TO SHEET METAL FORMING

K. YOSHIDA and K. MIYAUCHI

The Institute of Physical and Chemical Research, Japan

ABSTRACT

The material behavior in sheet metal forming has been investigated widely in the past several years in order to better assess press performance. This has resulted in a rapidly changing concept of press formability.

In addition to the conventional tensile and simulative tests, the stress-strain relation and workhardening behavior for various modes of deformation are analytically examined. New tests, such as the X value (ratio of deformation resistance under uniaxial and biaxial loading) are proposed. Strain analyses, including the forming limit analysis, have also been employed, using the scribed circle test, for more accurate understanding of complex shape forming in the press shops. Further, the deformation path is now known to strongly govern the forming limit curve.

The metal flow analysis is also attracting great attention as a possible means of extending the strain analysis. Formability is growing into a wider concept depending on such multi-factors as fracture, wrinkling, surface roughening, galling and geometrical accuracy.

In the near future, it will probably be necessary to establish a total assessment system which takes into account forming defects, material properties and chemical composition.

INTRODUCTION

The behavior of sheet metals in press forming has long been investigated from various points of view and the topic has already grown very large in the wider sense. It was probably first experimentally examined for explaining a difference in the forming limit and for finding the material factors governing the press formability of sheet metals. Then, some fundamental tests such as the tensile test and the hardness test were investigated in comparison with actual press performance as a means of predicting the formability.

References pp. 48–49.

It is now impossible to mention all the theoretical and experimental investigations relating to material behavior in press forming which have been published in many countries since World War II. As a consequence of the number of such fruitful studies, however, we can now systematically understand the multiplex relations among sheet metal properties, geometry of a pressed part and forming techniques including lubrication, if the forming is limited to axisymmetrical problems.

In asymmetric press forming, however, the number of geometrical factors increases abruptly, resulting in a considerable change in the role of sheet metal properties and press forming techniques. Since the systematization of complex shape forming has not yet been achieved, material behavior is discussed here mainly in relation to the press forming of complex parts such as rectangular and elliptic shells.

For the first time at the International Deep Drawing Research Group (IDDRG) Colloquium in Liege, in 1966, many papers were presented relating to complex-shape forming operations. Some examples: comparison of press performance with 1) the tensile test values [1]–[3], 2) the Swift Cup test [4], and 3) deformation and fracture analyses of rectangular shells [5], [6] which dealt with the wall breakage at the shell corner occurring near the end of forming.

The formability limit analysis according to the scribed circle test for sheet metal forming problems in the press shop was discussed in the 4th IDDRG Colloquium in Turin, 1968 [7]. The fracture behavior was dealt with in several papers in the same Colloquium, where the effect of deformation path on the forming limit was probably first discussed with respect to punch stretching and stretch flanging [8]. The deformation (constitutional) diagram, having the same coordinates as the forming limit diagram, was proposed for showing the deformation path and distribution of combined strains, and for providing the proper direction for lowering the severity of forming or to bring out the highest formability in sheet metals.

In the 5th IDDRG meeting in Tokyo, 1970, the analysis of deformation behavior in press forming of autobody parts was advanced further by using the deformation diagram and the strain composition diagram [9]–[11]. At that time, the different types of fracture in U-shaped stretch flanging were investigated experimentally and systematized depending on the blank geometry [12]. In the next IDDRG meeting, in Amsterdam, 1972, a number of interesting papers on strain analysis and complex shape forming were presented both in the open and closed sessions [13]. There were also impressive new approaches such as assessment of failure sensitivity by torsional testing of rings [14] and metal flow analysis in irregular shape shells [15].

In the two following IDDRG Colloquia in Gothenburg and Ann Arbor, except for wrinkling behavior [16], [17] and deformation path [18], no further investigation into the material behavior in complex shape forming was made in particular. On the other hand, the social demand for energy saving and weight reduction resulted in a remarkable increase of papers on the formability and galling problems in the press forming of high strength low alloy steel sheets.

Thus, a considerable number of papers on the material behavior in complex shape forming have been published in the IDDRG meetings over the past ten years. But not enough is known as yet to establish an assessment method of the press performance of large, complex panels. The Japan Sheet Metal Forming Research Group (originally The Conical Cup Test Research Group in Japan) consisting of scientists and engineers from the non-commercial institutes and steel and car makers in Japan, started its cooperative activities in 1957. It has participated in the IDDRG Colloquiums as the Japanese representative since 1960.

Recently, in addition to fracture limit analysis, there has been underway a new

TABLE 1

Factors Affecting the Material Behavior in Press Forming

SHEET METAL ──────────────┬──── Tensile Properties
                          │   n, r, X, strength, elongation
                          │
                          ├──── Hardness
                          │
                          ├──── Thickness
                          │
                          └──── Roughness

                    Simulative tests
                       ↗
              Classification (Simplified models)
                   ↗
DEFORMING BEHAVIOR ──────────── FORMING TECHNIQUES
                          ┌──── Type of forming
                          │   (Deformation mode)
                          │
                          ├──── Forming geometry
                          │
                          ├──── Lubrication
                          │
                          ├──── Forming speed
                          │
                          └──── Deformation path

FORMING LIMIT ──────────────┬──── Deforming limit
                            │   (Fracture limit)
                            │
                            ├──── Geometrical defects
                            │
                            ├──── Roughness limit
                            │
                            ├──── Galling
                            │
                            ├──── Panel stiffness
                            │
                            └──── Fatigue strength

program of deformation analysis of shallow and large complex panels. This analysis covers everything from the setting of a blank in the die to the taking-out of the pressed part from the die cavity, especially in relation to various geometrical defects resulting from unbalanced stress and strain distributions in the very small strain range, where no serious fracture problem takes place under ordinary stamping conditions. Such geometrical defects may also be regarded as a kind of forming limit without fracture. The concept of press formability is therefore still growing in its nature and extent.

FACTORS AFFECTING MATERIAL BEHAVIOR

There are various well-known factors governing the material behavior in press forming, as shown in Table 1. However, no factor can be related individually with the press formability or deformation behavior of sheet metals. A relation between the forming limit and a test value is usually upset by changing the lubrication conditions in punch-stretching. Although it may be impossible to describe such complicated relations among many factors even roughly, this paper tries to deal with the recent important topics relating to the material behavior as much as possible.

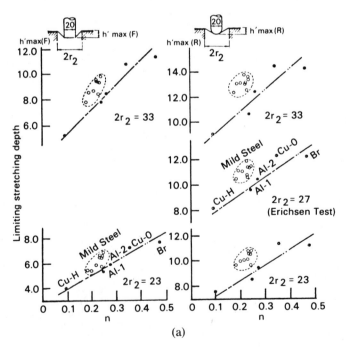

(a)

Fig. 1.   Relation between limiting stretching depth and n value: (a) Combined punch-stretching, and (b) Pure punch-stretching. All dimensions in mm.

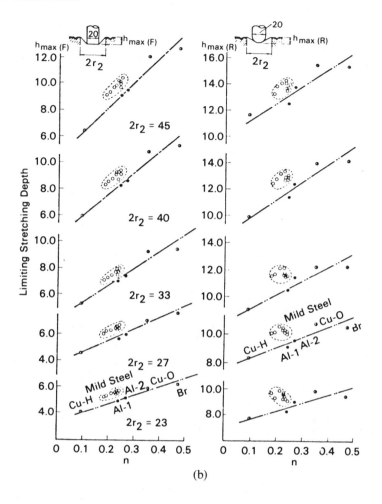

(b)

TENSILE TEST VALUES

Tensile test values have been used for assessing the deformation behavior and press formability of sheet metals, on the assumption that there is no significant change in the stress-strain relation for various modes of deformation. These test values have been working well for the rough, first-order assessment of formability. The recent strong demand, however, for selecting sheet metal and forming condition more effectively to save the cost of production has resulted in the more accurate understanding of tensile test values in relation with the different deformation modes.

The punch-stretchability of sheet metals is the ability to biaxially elongate in the die cavity until a crack occurs on the blank. This property is strongly affected by the work-hardening coefficient (n value) governing the necking strain and strain distribution. As the strain distribution depends most strongly on the ge-

References pp. 48–49.

ometry of forming and the forming limit is remarkably increased by the presence of draw-in (a component of deep drawability), a relation between the limiting stretching depth and the n value is considerably changed with the varying conditions shown in Fig. 1. The weakened relation is caused by the draw-in and also by the effect of forming geometry and friction (or lubrication) on the strain distribution, as is schematically explained in Fig. 2 [19]. In the round bottomed

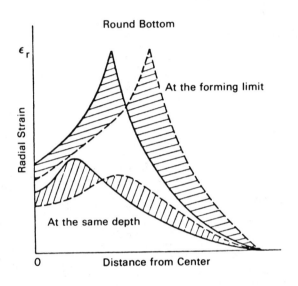

Solid line: low workhardenability
Broken line: high workhardenability

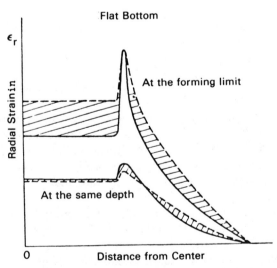

Fig. 2.   Radial strain distribution models for different workhardening characteristics.

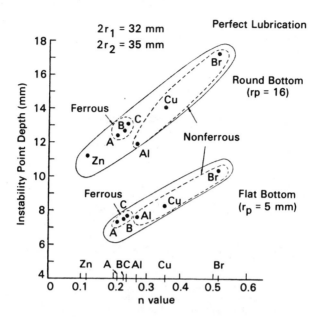

Fig. 3(a).   Relation between the depth at instability point and *n* value. r_1: Radius of punch, r_2: Radius of die opening, r_p: Punch profile radius.

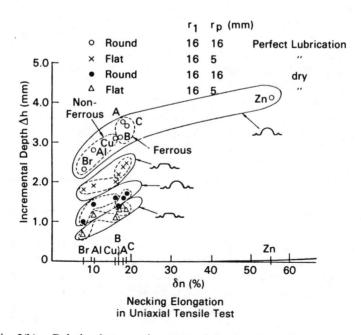

Fig. 3(b).   Relation between incremental depth and necking elongation.

References pp. 48–49.

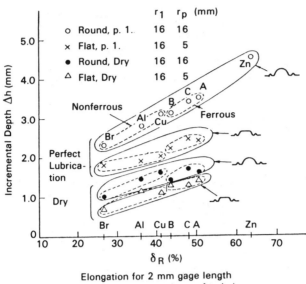

Fig. 3(c).   Relation between incremental depth and elongation of a notched specimen.

punch-stretching under low lubrication, the strain level in the center area surrounded by the hoop fracture is lowered as the n value increases. This is because, for the lower n material the center area deforms more easily, being affected by the stress gradient in the earlier stages of forming.

The effect of n value on the punch-stretchability is therefore greatest when the fracture is observed at the top of a round-bottomed punch under perfect lubrication. Deformation behavior in flat bottom and overhang regions is always in close relation with n value.

In addition to the effect of n value, the deformability of sheet metals after the onset of necking is also a factor affecting the punch-stretchability. Fig. 3 shows relations between the cup depth at the onset of instability and n value. It also shows relations between the incremental depth from the onset of instability to fracture and the necking elongation in the ordinary tensile test (or in the subsequent tensile test of a notched specimen obtained from the material subjected to uniform elongation.) The simulating effect of deformation mode makes the subsequent test much closer to the subsequent formability beyond the onset of instability in punch-stretching [20].

Punch-stretching accompanied by the flange draw-in is referred to as combined stretching [19], while the pure stretching described above is provided with locking beads, to produce no draw-in from the flange area. The combined stretchability depends on the pure stretchability and deep drawability, that is, mainly on the n and r values. Fig. 4 illustrates some typical relations between the forming limit (as measured by stretching depth-h_{max}) and n or r value for the shells of different

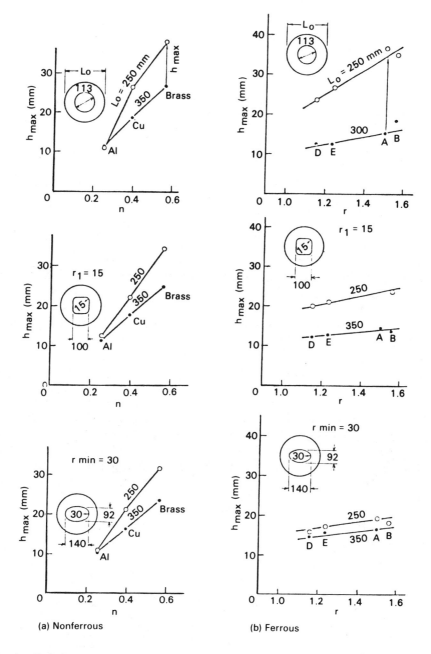

Fig. 4.   Relation between the stretching depth and n or r value for different forming geometry. All dimensions in mm.

References pp. 48–49.

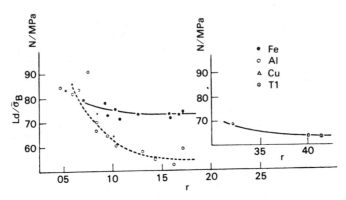

Fig. 5.   Relation between r and resistance to shrink flanging. L_d: Max. punch Load, $\bar{\sigma}_B$: Tensile strength.

geometry [21]. The relation between h_{max} and r in steel sheets may be seen to be fairly strong, though more or less differently affected by the forming geometry. In nonferrous materials which have almost the same r value, the forming limit (h_{max}) is closely related to n value for all the different geometries of forming. The different sensitivity to blank size between r and n values is due to a difference in the mechanisms which affect stretchability and deep drawability. The effect of blank size is also seen to be different for different forming geometries in steel sheets, due to different contributions to the forming limits in stretching and in drawing.

The effect of r value on deep drawability is widely known, but sometimes very weak in certain groups of sheet metals. It can be separated into two effects: one on the resistance to shrink flanging and another on the fracture strength under biaxial stretching load, as shown in Figs. 5 to 7 [22]. The resistance to shrink flanging decreases remarkably with an increase in r value in aluminum and copper sheets only. In contrast to this, the fracture strength compared with the uniaxial

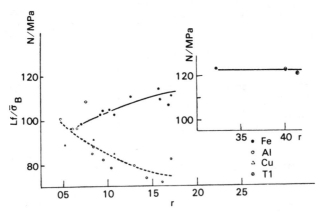

Fig. 6.   Relation between r and fracture strength. L_f: Punch load at fracture, $\bar{\sigma}_B$: Tensile strength.

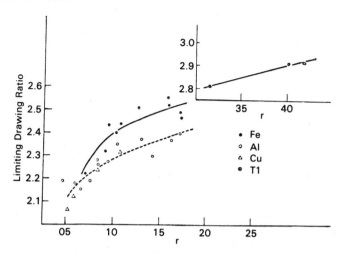

Fig. 7. Relation between limiting drawing ratio and r.

tensile strength is increased by increasing r value in ferrous materials, while it decreases in a series of aluminum and copper alloys. Both the effects of r value are indistinct in titanium sheets. The total effect of r value on the limiting drawing ratio in a cylindrical cup test is considerably varied for different groups of sheet metals, as shown in Fig. 7.

An example of a quantitative expression for the varying effect of r value corresponding to a change in the mode of deformation is the X value which expresses the ratio of uniaxial to biaxial flow stress [23]. Fig. 8 shows a relation

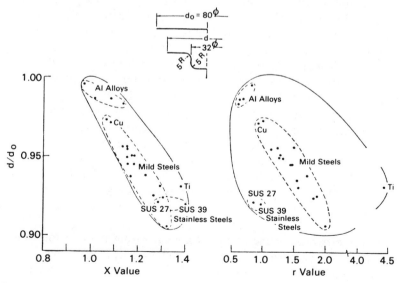

Fig. 8. Relation of peripheral shrinkage at fractue with X or r value. ϕ: Diameter and R: Punch or die profile radius in mm.

References pp. 48–49.

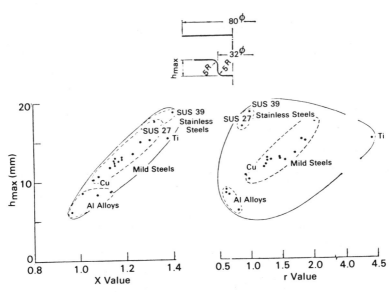

Fig. 9.   Relation of the limiting depth with X or r value. ϕ: Diameter and R: Punch or die profile radius in mm.

between X or r value and the peripheral shrinkage of a fractured blank, and Fig. 9 a relation between X or r-value and the limiting depth, for various kinds of sheet metals. Such formability characteristics are more strongly related with X value than r value. It is thus made clear that r value derived from the uniaxial tensile test is not a sufficient index to reveal all the anisotropic natures of sheet metals under the different modes of deformation. Besides, this different effectiveness of r value for different materials is obviously in close relation with their crystal structures, which suggests that this difference results from the different natures in the active and latent slip systems. These differences relate to the active number of slip system, angle between the corresponding slip systems, and the lowest energy system of the given mode of macro-deformation.

For the assessment of press formability, especially punch-stretchability, the stress-strain relation characterized by n value is very important and complicated as shown in Fig. 10. The stress-strain curves for different materials and deformation modes differ considerably from each other, and suggest the low possibility of single n value to express the workhardening property for the entire measured strain range. Although we have usually used the averaged n value, the n value derived from a small range of strain may be required for more accurate prediction of the instability point and punch-stretching limit. Fig. 11 shows the n^* value according to the two point method over a small strain interval for various sheet metals. This kind of information about the change of n value has already been successful in explaining the actual press performance, using the corresponding n^* for the given strain in a pressed part.

COMPLEX SHAPE FORMING

As previously described [5], complex shape forming has brought out new problems in comparison with the cylindrical forming; for example, wall failures in the deep drawing of elliptic and square shells, shown in Fig. 12. In the punch-stretching of complex shape, there is present the effect of strain accumulation on the stretching limit, as is explained by the schematic models of stretch deformation mode in Fig. 13 [24]. As the fracture takes place at the corner of a flat bottomed square punch, the deformability of sheet metals in the diagonal direction corresponds to the punch-stretching limit of a square shell. The total elongation of the deformed diagonal length between the point of non-displacement and the punch profile region depends on the external radial force in the diagonal direction

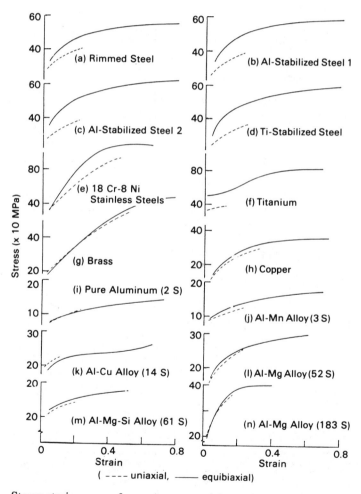

Fig. 10. Stress-strain curves for various materials on the assumption of isotropy.

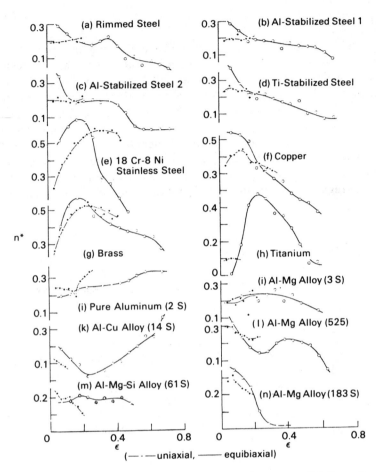

Fig. 11.   Dependence of n^* on the mode and amount of deformation.

and the other external eccentric forces transmitted from the adjacent straight punch profile regions. In other words, the total elongation of the deformed length can be increased by increasing the said external eccentric forces without increasing the radial tension at the punch corner. The effect of such strain accumulation due to the external eccentric forces on the forming limit reveals that the severity of press forming does not always depend on the local size of the most severely deformed region.

The strain distributions in the flat bottom region of square and elliptic punches are shown in Fig. 14 [25]. In the center portion of square punch-stretching, there is a state of equibiaxial stretch deformation the same as in the bottom of a cylindrical punch. The straight peripheral regions are subjected to biaxial tension, where the radial strain is predominant, while the deformation pattern in the corner region is somewhat similar to what happens in cylindrical punch-stretching, except that there is present the strain gradient, as schematically shown in

(a)

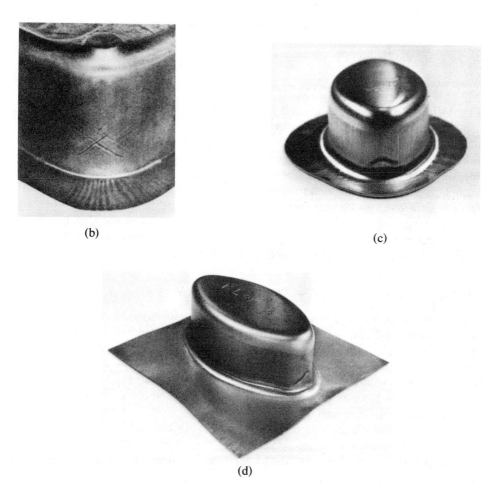

(b)

(c)

(d)

Fig. 12. Photographs of wall breakages: (a) in square shell [6], (b) X type in square shell [5], (c) in elliptic shell and (d) from a square blank.

References pp. 48–49.

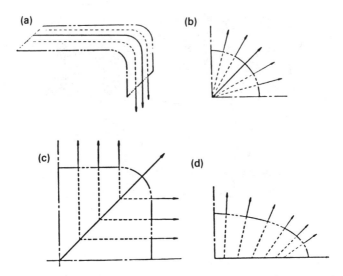

Fig. 13.   Schematic models of stretch deformation mode.

Fig. 15 [19]. In larger square shell forming, however, there is a very interesting size effect on the deformation behavior, accompanying the well-known size effect on the forming limit. Fig. 16 shows the strain distribution in the flat square bottom and the growth of strain at five points according to the strain gage measurement, where the peripheral strain is larger than the radial strain in the straight region near the punch profile [26]. This peculiar behavior probably results from a difference in the resistance to shrink flanging between the corner and straight flange regions. According to Figs. 14 to 16, we can see that the deformation behavior is very delicately varied with various forming conditions, especially in the press forming of shallow complex shells.

FORMING LIMIT ANALYSIS—SCRIBED CIRCLE TEST

In more complex, large-sized parts as compared with square and elliptic panels, it is extremely difficult to establish a total assessment system of forming severity. The scribed circle test of the plastic instability and fracture in sheet metal forming was proposed in 1963 [27] and is now being applied in press shops throughout the world on various problems. The major thrust is the improvement of press performance for large-sized parts of very complex geometry, which is governed by many different forming factors in various ways.

From this point of view, the scribed circle test has been one of the main themes in the IDDRG activities for many years and is now the most familiar means for assessing the forming severity and defects in the press forming of large-sized autobody panels.

In Japan, the scribed circle test is used not only for measuring the fracture

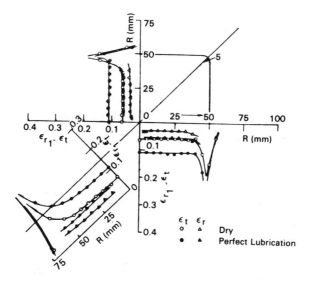

(a)   Square shell

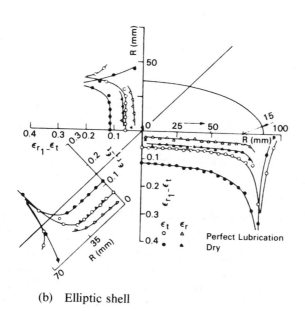

(b)   Elliptic shell

Fig. 14.   Strain distributions in the flat bottom region of square and elliptic shells. ϵ_r: Radial strain, ϵ_t: Thickness strain.

K. YOSHIDA, K. MIYAUCHI

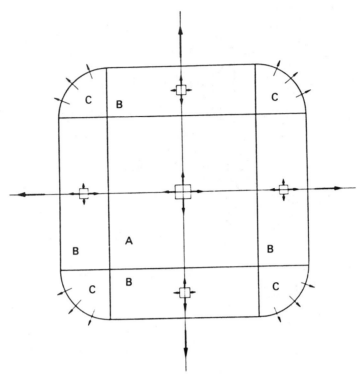

Fig. 15.   Schematic representation of deformation behavior in the flat bottom region of a square shell.

strain but also for obtaining the deformation diagram [8] which shows the combined strain distribution and deformation path and makes it easier to see the whole deformation behavior in the pressed panels. Good examples are shown in Figs. 17 and 18 [9].

DEFORMATION PATH AND PRESTRAIN

During experimental investigations using strain analysis in relation with the wall failures, the effect of deformation path and prestrain on the subsequent stress-strain relations, elongation and press formability has been revealed as a major problem. The forming limit according to the scribed circle test varies remarkably with a change in the deformation path [7].

In the tensile test of prestrained sheet metals, the shape of load-elongation curve is strongly affected by the test direction as shown in Fig. 19. That is, it depends on the variation of active slip systems from the primary to secondary deformation processes. The r value is also changed with the prestrain, and its planar directionality in the prestrained sheet metal is closely connected with the planar directionality of previous deformation, Fig. 20. Although the *r* value is not necessarily stable in the ordinary sheet metals, it seems to be more unstable in the prestrained than in the virgin material, as shown in Fig. 21.

As the stretch-flangeability is mainly governed by the ultimate ductility and the planar anisotropy, it is therefore strongly affected by the mode and amount of previous deformation [8]. Figure 22 illustrates the punch-stretching limit of

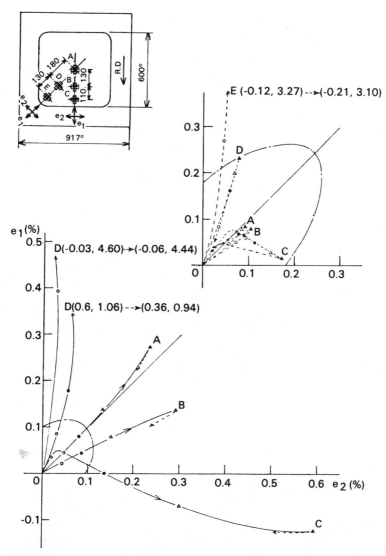

Fig. 16.   Deformation behavior at very small strain level in the flat bottom region of a large square shell. All dimensions in mm.

References pp. 48–49.

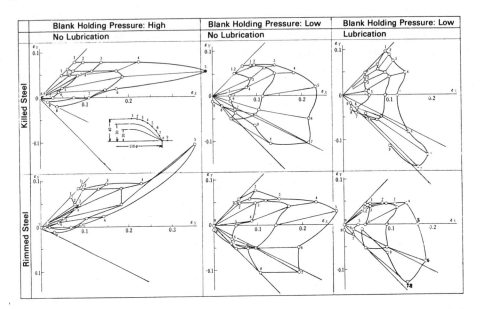

Fig. 17.   Variations in the deformation diagram due to changes in forming conditions.

S.C.D : 10 mm, Sheet Thickness: 0.7 mm
Killed Steel
Decarburized Rimmed Steel

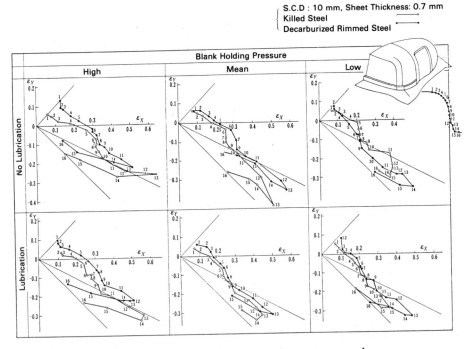

Fig. 18.   Deformation diagram of a quarter panel.

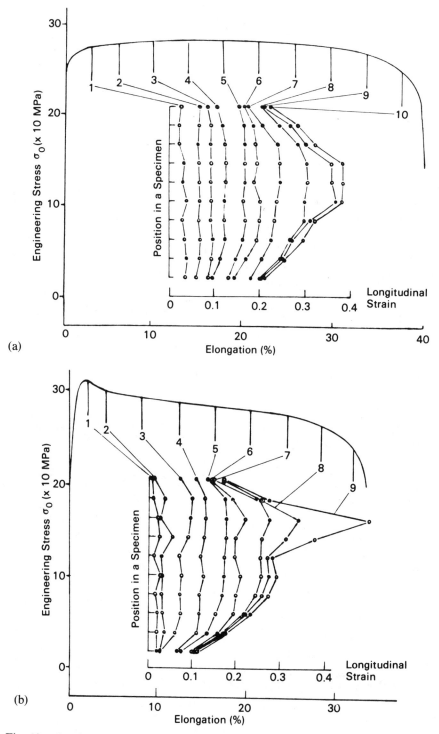

Fig. 19.   Load-elongation curve and longitudinal strain distribution in 15% cold rolled decarburized rimmed steel: (a) Rolling direction, (b) 45° and (c) 90° to rolling direction.

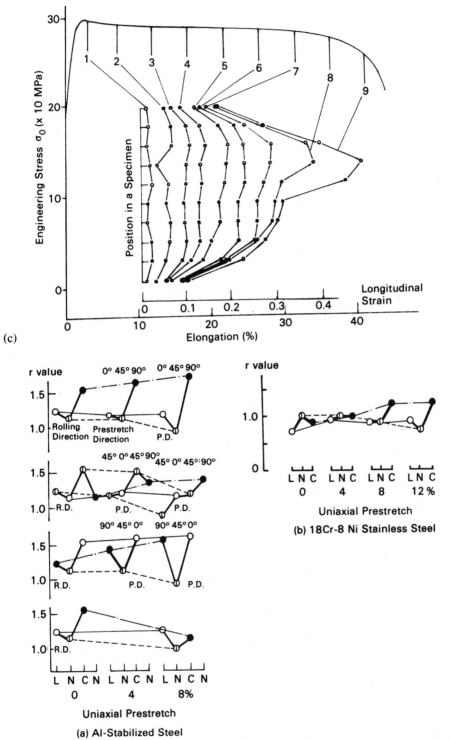

Fig. 20.   Change in r value due to increase in prestrain: (a) Al-stabilized steel, (b) 18 Cr-8 Ni stainless steel.

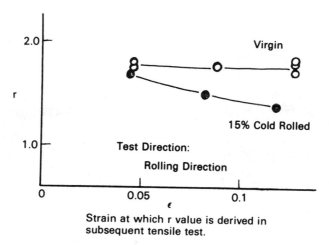

Fig. 21.   Effect of prestrain by cold rolling on r value.

prestrained sheet metals. Aluminum is very sensitive to a change in the deformation mode, probably because of its few slip systems.

In deep drawing, the forming limit is also in good agreement with the r value, though it is difficult to measure the r value [28]. However, the formability in complex shape forming has not yet been clarified for various conditions. For

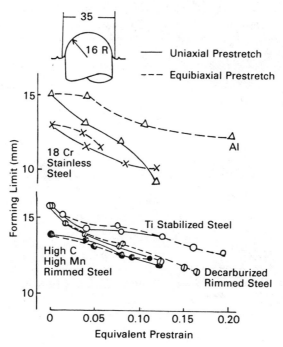

Fig. 22.   Punch-stretching of prestrained sheet metals. ϕ: Diameter and R: Punch profile radius in mm.

References pp. 48–49.

Virgin 30% cold rolled

Fig. 23. Flange wrinkling behavior.

Virgin

20% cold rolled

Fig. 24. Different shrink flanging behavior.

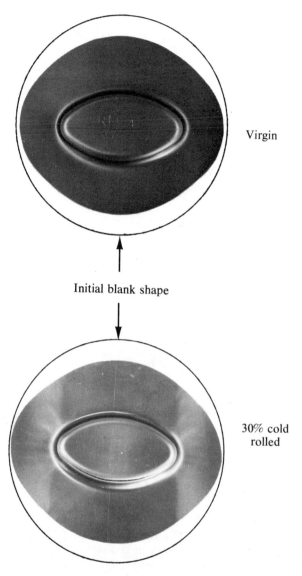

Virgin

Initial blank shape

30% cold
rolled

Fig. 25.   Different shrink flanging behavior of a circular blank.

example, wrinkling behavior depends strongly on the prestrain in sheet metal, as shown in Fig. 23, and becomes a significant factor affecting the forming limit. The wrinkling resistance depends on the planar anisotropy [29], workhardenability (prestrain), lubrication, and forming geometry including the blank size and shape.

Figs. 24 and 25 show photographs of formed cups from circular and rectangular

References pp. 48–49.

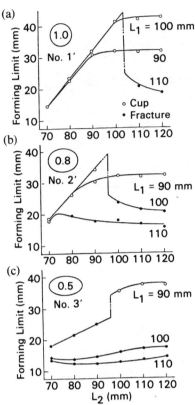

Fig. 26.   Geometrical effect of a rectangular blank on the forming limit in deep drawing of elliptic shells. Ellipticity = (a) 1, (b) 0.8 and (c) 0.5.

blanks in deep drawing of an elliptic shell. In the virgin material, the shrink flanging behavior is produced by plastic deformation all over the flange region except the corners of a square blank, while the prestrained sheet is deep drawn mainly by localized plastic deformation in the elliptic corner and by rigid-body displacement in the other remaining regions. The patterns of deformation in the square and circular blanks are different.

METAL FLOW ANALYSIS

In complex shape forming, principal strain axes may rotate in many deforming regions. Some difficulty of measuring the strains may accompany this shear strain. Fig. 26 shows a strange phenomenon with respect to the blank size effect in elliptic shell forming [15]. The limiting depth of elliptic shells is increased by increasing the blank size in the direction of minor axis (L_2). This additional flange decreases the shrink flanging of its adjacent region and consequently the local flange region along the major axis is merely compressed in the circumferential direction by such metal flows as shown in Fig. 27, while the increasing severity of local shrink flanging tends to lower the severity of forming.

TOTAL ASSESSMENT SYSTEM OF PRESS PERFORMANCE

The press performance of large-sized complex autobody panels is governed by many different factors compared with cylindrical forming. As a way of approaching this problem, Fig. 28 suggests the classification of the forming geometry and the corresponding material properties into three levels. Even complex parts such as autobody panels, may be considered at three levels of intricacy, namely, the localized region, the component region and the whole part. The forming severity of localized region is mainly governed by its geometry and the ultimate ductility

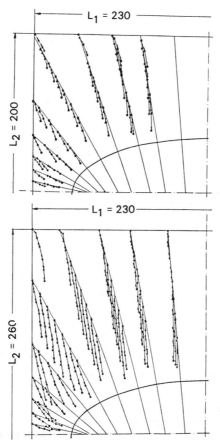

Fig. 27.   Geometrical effect of a square blank on metal flow pattern in the shrink flange portion of an elliptic shell.

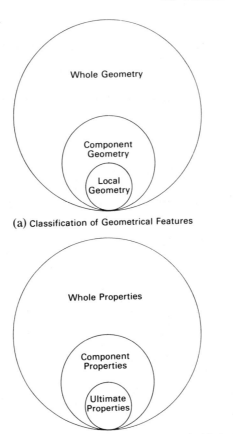

(a) Classification of Geometrical Features

(b) Material Properties corresponding to geometrical features

Fig. 28. Classification models for assessing factors in sheet metal forming: (a) Geometry features and (b) Corresponding material properties.

of sheet metal. The forming characteristics of a component region, while depending on its geometry, are closely related with forming techniques (such as the strain-accumulation-effect, and the strain-relaxation-effect in asymmetric shrink flanging) and also with some material properties (such as the ability to uniformly distribute the strain). The forming characteristics of the whole part may be treated as the averaged characteristics of the different forming behaviors in the various component regions taking into account the mutually interactive effects among them.

The material properties may also be classified corresponding to the above classification of the forming geometry and forming behavior.

From this point of view, it is also important to investigate the deformation and wrinkling behavior during a stamping operation, as shown in Fig. 29. Although many wrinkles occur in the earlier stages in the unsupported region, the wrinkles are gradually absorbed in the final geometry of the pressed part. If it were

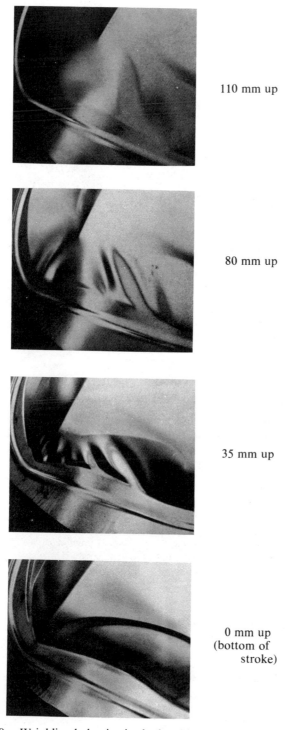

110 mm up

80 mm up

35 mm up

0 mm up
(bottom of
stroke)

Fig. 29.   Wrinkling behavior in the head lamp portion of a fender panel.

References pp. 48–49.

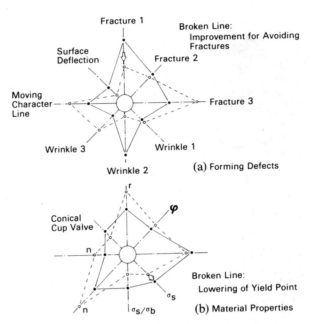

Fig. 30. Multiparametric diagrams of various forming factors in sheet metal forming: (a) Forming defects and (b) Material properties.

possible to predict the residual wrinkles in relation to identifiable forming factors, complex shape forming could be carried out with much higher efficiency.

As shown in Fig. 30, multiparametric diagrams of the constitutive factors offer some possibility of understanding the general features of each complex shape forming and of giving the general direction of improving the forming techniques and material properties. The goal is, however, still remote from where we stand now.

REFERENCES

[1] H. Josselin, Proc. 4th IDDRG Colloquium, Liege, (1966) Communication I.
[2] G. Pomey, M. Grumbach and R. Guillot, ibid., Communication II.
[3] M. Wybo, ibid., Communication XVIII.
[4] H. T. Coupland and R. R. Arnold, ibid., Communication XV.
[5] T. Okamoto and Y. Hayashi, ibid., Communication III.
[6] R. H. Heyer and J. R. Newby, ibid., Communication IV.
[7] G. M. Goodwin, la metallurgia italiana, 60 (1968), 767.
[8] K. Yoshida, K. Abe, K. Miyauchi and T. Nakagawa, ibid., 685.
[9] T. Kobayashi, H. Iida and S. Sato, Proc. Int. Conf. Sci. Tech. Iron and Steel, Tokyo, (1971) 837.
[10] T. Kobayashi, K. Murata, H. Ishigaki, T. Abe and R. Mio, ibid., 841.
[11] H. Josselin, M. Tong-Cuong and R. El Haik, ibid., 845.
[12] T. Matsuoka, C. Sudo and M. Kojima, ibid., 809.
[13] For example, M. J. Painter and R. Pearce, Proc. 7th Biennial Congress IDDRG, Amsterdam, (1972) 1.1.; S. S. Hecker, ibid., 5.1; T. Kobayashi, H. Ishigaki and T. Abe, ibid., 8.1.

[14] Z. Marciniak and J. Kolodziejski, ibid., 6.1.

[15] K. Yoshida, K. Yoshii, H. Komorida and K. Miyauchi, ibid., 9.1.

[16] K. Yoshida, H. Hayashi, K. Miyauchi, Y. Yamamoto and K. Abe, Proc. 8th Biennial Congress IDDRG, Gothenburg, (1972), 258.

[17] K. Yoshida, H. Takechi, S. Kuriyama and H. Hayashi, Proc. Biennial Congress IDDRG, ASM, Ann Arbor, (1976), 151.

[18] A. K. Ghosh and J. V. Laukonis, ibid., 167.

[19] K. Yoshida and K. Miyauchi, Sci. Papers Inst. of Physical and Chemical Research (Tokyo), 60 (1966) 5; Proc. 4th IDDRG Colloq., Liege, (1966) Communication V.

[20] H. Komorida, K. Miyauchi, Y. Numazawa and K. Yoshida, Reports, Inst. of Physical and Chemical Research, Tokyo, 44 (1968), 43.

[21] K. Miyauchi, C. Sudo and K. Yoshida, ibid., 45 (1969), 55.

[22] T. Oki, M. Miyahara, Y. Yamauchi, I. Kokubo and H. Hirano, Proc. Int. Conf. Sci. Tech. Iron and Steel, Tokyo, (1970), p. 962.

[23] K. Yoshida, K. Abe, K. Miyauchi and K. Yoshii, ibid., p. 968; K. Yoshida, K. Yoshii, H. Komorida, M. Usuda and H. Watanabe, Sheet Met. Ind., 48 (1971), 772.

[24] K. Yoshida and K. Miyauchi, Sci. Papers, Inst. of Physical and Chemical Research, 60 (1966), 137.

[25] K. Miyauchi, H. Tanaka, S. Honda, K. Hosono and K. Yoshida, Reports, Inst. of Physical and Chemical Research, 44 (1968), 197.

[26] I. Aoki, et al., The Japan Sheet Metal Forming Research Group Report No. 9 (1977-9-7, 8), Tokyo; To be published.

[27] S. P. Keeler and W. A. Backofen, Trans. ASM, 56 (1963), 1; S. P. Keeler, Sheet Met. Ind., 42 (1965), 683; ibid., 48 (1971), Nos. 5–10; SAE Paper No. 680092 (1968), 1.

[28] K. Miyauchi, Trans. JSME, No. 710-18 (1971), 17.

[29] H. Naziri and R. Pearce, La Metallurgia Italiana, 60 (1968), 727.

DISCUSSION

J. J. Jonas (McGill University, Canada)

You have given us a very good survey of the effect of n, r, and X, on drawing limits, and of the change of n with strain and, in particular, its decrease in the biaxial case. Recalling Fig. 3b, I would like to ask you a question about your data which shows much larger necking elongation in the case of zinc. Would you care to comment on the role of rate sensitivity in the drawing process? Could you say something about the possible effect of the m factor (rate sensitivity) on quantities like incremental depth and overall elongation?

Miyauchi

Zinc and zinc alloy sheets are very sensitive to the strain rate. In this case, zinc means Zn-Al alloy sheet which is very easily affected by change in strain rate. I have tried to keep final deformation-to-fracture approximately equal between the tensile test and punch stretch. If we compare this figure with following figure (Fig. 3c), we can see that the saturated strain is eliminated, so that, in this experiment, the punch stretching and the tensile tests are almost comparable in strain rate. In such a case, I think Fig. 3c is much better for explaining incremental depth and material properties.

Jonas

Well, could I just comment on that? I think that if we look at both of those figures the question still arises: what role would you attribute to rate sensitivity in the necking, or loss of ductility, process? I think this general question applies to both of your diagrams.

Miyauchi

We measure, of course, the strain rate in the tensile test. We found there is a very great effect of strain rate on zinc materials; however, in this case, for practical use at that time, we were satisfied just to have a comparison of the tensile test and the press forming test. If we apply almost the same strain rate, we can use this figure.

S. S. Hecker *(Los Alamos Scientific Laboratory)*

I would like to address the question of strain-rate sensitivity. In our punch stretching studies, in trying to correlate the cup height to some parameter that is easily measured, we found that the only parameter which correlated well with the cup height was the total elongation in uniaxial tension. By taking a good look at what controls the cup height which is mainly the forming limit and uniformity of strain), we found that the two factors which are the most important are, first of all, the strain hardening (as you have shown) but, indeed, just as important, is the strain-rate sensitivity. And here the important part is not the particular rate at which you conduct the test (because we have found that the cup height does not change up to six orders of magnitude) but, the point that is important is the strain-rate sensitivity of the material. In other words, when there is a rate change, does the flow stress level tend to go up or down? We found that the strain-rate sensitivity is very important in promoting uniformity of deformation and a higher forming limit level. This is particularly demonstrated in alloys like zinc. One of the materials investigated was a dispersion hardened zinc alloy which had a $m = 0.05$, which is quite high for metals. And, it had an extremely high cup height even though its n value was only 0.05, which is very low for metals. So we found that the two most important factors are strain hardening and strain-rate hardening. And, the reason that total elongation correlated well was that both of those factors go into making up a high total elongation. That is n controls mostly what happens up to maximum load, and strain-rate sensitivity, as A. K. Ghosh has shown, plays a very important role in post-uniform deformation. So, I think it is the combined factors of strain hardening and strain-rate hardening that are the most important for cup height.

Miyauchi

I agree with you except for zinc and zinc alloy materials. According to my experience in tensile tests, the necking elongation is more sensitive to the strain rate in mild steel, aluminum, and copper sheets. The necking elongation is the major factor showing mild steel to be improved over all other materials.

J. W. Hutchinson *(Harvard University)*

Has the work you've been describing had any influence on the press shop work in Japan?

Miyauchi

That is a difficult question for me to answer. However, there is a Japanese research group, consisting of auto making, steel making, and all other companies, and called the Japan Sheet Metal Forming Research Group—the representative of the group is Professor Yoshida. In this group we can communicate very well among our institute scientists and our research engineers in car making companies and also the engineers and researchers in the steel making companies. These results have been employed and used in actual press shops for more accurate understanding or assessment of press performance.

W. G. Brazier *(Fisher Body Division, General Motors Corporation)*

I noted with interest the slide that showed the variation in n value for different materials and the variation in elongation. My question is—have you looked at various materials (groups of materials) such as A.K. steel and commercial-quality steel and examined the variability in n-behavior at various degrees of deformation for each group of steels?

Miyauchi

The measurement of strain is almost the same between rimmed and killed steels. However, although our data measured n, as strain value, during necking in tensile test, the strain propagation behavior just before and during necking suggests that the n aspect is probably greater in killed steels than that of rimmed steels. The strain propagation is very large in A. K. steels rather than in rimmed steels. I think the n aspect is larger in the necking elongation range.

J. L. Duncan *(McMaster University, Canada)*

On the Zn-Al alloy, we find that the rate sensitivity at room termperature and at the rates at which you carry out the tests is of the order of 0.05. The question I would like to ask refers to the X value which is the ratio of the biaxial to uniaxial strength. You show a very much better correlation between the X value and cup depth than between cup depth and r value. Do you have a convenient method for measuring the X value? Can it be measured as easily as the r value or is it only something to do research on?

Miyauchi

I regret to say that for the time being we have not been able to find an economical and easy method for determining such parameters.

W. F. Hosford *(University of Michigan)*

Are you using the bulge test to measure biaxial strain to determine the X value?

Miyauchi

Yes.

PLASTICITY ANALYSIS OF
SHEET METAL FORMING

P. B. MELLOR and A. PARMAR

University of Bradford, West Yorkshire, United Kingdom

ABSTRACT

The paper surveys the application of plasticity theory to sheet metal forming problems and attempts to highlight those areas where knowledge is lacking. From an engineering standpoint the most valuable analyses would seem to be those which predict trends in deformation behaviour. It is only rarely that the material properties and lubrication characteristics are known with sufficient accuracy to make correlations between theory and practice worthwhile in a particular case. There would, however, seem to be definite advantages in the same research team following both theoretical and experimental paths.

The processes considered are deep-drawing (with a flat-headed punch), stretch forming (by hydraulic pressure, with a hemispherical punch, and in-plane stretching) and general sheet pressing. The easiest theoretical problems to solve are those where the deformation is stable. The most interesting and difficult problems deal with failure which can take the form of wrinkling and puckering, tensile instability, limit strain or final fracture.

In engineering the load required for a given process is usually determined with sufficient accuracy from a simple approximate analysis and in carrying out a more detailed analysis we are primarily interested in the prediction of the strain distributions and in their correlation with experimental measurements of strain. This is one method of making certain that an appropriate yield criterion and its associated flow rule have been selected. It is impossible to ignore anisotropy when discussing sheet metal forming. The theory of orthotropic anisotropy describes the deformation of deep-drawing steels quite well but is unsatisfactory for most other materials. In particular when the r-value is less than unity we must look at other yield criteria and flow rules to describe the plastic flow of such materials.

INTRODUCTION

It is pertinent to enquire at the outset what impact the theory of macroscopic plasticity has had on practical sheet metal forming. So far as direct application to industrial processes is concerned the answer must surely be none, and it does not appear likely that it will have direct application in the future. There is no doubt that suitable mathematical and numerical techniques, and computers, are available and that they can be used to solve problems in which the constitutive equations for thin sheet metals will become more and more complex. The real difficulty lies in the sheer amount of experimental work that is required to measure the variables which have to be fed into the theoretical equations in order to give good correlation between theory and practice. The variables include not only the mechanical and geometrical properties of the sheet metal, but also the elastic properties of the tooling and, most difficult of all, the state of lubrication between tooling and workpiece. The production of a component from sheet metal is a unique event and it is too sanguine to hope that the values of all the relevant variables of this unique event can be known.

The above is not a plea to abandon all attempts at obtaining a basic understanding of the plastic forming of metals. Indeed, with the increasing complexity of sheet metal forming processes, there is a need to improve our knowledge of the basic mechanics of such processes and it will be necessary to check theoretical predictions against carefully controlled experiments. However, the most important contribution that the theory of plasticity can make to industrial processes is to provide a framework for the rational discussion of such processes and to predict the effect of variables on the trends in loads and strain distributions.

Before the age of the digital computer the researcher intent on applying the theory of plasticity to metal forming problems had necessarily to make approximations in order to obtain a solution. This disciplined him to seek a critical understanding of the process with which he was dealing. With the advent of the modern computer there is less need to make approximations and there has been a tendency for the experimentalist and theoretician to go their own ways, to the detriment of the subject as a whole. It is sometimes apparent that the author of a theoretical paper has never observed the forming process he is attempting to analyse and equally it is sometimes evident that an experimental programme would have been arranged differently if the researcher had had a better appreciation of the theory of plasticity. The issues dealt with in sheet metal forming are complex and a degree of specialization is inevitable, but there does seem to be a need for more direct co-operation between specialists in metallurgy, mathematics and engineering, even within the same institution.

In the following a review is given of the applications of the theory of plasticity to deep-drawing and to stretch-forming. An attempt will be made to assess the situation as it stands before the start of this symposium. Areas where further basic information seems to be required will be identified.

DEEP-DRAWING AND PRESSING

The deep-drawing of a circular blank into a cylindrical shell has been studied many times. In deep-drawing, Fig. 1, the outer annular zone, X, is held under normal compression by the blank-holder to prevent buckling under the circumferential compressive stresses which are induced in this region as the punch

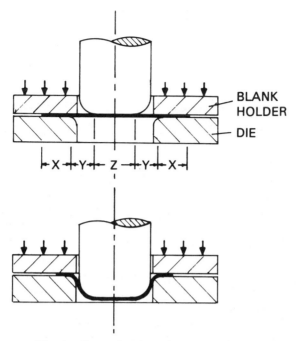

Fig. 1.   Deep-drawing of a cylindrical cup.

advances to pull the blank radially inwards towards the die opening. As the material from region X passes over the die it is subjected to bending and un-bending under tension. The material initially in region Y is bent and stretched either over the die profile or over the punch profile. Finally the material which is initially over the flat-face of the punch, region Z, is stretched under biaxial tensile forces. Fracture, when it occurs, is normally in material which is over the punch head profile radius. The limiting drawing ratio is defined as the ratio of the diameter of the largest blank which can be drawn successfully to the diameter of the punch. The maximum limiting drawing ratio that can be achieved in conventional drawing is approximately 2.2 and this will produce a cup of height approximately equal to the diameter of the punch.

The first phase of research work on deep-drawing can be thought of as starting with the work of Sommer [1] in 1925 and culminating in the work of Chung and Swift [11] in 1951. This early work has been summarized by Alexander [12]. In this period other significant contributions were made by Eksergian [2, 3], Siebel and Pomp [4], Sachs [5], Fukui [6], Swift [7, 8], Oehler [9], and Hill [10]. Hill analysed the pure radial drawing of an annular flange assuming a rigid plastic material. He considered two cases, plane strain (no change in the thickness of the blank) and the more realistic case of plane stress (where it was assumed that the mean pressure exerted on the blank by the blank-holder was small compared with the current yield stress of the material). Frictional forces were assumed negligible and the material was taken to be isotropic. The Tresca yield criterion was assumed together with the Lévy-Mises flow rule. Although this was known to be theoretically incorrect it was a device often adopted in pre-computer days to obtain strain distributions which were not inconsistent with expected boundary conditions and with experiment. Hill showed that the generalized strain could with little error be taken to be equal to the circumferential strain in the flange. He found that the positions of the particles in the deforming flange showed little dependence on the stress-strain properties of the material, but that there were greater differences in the thickness with the general tendency of work-hardening to reduce the variation in thickness across the annulus at any moment. Chung and Swift [11] used the same yield criterion and flow rule in their solution, which was extended to take in friction at the rim of the blank, and bending and un-bending under tension at the die. Their assumptions were justified by detailed comparison with their experimental results. This work led to a good understanding of radial drawing and to an appreciation of the thinning of the material as it passes over the die. They were not able to extend the solution to material being deformed over the punch head and they did not attempt to predict the limiting drawing ratio. They were however able to construct the punch load/punch travel diagram on the reasonable assumption that the contribution of the punch stretching to the total load was negligible.

Although the above theoretical work assumed that the material was isotropic it was well known that sheet material in particular could be strongly anisotropic. Lankford, Snyder and Bauscher [13] recognized the importance of the variation of r-value with orientation in the plane of the sheet for commercial low-carbon steel and indicated how this variation could be exploited in unsymmetrical pressings. It was this paper in particular which sparked off widespread interest in applications to metal forming.

Various theories of anisotropic yielding were developed from 1939 onwards and given open publication in the late nineteen forties. The theories which have proved most useful in engineering applications are based on modifications of the von Mises yield criterion and its associated flow rule. The theories developed independently by Jackson, Smith and Lankford [14], Hill [15], and Dorn [16], have several points in common. The theory proposed by Hill has the advantage of rigor within the basic assumptions adopted and it is perhaps the simplest to understand, with the consequence that it is most generally quoted today. The anisotropy is regarded as being due to preferred orientation and is assumed to have three mutually orthogonal planes of symmetry at every point. The lines of intersection of the planes are termed the axes of anisotropy. It is further assumed that the anisotropy is uniformly distributed in magnitude and direction which would appear to be a reasonable approximation for cold rolled sheet. The axes of anisotropy in this case lie in the direction of rolling, transversely in the plane of the sheet and normal to this plane. The theory assumes that there is no Bauschinger effect and that hydrostatic stress does not influence yielding. Hence linear terms are not included and only differences between normal stress components appear in the yield criterion. Jackson, Smith and Lankford developed the theory for a work-hardening material and assumed that the degree of hardening is a function only of the total plastic work and is otherwise independent of the strain path. Again this was an extension of previous work for isotropic materials. Hill [10], has given expressions for generalized stress and strain increment.

To apply this theory to sheet metal forming it is necessary to carry out tensile tests on specimens cut in directions at 0°, 45° and 90° to the direction of rolling and to measure the r-values for each direction. It is implicit in the theory that the strain ratios do not vary with the amount of plastic strain. In comparing experiment and theory it is usual to adopt an empirical expression to the experimentally determined stress-strain curve. This often has the form $\bar{\sigma} = K\bar{\epsilon}^n$ where $\bar{\sigma}$ is the generalized stress, $\bar{\epsilon}$ the generalized strain, K a constant, and n the well known work-hardening index. The value of n is, according to this theory, independent of the applied stress system. It can be appreciated that the application of the theory to real metals is very circumscribed. Nevertheless the concepts of n-value and r-value have proved very useful in what might be termed the second phase of the application of plasticity theory to deep-drawing.

In 1960 Whiteley [17] published results which showed that preferred orientation is the most important material variable influencing the performance of ordinary ductile metals in cylindrical cup-drawing with a flat-headed punch. He showed that the limiting drawing ratio increased as the average r-value of the material increased. Other investigators, Fukui, Yoshida and Abe [18], and Warwick and Alexander [19], found low correlation between limiting drawing ratio and material properties. Experiments by Lilet and Wybo [20], provided results in qualitative agreement with those of Whiteley.

There have been several theoretical studies of the dependence of the limiting drawing ratio on the n-value and r-value. In most cases failure was assumed to occur by necking under plane strain tension where the punch profile radius joins the straight punch stem. The maximum loads in pure radial drawing were calculated for various n- and r-values and various drawing ratios. At the limiting drawing ratio the maximum radial drawing load is equal to the load necessary to cause necking in plane strain tension in material on the punch stem. This criterion was used by Yamada [21] and Moore and Wallace [22]. More detailed calculations were reported by Chiang and Kobayashi [23] and by Budiansky and Wang [24]. Similar computations were carried out by El-Sebaie and Mellor [25, 26] who used a method suggested by Yamada and Koide [27] to eliminate an indeterminacy at the beginning of the calculations resulting from the use of the empirical equation $\bar{\sigma} = K\bar{\epsilon}^n$. All these results are, of course, based on a partial solution to the problem. The solutions neglect any possible weakening effect due to stretching of the sheet over the punch-head, and the presence of a die ring and its effect on the deformation is ignored. Nevertheless, the theoretical results give a useful guide to the effect of n and r on the limiting drawing ratio when deep-drawing with a flat-headed punch. The broken lines in Fig. 2, [26], agree with experimental findings that the n-value has little effect on the limiting drawing ratio whereas increasing r-values give increasing limiting drawing ratios. The theoretical curves overestimate the limiting drawing ratios that might be expected in practice mainly because of two reasons. Firstly, frictional forces in the flange are neglected and secondly necking will only occur at the join of the punch profile radius with the

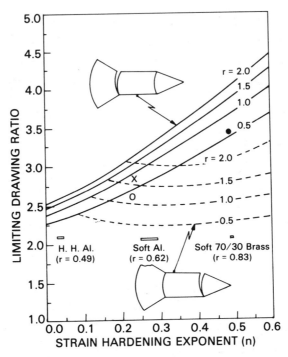

Fig. 2. Variation of theoretical limiting drawing ratios with *n*-values and comparison with experimental results.

Experimental results:
- ☐ Conventional deep-drawing (ref 25)
- ○ Soft aluminium, pressure assisted deep-drawing (ref 28)
- × Mild steel, pressure assisted deep-drawing (ref 29)
- ● Soft 70/30 brass, pressure assisted deep-drawing (ref 26)

punch stem if there are high frictional forces over the punch profile radius. This latter state can be approached if the blank is drawn into high pressure fluid. Given optimum conditions the necking in plane strain at the punch stem will be suppressed and failure will then take place by necking under simple tension in material emerging from the radial drawing region. Theoretical results for this criterion are shown by the full lines in Fig. 2, [26], where it is now clear that the *n*-value and to a lesser extent the *r*-value both have an important effect on the limiting drawing ratio. Experimental results by El-Sebaie and Mellor and by Y. Kasuga et al [28, 29] substantiate the trends of the theoretical results. In pure deep-drawing the effect of *r*-value is all important since this gives increased strength in plane strain tension relative to the radial drawing load. In general pressing both the *n*-value and *r*-value are of importance. Similar principles have been discussed by Keeler and Backofen for isotropic materials [30].

The Schuler Hydro-mechanical Process, Fig. 3, [31], illustrates superb engineering know-how in overcoming most of the problems of sheet pressing, albeit on a relatively small geometrical scale. Deep cylindrical, prismatic, tapered and conical shells can be made in one stage. The blank is drawn into the pressure chamber where controlled fluid pressure holds the deforming blank tightly against the punch thus restricting plastic deformation in that area. The pressure also deforms the blank upwards between the punch and the die, reducing the risk of puckering. Finally, the upward deformation means that the flange is drawn across the top surface of the die but is not bent over it. Limiting drawing ratios of 2.7 and over are common and since there is much less thinning in the final component compared with conventional deep-drawing it is possible to use thinner gauge materials. The Hydro-form Process, where the oil pressure is contained by a rubber diaphragm, also enjoys most of the same advantages. These particular processes are mentioned to emphasize the fact that when certain types of failure have been recognized they can sometimes be eliminated.

Relatively little work has been done on failure by wrinkling in the flange and even less on failure by puckering of unsupported material between punch and die. Geckeler [32] and Senior [33], have proposed theoretical relations for wrinkling in the flange. Senior concluded that while the critical flange dimensions are dependent on the material properties, there is little difference between the normally used work-hardening materials and that the critical diameter is affected to a far greater extent by the geometry of the drawing tools and the material thickness. Naziri and Pearce [34], concluded from experiments that a high r-value

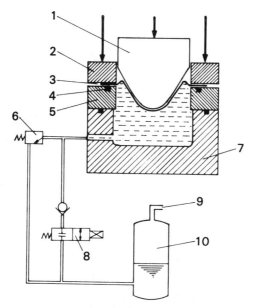

Fig. 3.   The Schuler Hydro-mechanical Process.
1. Punch, 2. Blankholder, 3. Partially formed pressing,
4. Plastic seal, 5. Drawing ring, 6. Pressure control valve,
7. Water chamber, 8. Solenoid operated control valve,
9. Compressed air 80 lbf/in², 10. Reservoir.

was beneficial for resisting wrinkling. With most pressings puckering is more of a problem than flange wrinkling and the blankholding force to suppress puckering is higher than that necessary to suppress wrinkling. In large pressings draw-beads are used to restrain the flow of the material into the die. In shallow pressings it is necessary to stretch the material to ensure that all the material has yielded and thus reduce the effect of elastic spring back. High strength materials, of relatively low ductility, are therefore particularly difficult to deal with. Etheride and Marbey [35], have paid particular attention to the effect of drawing-in on the failure of a pressing.

There have been fewer attempts to apply plasticity theory to deep-drawing with a hemispherical headed punch. Woo [36] has given a numerical solution for a material exhibiting normal anisotropy. He obtained the stress and strain distribution disregarding die profile effects. More recently Wifi [37] has used an

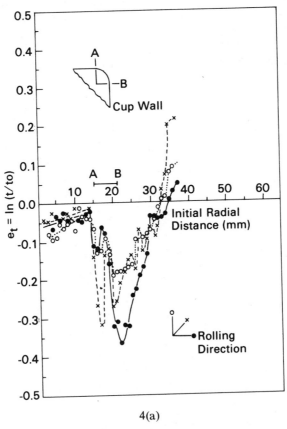

4(a)

Fig. 4.   Measured thickness strains in deep-drawn cups at 0°, 45° and 90° to direction of rolling.
(a) Soft aluminum; $r_0 = 0.97$, $r_{45} = 0.41$, $r_{90} = 0.68$.
(b) Soft 70/30 brass; $r_0 = 0.86$, $r_{45} = 0.85$, $r_{90} = 0.78$.

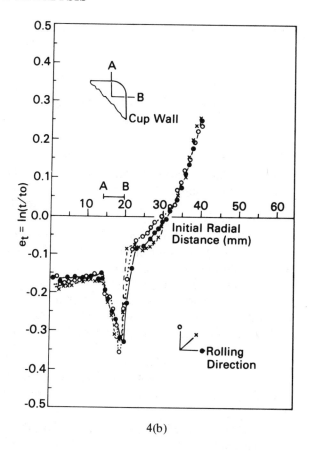

4(b)

incremental variational method to analyze axisymmetric elastic plastic solids at large strains including deep-drawing with a hemispherical headed punch. He has allowed for the die profile but in his example he has assumed the material to be isotropic. He stated that such problems are computationally time consuming.

In practice planar anisotropy leads to earing in the flange and to nonuniformity in the thickness of the finished component. Lee and Kobayshi [38] have used a matrix method of analysis for rigid-plastic materials to solve the problems of plane-stress bore expanding and flange drawing. The analysis was performed for two materials; one in which the anisotropy exists only in the direction of sheet thickness (normal anisotropy) and the other possessing the two mutually perpendicular axes of anisotropy in the plane of the sheet with the other axis in the direction of the thickness (planar anisotropy, or orthotropy). The results for planar anisotropy showed that in bore expanding the bore remained closely round even after considerable expansion but the thickness strains near the bore varied with orientation. In flange drawing the method enabled an analysis of ear formation to be made. The authors concluded that the nonuniform thickness distributions, caused by planar anisotropy, appear to have important implications for the occurrence of instability, such as necking in bore expanding and wrinkling in flange drawing. The importance of planar anisotropy on thickness distributions in a cylindrical cup drawn with a flat-headed punch is clear from the experimental results of Fig. 4 [25]. It is evident that the planar anisotropy of this particular aluminium sheet is much greater than for the brass and this is reflected in the thickness strain distributions of the drawn cup.

PURE STRETCH FORMING

The deformation of a circular metal diaphragm by a uniform lateral pressure has conceivably attracted more attention from researchers than any other problem in the plastic deformation of sheet, but it seems correct to state that there is as yet no reliable theoretical guide to the instability strain for this process. Interest in this process is now really centered on its use as a method of determining the work-hardening characteristic up to larger plastic strains than can be achieved in simple tension. An early solution by Gleyzal [39] for small strains was based on the total-strain theory of plasticity. Hill [40] developed a more general solution for small strains based on the Mises theory, but his method of successive approximations is only valid for sufficiently work-hardened materials. A special solution by Hill based on the assumption that particles in the deforming diaphragm at each stage move normally to the momentary profile of the bulge is of greater practical interest. Various numerical solutions, based on the Hencky theory, have been obtained by Weil and Newmark [41], Woo [42] and Storåkers [43]. Ross and Prager [44] have solved the problem using the Tresca yield criterion and the associated flow rule with the assumption that the thickness of the diaphragm is uniform at each stage. Chakrabarty and Alexander [45] used the Tresca yield criterion and the associated flow rule together with the assumption that the diaphragm has a spherical shape. The results of Hill and of Ross and Prager are contained as special solutions. Iseki, Jimma and Murota [46] have developed a large displacement, elastic-plastic finite element analysis for the hydrostatic bulging of axisymmetric and non-axisymmetric metal diaphragms. The solution was found to be in good agreement with experimental results for a circular diaphragm. Yamada and Yokouchi [47, 48] have given solutions for rigid-plastic and elastic-plastic materials. The rigid plastic solution was for the bulging of a circular diaphragm assuming normal anisotropy. Wang and Shammamy [49] have given the most comprehensive set of results for a material exhibiting normal anisotropy.

The variations of some theoretical instability thickness strains with the n-value are compared in Fig. 5 with some experimental instability strains obtained by Mellor [50]. The theoretical instability strains given by Wang and Shammamy appear to overestimate the actual instability strain achieved in practice. Although the anisotropic parameters were not measured in the experimental work it is now obvious that all the materials were anisotropic.

The stretch-forming of a circular blank by a hemispherical punch is a similar problem. In this case however the geometry of the sheet metal is known once it is in contact with the punch. Perhaps the greatest difficulty here, as in deep-drawing, is to characterize the frictional conditions in a realistic manner. Theo-

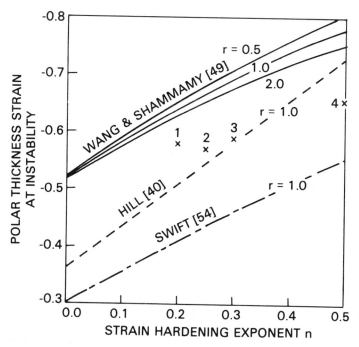

Fig. 5.   Theoretical predictions of polar thickness strains at instability for an isotropic material.

   Experimental values:
   1. Killed steel
   2. Soft aluminum
   3. Soft copper
   4. Softened stainless steel

retical solutions for isotropic sheet have been obtained by Woo [51], Chakrabarty [52] and Wifi [37]. Stretch-forming of anisotropic sheet has been considered by Wang [53].

The end of useful straining in a stretch-forming operation is difficult to define. Certainly with annealed material a maximum in the fluid pressure in the diaphragm test may signal the instability condition but with most materials further useful straining is obtained under falling pressure. The general instability conditions were discussed by Swift [54] for isotropic materials and extended to anisotropic materials by Moore and Wallace [22]. Dillamore, Mella and Hazel [55] pointed out an error in ref [22] which had resulted in a prediction that the r-value had an appreciable effect on the instability strain. In fact the instability strain shows only a small dependence on the r-value. Interest in the end of useful straining was stimulated by Keeler's definition of a limit strain [56, 57]. Since Keeler's original work there have been many experimental investigations to obtain forming limit diagrams. Here, we wish to restrict discussion to the attempts that have been made to understand the mechanics of the plastic deformation leading to the appearance of a local neck. Marciniak and Kuczyński [58, 59],

References pp. 72-74.

have devised an ingenious method for the experimental study of frictionless in-plane stretching and they also gave a theoretical analysis for the growth of an existing inhomogeneity into a localized neck. The analysis and the type of initial inhomogeneity have been discussed by Sowerby and Duncan [60], and by Minh, Sowerby and Duncan [61]. Azrin and Backofen [62] designed a different in-plane stretching test and as a result of experiments they reached the conclusion that the size of any initial inhomogeneity existing in commercial sheet was much smaller than that required by the theory of Marciniak and Kuczyński. Azrin and Backofen assumed that the initial inhomogeneity took the form of local thinning; Minh, Sowerby and Duncan [63] have suggested that the inhomogeneity can take the form of internal defects. The assumption that deformation in stretch-forming is divided into a stable phase and a diffuse necking phase, and that it is during this latter phase that a localized neck develops is supported by the detailed observations of strain development by Painter and Pearce [64]. Tadros and Mellor [65] have suggested a modification to the original Marciniak and Kuczyński analysis to allow for uniform plastic deformation up to instability. This was in response to the fact that the measured limit strain in plane strain tension is always equal to or greater than the instability strain in plane strain tension whereas the Marciniak and Kuczynski analysis predicts a lower value. There is a real need for a detailed experimental investigation to establish whether or not the proposed inhomogeneity exists. Yamaguchi and Mellor [66] and Parmar, Mellor and Chak-rabarty [67] have put forward models based on the increase in surface roughness with plastic strain. It is however clear from the experimental work of Tadros [68] that the surface roughness is superimposed on the developing neck. In other words the final local neck does not develop from a trough of surface roughness.

The Marciniak and Kuczyński analysis assumes orthotropic behaviour and predicts that for a constant n-value the limit strain increases as the r-value decreases. This is contrary to the experimental findings of Woodthorpe and Pearce [69].

The effect of strain-rate sensitivity is considered in ref [59]. This increases the value of the limit strain and influences the shape of the forming limit curve. However the theory still underestimates the limit strain in plane strain tension.

STATUS OF THE ANISOTROPIC THEORY

The deformation of real materials is extremely complex and it is hardly sur-prising that the simple macroscopic theories used for isotropic and orthotropic behaviour are inadequate. The nature of preferred orientation has been reviewed by Dillamore and Roberts [70] and the possibility of texture strengthening and the relation to the orthotropic theory have been discussed by Backofen et al [71], Hosford and Backofen [72] and Hosford [73]. Wilson [74, 75] has discussed the plastic anisotropy in sheet metals and Sowerby and Johnson [76] have recently given a review of texture and anisotropy in relation to metal forming.

Only detailed experimental work will confirm whether or not a particular macroscopic theory is satisfactory. Bramley and Mellor [77] have carried out experiments in simple tension and balanced biaxial tension to check the theory of orthotropic anisotropy for four killed steels from different rolling mills. The r-values were measured from tensile specimens cut at every 10 degrees to the rolling direction. The r-value for a particular orientation was found to be constant up to the maximum measured strain of 0.2. Using the work-hardening characteristics obtained in simple tension, the theory of orthotropic anisotropy was used to predict the stress-strain curve in balanced biaxial tension. This was then compared with an experimental curve obtained from the diaphragm test. The correlation for one of the killed steels is shown in Fig. 6. The theory predicts correctly that the biaxial curve will be higher than the uniaxial curve and the correlation is considered to be good. Average r-values and an average uniaxial stress-strain curve were used in the correlation. If σ_b is the polar stress and ϵ the thickness strain then

$$\sigma_b = \left[\frac{1 + r}{2} \right]^{1/2} \bar{\sigma}, \qquad \epsilon = \left[\frac{2}{1 + r} \right]^{1/2} \bar{\epsilon}, \tag{1}$$

where r is the average r-value, $\bar{\sigma}$ is the average uniaxial stress and $\bar{\epsilon}$ the average uniaxial strain.

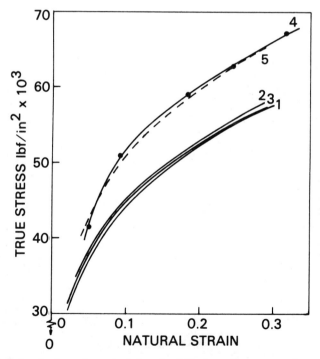

Fig. 6. Work-hardening characteristics for killed steel: measured average r-value is 1.42. Curves 1, 2, 3 are experimental curves in simple tension at 0°, 45° and 90° to rolling direction respectively. Curve 4 is experimental curve for balanced biaxial tension. Curve 5 is balanced biaxial curve predicted from average r-value and corresponding work-hardening characteristic.

References pp. 72-74.

Pearce [78] carried out similar experiments and made the same type of corre-lation for several materials including an annealed rim-steel having an average r-value of 0.38, see Fig. 7. It is seen that the experimental curve in biaxial tension is above the uniaxial curves whereas the predicted curve falls well below the uniaxial curves. The difference between the two curves is far too great to be explained by experimental error, although the small diameter (112 mm) of the diaphragm testing machine might have given rise to a somewhat higher biaxial curve than if a larger machine had been used, see the work of Horta, Roberts and Wilson [79]. It is now clear that the theory of orthotropic plasticity does not explain the behaviour in biaxial tension where the r-value is less than unity. Dillamore [80] arguing from crystal plasticity has concluded that the orthotropic theory is only likely to give reasonable correlation for r-values between 1 and 2.

Bramley and Mellor [81] also examined the behaviour of titanium 115 in bal-anced biaxial tension. This material had an average r-value of 2.85. The predicted balanced biaxial curve gave reasonable correlation despite the fact that the n-value in simple tension showed large variations with orientation. Rogers and Roberts [82] made a crystallographic study of the plastic deformation of the same material and found that twinning occurred to a much greater extent under bal-anced biaxial tension.

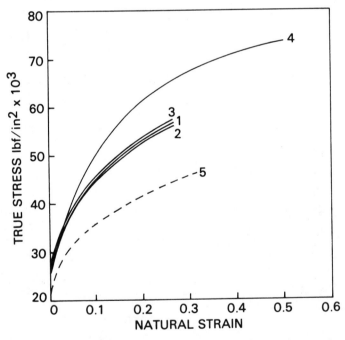

Fig. 7. Work-hardening characteristics for annealed rim-steel: measured average r-value is 0.38. Curves 1, 2, 3 are experimental curves in simple tension at 0°, 45° and 90° to rolling direction respectively. Curve 4 is experimental curve for balanced biaxial tension. Curve 5 is balanced biaxial curve predicted from average r-value and corresponding work-hardening characteristic.

Yoshida et al [83, 84] also found poor correlation between simple tension and balanced biaxial tension and suggested the inclusion of a new parameter called the X-value which is the ratio of the measured stress in balanced biaxial tension to the stress in simple tension.

A NEW YIELD CRITERION

Hill [85] has proposed a new yield criterion, which might accommodate the results for balanced biaxial tension, and which is of particular interest for those materials having r-values less than unity. The criterion for plane stress, assuming that there is no planar anisotropy, is,

$$(1 + 2r)|\sigma_1 - \sigma_2|^m + |\sigma_1 + \sigma_2|^m = 2(1 + r)Y^m \qquad (2)$$

where Y is the yield stress in uniaxial tension, r is the usual strain-ratio measured in uniaxial tension and m is a new parameter which must also be determined experimentally. For the yield locus to be convex $m \geq 1$. When $m < 2$ the locus is elongated in the direction of balanced biaxial tension. The effect of the parameter m is shown in Fig. 8. $m = 2$ gives Hill's original yield criterion [15].

The value of m for a particular material can be determined by comparing the experimental work-hardening characteristics for uniaxial tension and balanced biaxial tension. To predict the balanced biaxial curve from the uniaxial results requires the definition of a generalized stress and a generalized strain increment. We are proposing here that the generalized strain increment be defined using the principle of equivalence of plastic work. This follows naturally from previous work on isotropic and anisotropic materials and of course will lead to the same restrictions on the use of the theory. From the yield criterion the generalized stress $\bar{\sigma}$ is defined by,

$$\bar{\sigma} = \left[\frac{1}{2(1 + r)} \{(1 + 2r)|\sigma_1 - \sigma_2|^m + |\sigma_1 + \sigma_2|^m\} \right]^{1/m} \qquad (3)$$

and from the assumption of equivalence of plastic work it follows that the generalized strain increment,

$$d\bar{\epsilon} = \frac{[2(1 + r)]^{1/m}}{2} \left\{ \frac{1}{(1 + 2r)^{1/(m-1)}} |d\epsilon_1 - d\epsilon_2|^{m/(m-1)} + |d\epsilon_1 + d\epsilon_2|^{m/(m-1)} \right\}^{(m-1)/m} \qquad (4)$$

Following previous work [77] it is reasonable in the diaphragm test to assume rotational symmetry about an axis normal to the sheet. Hence, using the above equation it is possible, assuming an average r-value and an average stress-strain curve in uniaxial tension, to predict the variation of polar stress with polar thickness strain in the diaphragm test. Thus putting $\sigma_1 = \sigma_2 = \sigma_b$ in equation (3) the polar stress σ_b is given by,

$$\sigma_b = \frac{[2(1 + r)]^{1/m}}{2} \bar{\sigma} \qquad (5)$$

where r is the average r-value and $\bar{\sigma}$ is the average stress in uniaxial tension. Similarly, putting $d\epsilon_1 = d\epsilon_2 = \dfrac{d\epsilon}{2}$ where $d\epsilon$ is the thickness strain increment at the pole,

$$d\epsilon = \frac{2}{[2(1+r)]^{1/m}} \, d\bar{\epsilon} \qquad (6)$$

and since there is no change in the strain ratio at the pole throughout the deformation,

$$\epsilon = \frac{2}{[2(1+r)]^{1/m}} \, \bar{\epsilon} \qquad (7)$$

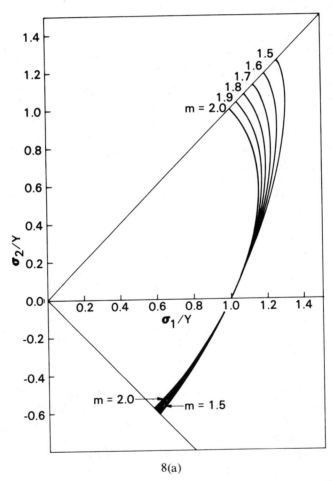

8(a)

Fig. 8.   Yield loci based on new anisotropic yield criterion proposed by Hill.
(a) $r = 1.0$
(b) $r = 0.5$

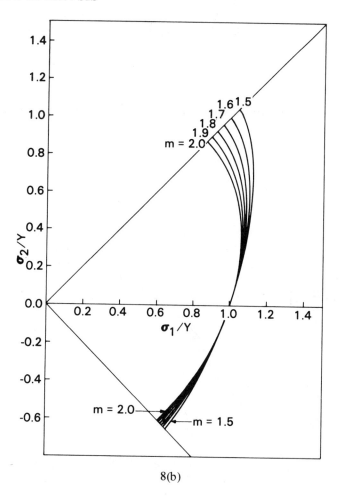

8(b)

Fig. 9 shows experimental results for uniaxial and balanced biaxial tension obtained by Taghvaipour [86]. For the steel in Fig. 9(a) the average r-value was 0.44 (this is designated Steel F in Taghvaipour's thesis). Despite the low r-value it was found that the balanced biaxial stress-strain curve fell slightly above the average of the uniaxial stress-strain curves. The original anisotropic theory, m = 2, predicts that the balanced biaxial curve falls well below the uniaxial curve, Fig. 9(a). By taking different m-values and using the above equations it has been found that the best correlation between theory and experiment is achieved when $m = 1.5$. For soft aluminium, having an average r-value of 0.723, similar reasoning gives an m-value of 1.8, Fig. 9(b).

The m-value has its greatest effect for balanced biaxial tension and it is of interest to note that Taghvaipour [86, 87] found reasonable correlation between simple tension and plane strain compression results on the basis of the original anisotropic theory. It is clear, however, that the original anisotropic theory leads

References pp. 72-74.

to erroneous results, especially for low r-value materials deformed under biaxial tension and there is a need to reconsider the effect of r-value on the theoretical prediction of limit strains. It will also be necessary to consider the effect of the new proposal on the limiting drawing ratio in the deep-drawing of a flat-ended cup, although the effect is likely to be less in this case because the deformation being considered goes from simple compression to plane strain tension.

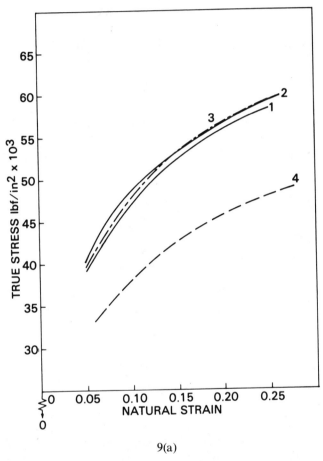

9(a)

Fig. 9. Estimation of m-values from work-hardening characteristics in simple and biaxial tension.
   (a) Steel 'F', $r_{av} = 0.44$
   (b) Soft Aluminum, $r_{av} = 0.72$
   Curve 1: Experimental curve, simple tension, based on average of curves along $0°$, $45°$ and $90°$ to direction of rolling.
   Curve 2: Experimental curve, diaphragm test.
   Curve 3: Balanced biaxial tension curve predicted from Curve 1, based on equations (5) and (7). For Steel F, $m = 1.5$, for Soft Aluminum, $m = 1.8$.
   Curve 4: Balanced biaxial tension curve predicted from Curve 1, based on average r-value and equation (1).

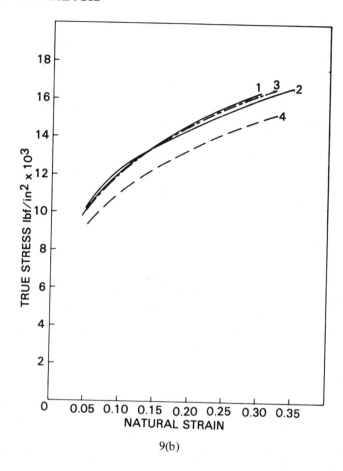

9(b)

It is difficult to obtain a full yield locus for sheet metal and therefore the experimental methods of Lee and Backofen [88] and of Tozawa, Nakamura and Shinkai [89] are of great interest. In determining an experimental yield locus at large plastic strains it must be remembered that any point on the locus should be at the same generalized strain. Since the generalized strain for a real material is unknown it is necessary either to make an assumption of equivalence of plastic work or to stretch a material up to a known plastic strain and then determine the re-yielding of that material under a number of stress systems.

CONCLUSIONS

The plastic deformation of metals is a complex subject and it is not to be expected that such deformation can be fully described by a simple yield locus and its associated flow rule. Any constitutive equations that are used in a metal forming problem will be the result of compromise and they may be applicable only for a certain stress system and only following a certain strain path. It is necessary to check the validity of any theory by careful experiments, paying particular attention to the correlation between theoretical and experimental strain distributions. It is easier to determine a yield locus that accounts for the Bauschinger effect and the different behaviour of a metal under tensile and compressive conditions than it is to incorporate such basic information into an analysis of a sheet metal forming process.

The macroscopic theory of plasticity as applied to sheet metals started out by describing isotropic behaviour. The work-hardening characteristic was characterized by the n-value. The introduction of anisotropy involved the measurement of r-value and it is now being suggested that there is a need to measure yet another parameter, the m-value. At the moment this necessitates the use of an ancillary diaphragm test. If the diaphragm test is adopted for this purpose there will be a need to standardize the geometry of the test and the test procedure.

REFERENCES

[1] M. H. Sommer, Maschinenbautechnik, 4 (1925), 1171.
[2] G. L. Eksergian, Trans. ASME, 48 (1926), 609.
[3] G. L. Eksergian, Metal Industry, 23 (1927), 483.
[4] E. Siebel and A. Pomp, Mitt. K. Wilh. Inst., Eisenforsch, 11 (1929), 139.
[5] G. Sachs, Proc. Inst. Automobile Eng., 29 (1934–35), 588.
[6] S. Fukui, Sci. Papers No. 849 and 885, Institute of Physical and Chemical Research, Tokyo, (1938–39), 1432 and 373.
[7] H. W. Swift, Proc. Inst. Automobile Eng., 34 (1939–40), 361.
[8] H. W. Swift, Engineering (Lond.), 166 (1948), 333 and 357.
[9] G. Oehler, Metallwistschaft, 16 (1939), 1059.
[10] R. Hill, The Mathematical Theory of Plasticity, Oxford Press (1950).
[11] S. Y. Chung and H. W. Swift, Proc. Inst. Mech. Eng., 68 (1951), 165.
[12] J. M. Alexander, Mat. Rev., 5 (1960), No. 19.
[13] W. T. Lankford, S. C. Snyder and J. A. Bauscher, Trans. ASM, 42 (1950), 1197.
[14] L. R. Jackson, K. F. Smith and W. T. Lankford, Met. Technol. Tech. Pub. No. 2440 (1948).
[15] R. Hill, Proc. Soc. Lond., A. 193 (1948), 281.
[16] J. E. Dorn, J. Appl. Phys., 20 (1949), 15.
[17] R. L. Whiteley, Trans. ASM, 52 (1960), 154.
[18] S. Fukui, K. Yoshida and K. Abe, Sci. Papers, Institute of Physical and Chemical Research, Tokyo, 54 (1960), 199.
[19] J. O. Warwick and J. M. Alexander, J. Inst. Met., 91 (1962–63), 1.
[20] L. Lilet and M. Wybo, Sheet Metal Ind., 41 (1964), 783.
[21] Y. Yamada, J. Japan Soc. Tech. Plasticity, 5 (1964), 183.
[22] G. G. Moore and J. F. Wallace, J. Inst. Met., 93 (1964), 33.
[23] D. C. Chiang and S. Kobayashi, J. Engr. Ind. (Trans. ASME, B), 88 (1966), 443.
[24] B. Budiansky and N. M. Wang, J. Mech. Phys. Solids, 14 (1966), 357.
[25] M. G. El-Sebaie and P. B. Mellor, Int. J. Mech. Sci., 14 (1972), 535.
[26] M. G. El-Sebaie and P. B. Mellor, Int. J. Mech. Sci., 15 (1973), 485.
[27] Y. Yamada and M. Koide, Int. J. Mech. Sci., 10 (1968), 1.
[28] Y. Kasuga, N. Nokaki and K. Kondo, Bull. JSME, 4 (1961), 394.
[29] Y. Kasuga and S. Tsutsumi, Bull. JSME, 8 (1965), 120.
[30] S. P. Keeler and W. A. Backofen in discussion of Ref [17].
[31] E. Buerk, Sheet Metal Ind., 44 (1967), 182.
[32] J. W. Geckeler, Z. Angew. Math. Mech., 8 (1928), 341.
[33] B. W. Senior, J. Mech. Phys. Solids, 4 (1956), 235.
[34] H. Naziri and R. Pearce, Metal. Ital., 40 (1969), 727.

[35] R. A. Etheride and F. J. Marbey, Proc. 12th Int. Mach. Tool Design and Res. Conf., Manchester (1971).

[36] D. M. Woo, Int. J. Mech. Sci., *10* (1968), 83.

[37] A. S. Wifi, Int. J. Mech. Sci., *18* (1976), 23.

[38] S. H. Lee and I. Kobayashi, Proc. 15th Int. Mach. Tool Design and Res. Conf., Birmingham (1974).

[39] A. Gleyzal, J. Appl. Mech. (Trans. ASME, E), *70* (1948), 288.

[40] R. Hill, Philos. Mag, *41* (1950), 113.

[41] N. A. Weil and N. M. Newmark, J. Appl. Mech. (Trans ASME, E), *77* (1955), 533.

[42] D. M. Woo, Int. J. Mech. Sci., *6* (1964), 303.

[43] B. Storåkers, Int. J. Mech. Sci., *8* (1966), 619.

[44] E. W. Ross and W. Prager, Q. Appl. Math., *12* (1954), 86.

[45] J. Chakrabarty and J. M. Alexander, J. Strain Anal., *5* (1970), 155.

[46] H. Iseki, T. Jimma and T. Murota, Bull. JSME, *17* (1974), 1240.

[47] Y. Yamada and Y. Yokouchi, Seisan Kenkyu, *19* (1967), No. 12.

[48] Y. Yamada and Y. Yokouchi, Seisan Kenkyu, *21* (1965), No. 11.

[49] N. M. Wang and M. R. Shammamy, J. Mech. Phys. Solids, *17* (1969), 43.

[50] P. B. Mellor, J. Mech. Phys. Solids, *5* (1956), 41.

[51] D. M. Woo, The Engineer, Lond., *220* (1965), 876.

[52] J. Chakrabarty, Int. J. Mech. Sci., *12* (1970), 315.

[53] N. M. Wang, J. Appl. Mech. (Trans. ASME, E), *37* (1970), 431.

[54] H. W. Swift, J. Mech. Phys. Solids, *1* (1952), 1.

[55] I. L. Dillamore, P. Mella and R. J. Hazel, J. Inst. Met., *100* (1972), 50.

[56] S. P. Keeler and W. A. Backofen, Trans. ASM, *56* (1963), 25.

[57] S. P. Keeler, SAE, Paper No. 650535 (1965).

[58] Z. Marciniak and K. Kuczyński, Int. J. Mech. Sci., *9* (1967), 609.

[59] Z. Marciniak, K. Kuczyński and T. Pokora, Int. J. Mech. Sci., *15* (1973), 789.

[60] R. Sowerby and J. L. Duncan, Int. J. Mech. Sci., *13* (1971), 217.

[61] H. Van Minh, R. Sowerby and J. L. Duncan, Int. J. Mech. Sci., *16* (1974), 31.

[62] M. Azrin and W. A. Backofen, Metall. Trans., *1* (1970), 2857.

[63] H. Van Minh, R. Sowerby and J. L. Duncan, Int. J. Mech. Sci., *17* (1975), 339.

[64] M. J. Painter and R. Pearce, Proc. 7th Biennial Congress, IDDRG, Amsterdam (1972).

[65] A. K. Tadros and P. B. Mellor, Int. J. Mech. Sci., *17* (1975), 203.

[66] K. Yamaguchi and P. B. Mellor, Int. J. Mech. Sci., *18* (1976), 85.

[67] A. Parmar, P. B. Mellor, J. Chakrabarty, Int. J. Mech. Sci., *19* (1977), 579.

[68] A. K. Tadros, PhD Dissertation, University of Bradford (1976).

[69] J. Woodthorpe and R. Pearce, Sheet Metal Ind., *46* (1969), 1061.

[70] I. L. Dillamore and W. T. Roberts, Met. Rev., *10* (1965), 271.

[71] W. A. Backofen, W. F. Hosford and J. J. Burke, Trans. ASM, *55* (1962), 264.

[72] W. S. Hosford and W. A. Backofen, Proc. 9th Sagamore Conf., Raquette Lake, N.Y. (1962).

[73] W. F. Hosford, Met. Eng. Quart., ASM (1966), 13.

[74] D. V. Wilson, J. Inst. Metals, *94* (1966), 84.

[75] D. V. Wilson, Met. Rev., *139* (1969).

[76] R. Sowerby and W. Johnson, Mat. Sci. and Eng., *20* (1975), 101.

[77] A. N. Bramley and P. B. Mellor, Int. J. Mech. Sci., *8* (1966), 101.

[78] R. Pearce, Int. J. Mech. Sci., *10* (1968), 995.

[79] R. M. S. B. Horta, W. T. Roberts and D. V. Wilson, Int. J. Mech. Sci., *12* (1970), 231.

[80] I. L. Dillamore, J. Phys. D: Appl. Phys., *7* (1974), 979.

[81] A. N. Bramley and P. B. Mellor, Int. J. Mech. Sci., *10* (1968), 211.

[82] D. H. Rogers and W. T. Roberts, Int. J. Mech. Sci., *10* (1968), 221.

[83] K. Yoshida, Report, Institute of Physical and Chemical Research, Tokyo, *45* (1969), 157.

[84] K. Yoshida, K. Yoshii, H. Komorida, M. Usuda, H. Watanabe, Sci. Papers, Institute of Physical and Chemical Research, Tokyo, *64* (1970), 24.

[85] R. Hill, Private communication of 18 January 1977.

[86] M. Taghvaipour, Ph.D. Dissertation, University of Birmingham (1970).
[87] M. Taghvaipour and P. B. Mellor, Proc. Inst. Mech. Engrs., *185* (1970–71), 593.
[88] D. Lee and W. A. Backofen, Trans. Met. Soc. AIME, *236* (1966), 1077.
[89] Y. Tozawa, M. Nakamura and I. Shinkai, Trans. Iron Steel Inst. Jpn., *11* (1971), 936.

DISCUSSION

C. A. Magee *(Ford Motor Company)*

In order to use this new flow criterion, is it necessary to measure both r and m (in your notation)?

Mellor

Yes! This is an addition to what we've already got to do.

Magee

Would the X value actually be simpler for empirical studies?

Mellor

Here it depends on how you are going to use it. If you are going to use it, not in an analysis, but to help you understand, then the X value is O.K.

S. S. Hecker *(Los Alamos Scientific Laboratory)*

It is interesting that, just when Hill gave up on his yield criterion for r values less than unity, we found some pretty good confirmation for $r \approx o$. I did some biaxial tube testing of heavily cold worked aluminum in which the r value was indeed very close to zero. It was very interesting to find that the yield locus, instead of looking like the isotropic von Mises, became a circle so that there is biaxial softening instead of biaxial strengthening. These results are very close to what Hill would have predicted for $r \approx o$.

Mellor

This was a cold worked tube? So the strains were very small. Were you loading and unloading? How do you define a yield criterion? Were you following a constant-strain path or were there reversals?

Hecker

Actually the strains were determined by both loading and unloading techniques. We determined the very small plastic strain offset yield surface and just loaded along constant stress ratios until failure to determine the larger plastic strain offsets. In both cases, the yield loci were quite circular instead of elliptical.

Mellor

Well, I must say I defended the original concept of anisotropy for a long time and there are now crystallographic reasons (put forward by Dillamore) why the simple anisotropic theory will only be correct between r values of 1 and 2. I'm putting that forward as a statement, I don't quite understand his reasoning. But he is quite confident about that. You must be very careful how you measure stress from bulge test. You must measure the radius of curvature which is quite difficult to do accurately. It was only after we spent several months making measurements that our experimental results were to be trusted. We then found out that the yield criterion fitted our results perfectly. We also have stretched (in-plane stretching), under a balanced biaxial field, a sheet (aluminum, brass, steel) with a small hole in it. This represented a void. This now turned into a pure plasticity problem in which we can expand the hole using the Marciniak technique. We now get a strain range from simple tension at the hole to pure balanced biaxial tension. So we are going, in one test, from simple tension to plane strain tension to balanced biaxial tension. And Hill's new yield criterion fits all our results perfectly for all the common materials.

A. K. Ghosh (Rockwell International)

I have a comment on Hecker's comment on the r-value. For heavily cold-rolled aluminum, how is a good measurement of r possible? r is the *ratio* of plastic strains and the plastic strains are very close to zero.

Hecker

There is no problem with the actual measurement of r because one uses strain gages. There is no difficulty in making sure one has a precise measurement, but you certainly have difficulty as to how representative a small strain is to what we call the r-value. That difficulty is certainly there.

Mellor

Are you satisfied with the answer, Dr. Ghosh?

Ghosh

Well, maybe the measurements are accurate, but they include elastic plus plastic strains and at those strain levels—mostly elastic!

Mellor

I will add that we have measured the r-values in titanium from the elastic to the plastic and put forward a theory incorporating the elastic deformation. In titanium, we found that the r-value increased as you went from the elastic to the plastic. We went up to a strain of 0.05. When we cold-rolled some titanium, we found there was a very big drop in r-value, even when we rolled it only 10%. We have not published these results.

U. F. Kocks *(Argonne National Laboratories)*

You've made repeated reference in your talk to path dependence. It seems to me that when you have observed a path dependence there will be no end to the number of parameters you must introduce to describe behavior. It seems to me an indication that one must go away from history descriptions of the behavior of a strain hardening material to state parameter descriptions. Unfortunately, that means that the first parameter that everyone seems to be using, (namely, the n-value) is the exponent of strain which is not a state parameter. So it seems to me that if one wanted to have any hope of having fewer parameters, of finding the behavior over a large range of applications, one has to start from scratch. Sorry about that!

Mellor

Well, if anyone comes up with the equations, I'll try to fit them to my experimental results.

S. Nemat-Nasser *(Northwestern University)*

I ask for your comments about reduced necking leading to bulk-wise necking. Has there been any experimental work prior to the localization (or with a machined diffuse neck)? Were tests pursued to understand the mechanisms which might lead to localization?

Mellor

The growth of a neck (that is, neck formation) is under active investigation in a lot of areas. I might just say that the strain-rate effect was mentioned and, while I realize the importance in superplastic alloys, I am not convinced yet, that it has a big effect for common materials on the timing of the neck formation. Now, Marciniak and Kuczyński took this into account. They were using what is now called the old theory of anisotropy and that predicts that if you have a very low r value, you get a very large strain. Now we all know that the r value doesn't have any effect on the limit strain and this new theory of Hill's does put that into perspective. In fact, the Hill's m value brings the strain down again. And, if you take a piece of soft aluminum, soft brass, or deep-drawing steel, the limit strain is almost identical in all three.

M. A. Bragard *(Centre de Recherche Metallurgiques, Belgium)*

How do you measure Hill's *m*-value?

Mellor

At the present time the most obvious method of getting Hill's *m*-value is by direct comparison of the biaxial curve, from the bulge test, with your simple tensile test results. From the research point-of-view, though, our experiments with the expansion of a hole combined with a theoretical analysis (which we have) allows us to determine an *m*-value. It is like an independent check which we use to predict what the bulge test should be. And it is giving reasonable results at the moment. But, as I say, any of these simple theories can only be applied to certain materials and then only over a restricted range of strains and strain paths.

H. A. Armen *(Grumman Aerospace Corporation)*

Do you know of any information that describes the size, shape and orientation of the newly defined yield surface when the material undergoes subsequent proportional or non-proportional loading beyond initial yielding?

Mellor

I do not know of any reliable information on yield surfaces subsequent to initial yielding, even for the simpler case of proportional loading. The yield surface should give values of stress at the same generalized strain and of course for a real material the generalized strain at this stage is unknown. For sheet metals, I would suggest that the only way of obtaining reliable information is to strain the material under biaxial tension in its plane to different strain levels and then determine the "new" initial yield locus at each strain level. Then at least the yield locus represents the same condition of the material. I will not comment on the difficulties of obtaining the initial yield surface.

SESSION II

CONSTITUTIVE RELATIONS
FOR SHEET METAL

Session Chairman
E.H. LEE

Stanford University
Stanford, California

PLASTIC DEFORMATION BEHAVIOR
UNDER CONDITIONS OF COMBINED STRESS

Y. TOZAWA

Nagoya University, Japan

ABSTRACT

Experiments are carried out by the biaxial compression method using a rectangular specimen prepared by stacking and gluing metal sheets. In order to represent the results systematically, a yield surface in the σ_X, σ_Y, σ_Z, $3\tau_{XY}$ stress space for large plastic strain is used. The work-hardening characteristics for both isotropic and anisotropic materials are discussed, and the rules for anisotropy induced by prestrain and for work-hardening caused by subsequent strain are determined. An equation which represents the yield surfaces for large subsequent strain of prestrained materials is proposed.

INTRODUCTION

The deformation behavior of sheet metal is usually expressed in terms of the characteristic values obtained by the tensile test, viz., yield point, tensile strength, elongation, n-value, r-value, etc. The tensile test is very basic and simple, and tensile values indicate to a certain degree the deformation characteristics of the sheet material. Although the tensile test is valuable for the measurement of mechanical properties of sheet metal, it offers information only under a very limited condition, i.e., monotonic uniaxial tension. The strain states in sheet metal during forming, on the other hand, are generated by combined stresses which follow necessarily from equilibrium of the internal stresses in the sheet and the external press loads. Such strain states are obviously not simple: the ratio of the principal strains at a material point is unlimited and need not remain constant during forming; moreover, the axes of the principal strains can rotate with respect to the sheet plane.

The deformation behavior under conditions of combined stress may be calculated from the measured tensile data by using theories of plasticity. Such calculations are, of course, meaningful only if the theory applied has been experimentally confirmed. To date there exists no theory which describes the deformation behavior of real materials with sufficient accuracy. Indeed, calculations based on an unsatisfactory theory may lead to wrong conclusions, causing misunderstanding and confusion.

The idea of characterizing formability of sheet metal for press forming on the basis of tensile data was predominant about ten years ago, but now that approach has been reconsidered and better approaches are being sought. An example of this trend is the X-value proposed by K. Yoshida [1].

To better understand the role of material in sheet metal forming, the deformation behavior of sheet metal under conditions of combined stress must be investigated. Results of such an investigation could be expected to explain the discrepancies in the existing plasticity theories, to establish the basis for a more satisfactory theory, and to provide guidelines for the development of sheet metal materials and the improvement of stamping technology.

SOME BEHAVIOR IN FORMING

Bending [2]—The directional dependence of the flow stresses in tension, compression and bending of 60/40 brass are shown in Fig. 1. From this figure, it

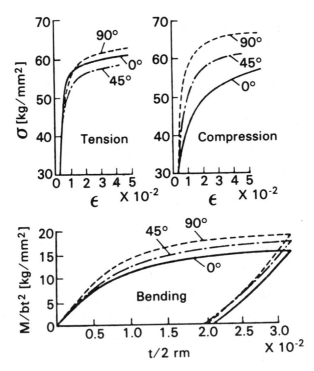

Fig. 1. Dependence of the deformation resistance in tension, compression and bending on the direction. (60/40 brass sheet 0.6mm in thickness prestrained by cold rolling to 40% reduction.)

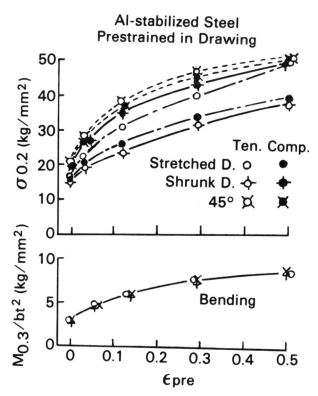

Fig. 2. Flow stresses in tension, compression and bending of aluminum-stabilized steel sheet, 0.8mm in thickness, prestrained by drawing to various strains ϵ_{PRE}.

is clear that the directional dependence in bending cannot be correlated with that in uniaxial tension. But, this phenomenon can be reconciled easily by taking into account the results of the compression test; since, in bending the inner region of the bent sheet is in compression.

Fig. 2 shows the flow stresses in tension, compression and bending of a predrawn, aluminum-stabilized, sheet steel. The flow stresses in bending are seen to be independent of the direction of the sheet specimen. The non-directionality in either 0° or 90° to the rolling direction is explainable by the fact that the flow stress in tension is high in the same direction as the flow stress in compression is low, and vice versa. At 45° to the rolling direction, however, the flow stresses are high in both tension and compression. This example shows that the uniaxial test results fail to correlate with those obtained in a slightly different deformation mode.

There is a further complication. During bending of a sheet with a large width-to-thickness ratio, the central portion of the width deforms under a plane strain condition. Thus, if the directional dependence of the flow stress under the plane

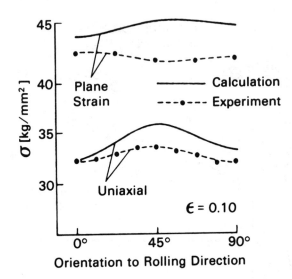

Fig. 3.   Planar anisotropy of the flow stress under the uniaxial tension and the plane strain condition. (Aluminum-stabilized steel sheet, 0.8mm in thickness, as-received.)

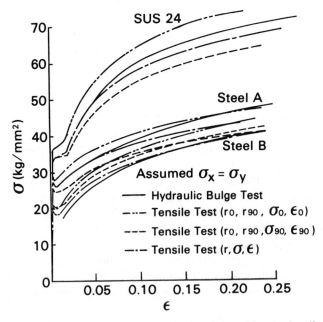

Fig. 4.   Through thickness stress-strain curves predicted from hydraulic bulge test and tensile test on the balanced biaxial stress assumption. (Stainless steel SUS24, 0.6mm thick; and low carbon steels A and B, 0.8mm thick.)

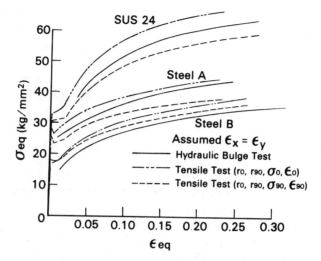

Fig. 5. Equivalent stress-strain curves predicted from hydraulic bulge test and tensile test on the balanced biaxial strain assumption.

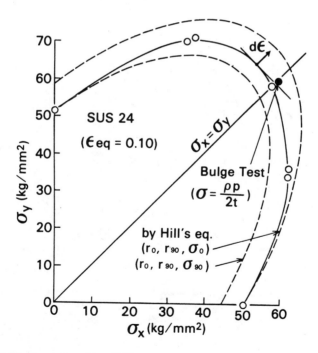

Fig. 6. Yield loci predicted by Hill's anisotropic theory compared with that obtained by experiment.

References pp. 108–109.

strain condition is different from that under the uniaxial stress condition, it will be even more difficult to correlate the bending results with the uniaxial behavior.

Fig. 3 shows a comparison of the planar anisotropy of the flow stress under the plane strain condition with that under uniaxial tension for as-received steel. It is seen that the variations of the experimental flow stresses with orientation to the rolling direction are different in the two cases. In the same figure are plotted the calculated curves based on Hill's anisotropic theory [3]. In making these calculations, the r-values at 0°, 45°, and 90° to the rolling direction and the uniaxial flow stress in the rolling direction were used. Comparing the calculated results with the experimental data shows clearly that the theory is in variance with experiment in the plane strain case.

Hydraulic Bulging [4]—A number of studies of deformation behavior of aniso- tropic sheet metals in biaxial tension have been made using the hydraulic bulge test to compare to Hill's anisotropic theory. They are divided into two groups on the basis of investigation procedure. Authors in one group, Bramley and Mellor [5][6], Pearce [7], Woodthorpe and Pearce [8] and Horta et al [9], have assumed that the stress state at the pole of the bulge is one of balanced biaxial tension,

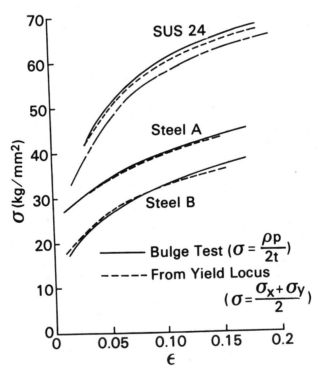

Fig. 7.   Interpretation of the stress-strain curves derived from measurements in hy- draulic bulge test. Chain line is uniaxial stress-strain curve obtained by through-thickness compressive test.

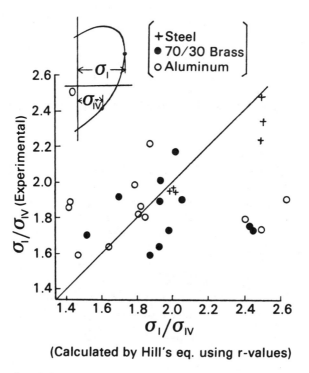

Fig. 8. The ratio of the stress components in two plane strain states, predicted by Hill's theory, compared with the experimental results.

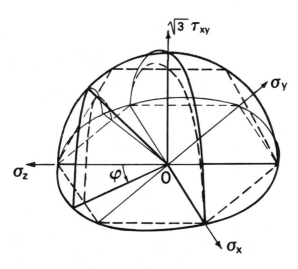

Fig. 9. Yield surfaces in σ_X, σ_Y, σ_Z, $\sqrt{3}\tau_{XY}$ stress space. The angle φ is a parameter representing the stress state.

i.e. $\sigma_X = \sigma_Y$, where σ_X and σ_Y are the stress components in the direction denoted by their subscripts, and X and Y represent the rolling- and transverse-directions. Kular and Hillier [10], in another group have assumed a state of balanced biaxial strain, i.e. $\epsilon_X = \epsilon_Y$, at the pole. The assumption of the balanced biaxial strain is supported by many experiments except one [11]. It is shown by Kular and Hillier [10] that the Hill's definition of equivalent stress and strain can be used to calculate the deformation behavior in biaxial tension using information from uniaxial tensile tests.

Figs. 4 and 5 show, respectively, the through-thickness stress-strain curves calculated on the basis of the balanced biaxial stress assumption and the equivalent stress-strain curves based on the balanced biaxial strain assumption. The apparently large discrepancies between the calculations and experimental data indicate that Hill's anisotropic theory fails to describe the real material behavior. As may be seen in Fig. 6, the calculated yield loci departs significantly from the experimental data.

The stress state at the pole of the bulge can be represented as a point on the yield locus using the rule of normality of the strain vector to the yield locus. The point obtained from the bulge test on the assumption of $\sigma_X = \sigma_Y$, naturally differs from the point determined by the strain state, but is near the tangent

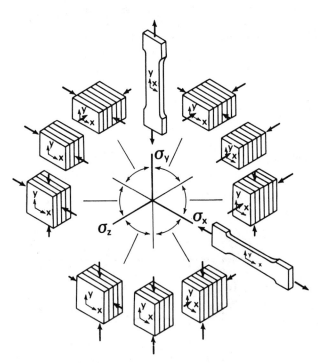

Fig. 10. Schematic representation of the experimental method to determine a yield locus in the normal stress plane.

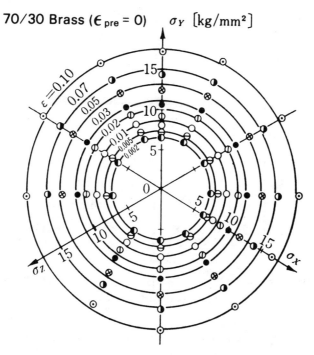

Fig. 11.   Yield loci at various strains for annealed 70/30 brass. X, Y and Z represent the rolling-, transverse-, and normal-directions, and σ_X, σ_Y and σ_Z are the stress components in the direction denoted by the subscripts.

through the point of $\epsilon_X = \epsilon_Y$ on the yield locus as shown in Fig. 6. The tangent is a line of constant mean stress, i.e. $(\sigma_X + \sigma_Y)/2 = c$. Therefore, the stress-strain curves calculated on the assumption of balanced biaxial tension from the observed value of pressure, polar radius of curvature and polar thickness are in good agreement with the curves in which stress is the mean value of stress components in the rolling- and transverse-directions as shown in Fig. 7. This suggests that for investigating the deformation behavior of sheet metal under biaxial tension, it is preferable to use the through-thickness compressive test instead of the bulge test, as discussed previously in [12]. In fact, the results obtained by the through-thickness compressive test were plotted in Fig. 7, and they compare favorably with the bulge data.

Ratio of Stresses in Plane Strain State—According to Hill's anisotropic theory, the ratio of the first (larger of the two) principal stresses corresponding to the two plane strain states in the first and fourth quadrants of the principal stress space can be determined by the r-values of sheet metal. Fig. 8 shows a comparison of the calculated stress ratios with experimental data for steel, 70/30 brass and aluminum sheets [2]. It is seen that while good agreement between theory and experiment is obtained for steel, the correlations for brass and aluminum are

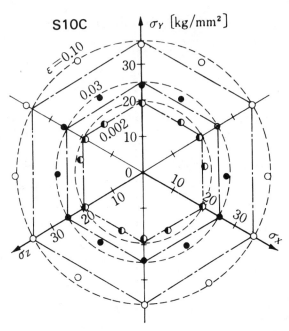

Fig. 12.   Yield loci in normal-stress plane for annealed low carbon steel (S10C). (circles: Mises law; regular hexagons: Tresca law.)

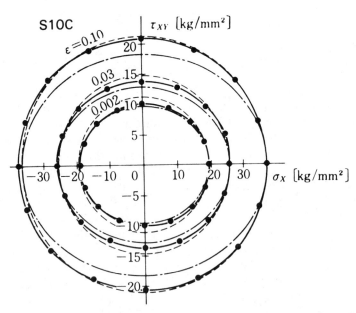

Fig. 13.   Yield loci in σ_X, τ_{XY} stress plane for low carbon steel (S10C). (circles: Mises law; ellipses: Tresca law.)

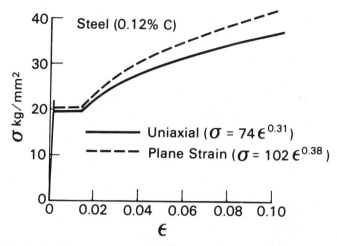

Fig. 14. Comparison of the strain-hardening characteristics in uniaxial and plane strain stress states.

rather poor. Since this stress ratio has been generally recognized as the most important material parameter influencing deep drawability of sheet metal, the result shown in Fig. 8 may explain why the correlation between the limiting drawing ratio and the r-value had not been successful for brass and aluminum sheets.

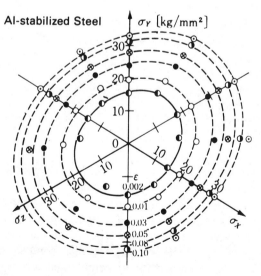

Fig. 15. Initial yield loci for aluminum-stabilized steel. Broken lines: equal ratio hardening loci based on a solid locus determined from experimental points at $\epsilon = 0.002$.

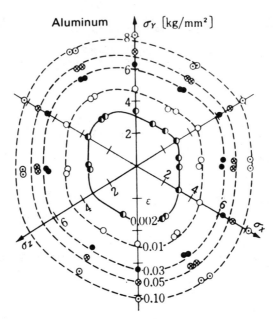

Fig. 16. Initial yield loci for aluminum. Broken lines: equal difference hardening loci based on a solid locus determined from experimental points at $\epsilon = 0.002$.

REPRESENTATION OF DEFORMATION BEHAVIOR AND EXPERIMENTAL PROCEDURE

Yield Surface in Stress Space—In order to describe the deformation behavior under combined stress systematically, a yield surface in the σ_X, σ_Y, σ_Z, $\sqrt{3}\tau_{XY}$ stress space has been proposed [13] as shown in Fig. 9. In this stress space, the plane of $\tau_{XY} = 0$ is referred to as the normal stress plane; the von Mises yield criterion is represented by a sphere and the Tresca criterion by a curved surface generated by revolution of a regular hexagon about the σ_Z-axis.

The length of stress vector in this space is equal to the effective stress σ conventionally defined for isotropic materials,

$$\sigma = \frac{1}{\sqrt{2}} \{\sqrt{(\sigma_X - \sigma_Y)^2 + (\sigma_Y - \sigma_Z)^2 + (\sigma_Z - \sigma_X)^2 + 6\tau_{XY}^2}\}. \qquad (1)$$

A given stress state (stress ratio) is a straight line through the origin. Thus, the constant stress state is represented by a conical surface which is defined by the conical angle ϕ. If there is planar anisotropy of the yield stress under the stress state ϕ, it will be shown by the intersection of a conical surface with the yield surface. The angle of rotation θ of the principal axes about Z-axis in the physical plane corresponds to a rotation of angle 2θ of the normal stress plane about σ_Z-axis in this stress space.

A single yield surface is constructed by the stress values corresponding to a constant plastic strain ϵ, defined by

$$\epsilon = \frac{\sqrt{2}}{3}\{\sqrt{(\epsilon_X - \epsilon_Y)^2 + (\epsilon_Y - \epsilon_Z)^2 + (\epsilon_Z - \epsilon_X)^2}\}, \qquad (2)$$

where ϵ_X, ϵ_Y and ϵ_Z are the principal strain components.
From these yield surfaces, the strain-hardening characteristics for a particular stress state are given by the distances between any two loci along the corresponding line.

Experimental Procedure—Experiments under combined stress were carried out by the biaxial compression method using the glued test pieces as has been discussed in a previous investigation [12].

Yield loci in the normal stress plane were constructed on the basis of tests under eleven different conditions as shown in Fig. 10, and the yield surfaces over the whole stress space were determined by yield loci in the normal stress plane

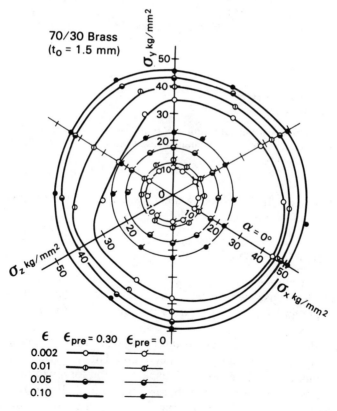

Fig. 17.　Yield loci after and before uniaxial tension in X-direction for 70/30 brass.

References pp. 108–109.

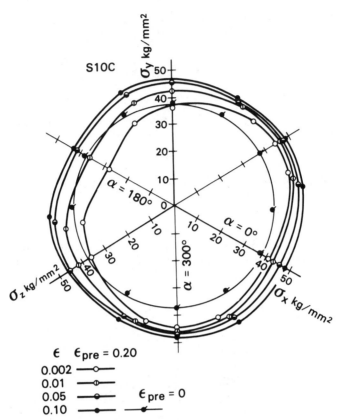

Fig. 18. Yield loci at various subsequent strains for the prestretched low carbon steel (S10C). α is the rotation angle of subsequent stress vector from the prestrain stress vector in the normal stress plane.

rotated about σ_z-axis. In determining the subsequent yield loci, specimens were prestrained, unloaded and then reloaded.

RESULTS AND DISCUSSION [14]-[17]

Strain-hardening Characteristics—Fig. 11 shows the yield loci in the normal stress plane for annealed brass, which are represented by circles. It is seen that the brass is isotropic and satisfies the Mises law quite well for plastic strain up to 0.1.

Yield loci in the normal stress plane and in the σ_x, τ_{xy} stress plane for annealed steel are shown in Figs. 12 and 13, respectively. This material is also isotropic, but its yield condition is closer to the Tresca criterion for small plastic strain and approaches the Mises criterion with increase in strain [13]. In other words, the strain-hardening characteristics under the plane strain condition differ from that under the uniaxial condition as shown in Fig. 14.

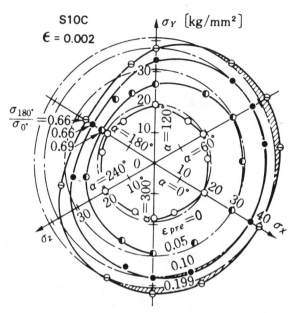

Fig. 19.   Yield loci at $\epsilon = 0.002$ for steel (S10C) prestretched by various strains.

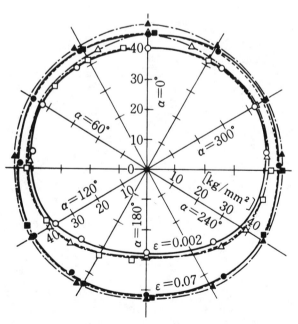

Fig. 20.   Congruence of yield loci for steel (S10C) prestrained in tension (○ ●), cold rolling (□ ■) and compression (△ ▲).

References pp. 108–109.

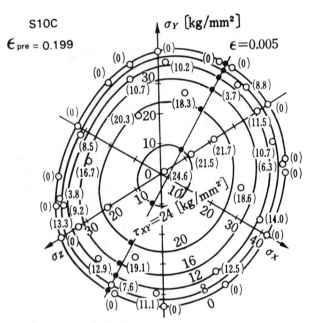

Fig. 21. Yield surface for steel (S10C) prestretched in X-direction shown by constant τ_{XY} lines. Numbers in the parentheses are values of τ_{XY} at the experimental points (○).

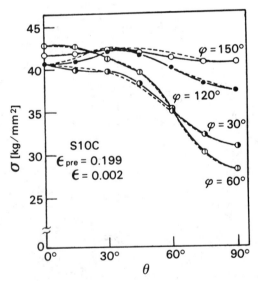

Fig. 22. Planar anisotropy of the yield stress at $\epsilon = 0.002$ under various stress states represented by φ (in Fig. 9) for steel (S10C) prestretched in X-direction. θ is a rotation angle of the principal axes about Z-axis. Broken lines are obtained from the approximate surface of revolution.

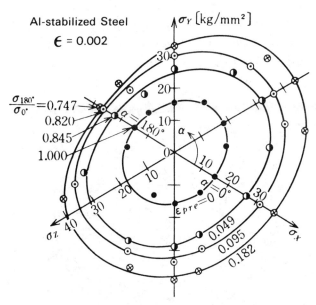

Fig. 23. Yield loci at $\epsilon = 0.002$ for aluminum-stabilized steel prestretched to various strains.

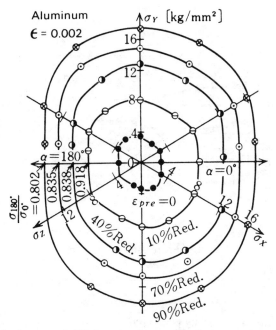

Fig. 24. Yield loci at $\epsilon = 0.002$ for aluminum prerolled to various reductions in thickness.

References pp. 108–109.

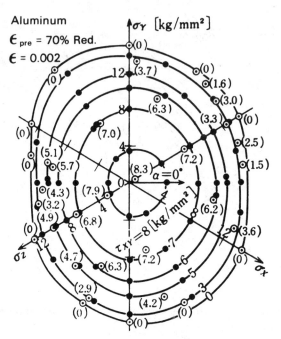

Fig. 25. Yield surface for prerolled aluminum shown by constant τ_{XY} lines. Numbers in parentheses are values of τ_{XY} at the experimental points ($\odot$).

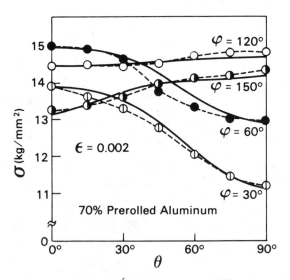

Fig. 26. Planar anisotropy of the yield stress at $\epsilon = 0.002$ under various stress states represented by φ (in Fig. 9) for prerolled aluminum. θ is a rotation angle of the principal axes about Z-axis. Solid lines are obtained by calculation.

Yield loci for as-received aluminum-stabilized steel, as shown in Fig. 15, are elliptical in shape (consistent with the measured $\bar{r} = 1.70$). The ellipses enlarge with increase in strain, keeping a shape very similar to the initial locus at $\epsilon = 0.002$.

Yield loci for annealed aluminum, as shown in Fig. 16, are hexagonal in shape rather than elliptical ($\bar{r} = 0.61$) and enlarge with increase in strain. The distance in radial direction from the $\epsilon = 0.002$ locus remains almost constant, independent of stress state.

These results suggest that for both isotropic and anisotropic materials, isotropic expansion of the yield loci can be divided into equal ratio hardening patterns and equal difference hardening patterns.

Yield Surface for Prestrained, Initially-Isotropic Materials—Fig. 17 shows the effect of prestrain on the yield loci for initially isotropic 70/30 brass. The loci are enlarged and distorted extremely by prestraining, and are further changed in shape with increase in subsequent strain. The subsequent yield stress after pre-stretching is lowest in the direction opposite to the prestrain vector due to the Bauschinger effect. The subsequent yield stress becomes highest in the same direction as the prestrain.

Fig. 18 shows the yield loci for prestretched steel. The subsequent yield behavior is similar to that shown in Fig. 17 except that the maximum hardening now occurs when the rotation angle of the subsequent stress vector from the prestrain stress vector α is about $60°$, not $0°$. The phenomenon that the strain-

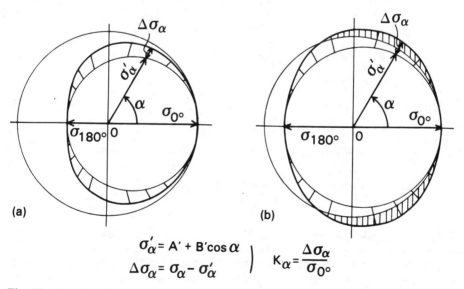

$$\sigma'_\alpha = A' + B'\cos\alpha$$
$$\Delta\sigma_\alpha = \sigma_\alpha - \sigma'_\alpha \qquad K_\alpha = \frac{\Delta\sigma_\alpha}{\sigma_{0°}}$$

Fig. 27. Explanation of excessive hardening factor K_α. Phenomenon of excessive hardening appears in (b).

References pp. 108–109.

Y. TOZAWA

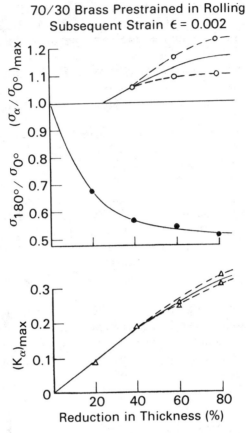

Fig. 28.   Appearance of phenomenon of excessive hardening depending on the Bauschinger effect and the excessive hardening factor.

hardening induced by prestrain becomes larger under some other subsequent
stress state than under the same one as the prestrain, we shall call excessive
hardening. The hatched area in Fig. 19 shows the extent of excessive hardening.
It occurs when the prestrain is beyond a certain value, depending on the material.
   Fig. 20 shows the effect of different kinds of prestrain on the yield loci and it
is seen that when restricted to the same magnitude of prestrain, these yield loci
are congruous with each other; even if the prestrain is not of the same kind.
   The yield surface for prestretched steel is shown by constant τ_{XY} lines in Fig.
21. It is seen that these lines are symmetric with respect to σ_X-axis (or the stress
vector of the prestrain) in the same manner as the yield loci in the normal stress
plane shown in Figs. 17–20. Furthermore, the farthest points from σ_X-axis on
each constant τ_{XY} line ($\bullet$) are on a line perpendicular to τ_X-axis. From these
results, we may assume that the yield surfaces for prestrained, initially-isotropic
materials are surfaces of revolution about the prestrain's stress vector. This

assumption was shown to be correct by further comparisons with experiment. As shown in Fig. 22, a comparison is made of the yield stress calculated using this assumption and the observed data taking into account planar anisotropy under various stress states. It is seen that the agreement is satisfactory.

Fig. 22 shows another interesting result. As the stress state is represented by ϕ as defined in Fig. 9, the uniaxial tension in X-direction and the uniaxial compression in Y-direction corresponds to $\phi = 120°$ and $\phi = 60°$ on $\theta = 0°$, respectively. It is seen that the directional dependence of the yield stress varies with the stress state. Thus, the planar anisotropy as determined by the tensile test offers only limited information.

Yield Surface for Prestrained, Initially-Anisotropic Materials—Fig. 23 shows the yield loci for aluminum stabilized steel prestretched in X-direction. The yield loci are enlarged and distorted with increase in prestrain. The shape becomes oblate in the direction opposite to the prestrain vector (which may be explained by the

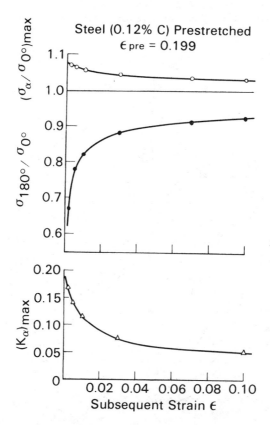

Fig. 29. Variations of phenomenon of excessive hardening, Bauschinger effect and excessive hardening factor with subsequent strain.

References pp. 108–109.

Bauschinger effect) and the major axis of yield loci rotates from σ_Z-axis towards the perpendicular to σ_X-axis.

Fig. 24 shows the effect of prestrain on the yield loci for aluminum. In this case, it is seen that the anisotropy decreases with increased thickness reduction, and the shape changes from hexagonal to elliptical. The major axis of the ellipse is perpendicular to the prestrain vector.

The yield surface for prerolled aluminum is shown by constant τ_{XY} lines in Fig. 25. These lines differ in shape and are not symmetrical with respect to the prestrain vector, because the anisotropy induced by the straining is superposed upon the anisotropy existing in the sheet before straining.

The measured yield surfaces for prestrained anisotropic materials may not be approximated by surfaces of revolution about the prestrain vector. As in the case of initially-isotropic materials, however, the results for anisotropic materials may be rearranged as follows. The degree of strain-hardening induced by prestrain is expressed by the ratio $\sigma_\alpha/\sigma_\alpha{}^\circ$ for the equal ratio hardening materials and by the quantity $(\sigma_\alpha - \sigma_\alpha{}^\circ)$ for the equal difference hardening materials. The symbols σ_α and $\sigma_\alpha{}^\circ$ denote the magnitudes of the stress vectors at a given strain in the α-direction after and before straining, respectively. When the above ratios or

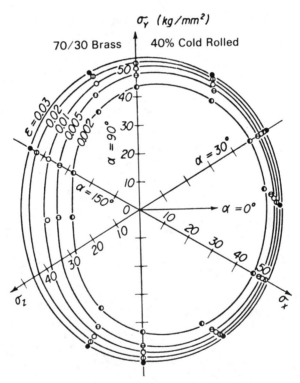

Fig. 30. Calculated results of yield loci at various subsequent strains for 70/30 brass prestrained by cold rolling. Substituting directions: $\alpha = 30°$, $90°$ and $150°$.

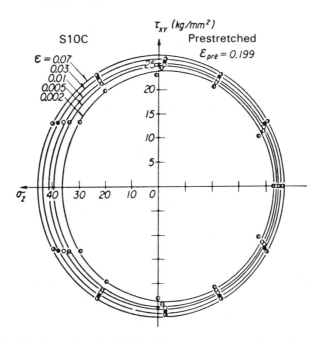

Fig. 31. Calculated results of yield loci in σ_Z, τ_{XY} stress plane for steel (S10C) pre-stretched in X-direction. Substituting directions: $\alpha = 0°$, 180° and 300° in the plane of $\tau_{XY} = 0$.

quantities are rearranged with respect to the rotation angle α, they become minimum at 180° and symmetric with respect to 180° regardless of the type of prestrain. Therefore, the strain-hardening anisotropy induced by prestrain depends on the angle between the subsequent stress vector and the prestrain stress vector regardless of the anisotropy existing in the sheet before straining. Thus, we assume that the ratio $\sigma_\alpha/\sigma_\alpha{}^\circ$ or the quantity $(\sigma_\alpha - \sigma_\alpha{}^\circ)$ is rotationally symmetric with respect to the prestrain stress vector in the yield surface for prestrained, initially anisotropic materials. Calculations based on this assumption agree fairly well with experiment [14]. An example of the results involving planar anisotropy is shown in Fig. 26.

Phenomenon of Excessive Hardening—As discussed earlier, excessive hardening occurs when the maximum hardening induced by prestrain occurs in a stress state different from the prestrain. It seems reasonable to assume that the yield stress in the different stress state from the prestraining is more or less influenced by the Bauschinger effect. In order to separate the influence of the Bauschinger effect from the phenomenon of excessive hardening, the influence of the Bauschinger effect is assumed to vary as

$$\sigma_\alpha' = A' + B' \cos \alpha \tag{3}$$

in the normal stress plane, where σ'_α is magnitude of the yield stress vector in the α-direction and A' and B' are constants determined by $\sigma_{0°}$, the yield stress at $\alpha = 0°$, and $\sigma_{180°}$, at $\alpha = 180°$. The yield locus for prestrained, initially-isotropic material is shown schematically in Fig. 27. The curve represented by Eq. (3) is similar in shape to a circle which is a representation for the kinematic hardening model. Now, the excessive hardening factor K_α is defined as

$$K_\alpha = (\sigma_\alpha - \sigma'_\alpha)/\sigma_{0°} \tag{4}$$

The phenomenon of excessive hardening is shown by hatched area outside of a circle with radius $\sigma_0°$ representing the isotropic hardening in Fig. 27(b). Therefore, when the Bauschinger effect is large (i.e., the ratio $\sigma_{180°}/\sigma_{0°}$ is small) and the factor K_α is small, the phenomenon of excessive hardening does not occur.

Fig. 28 shows the variations of the ratio $\sigma_{180°}/\sigma_{0°}$ and the maximum value of the factor K_α with prestrain. When the prestrain is large, the maximum value of the ratio $\sigma_\alpha/\sigma_{0°}$ becomes larger than unity, and the phenomenon of excessive hardening occurs. The variation of the excessive hardening with subsequent strain depends also on the Bauschinger effect and the excessive hardening factor as shown in Fig. 29.

Similar excessive hardening factors can be deduced for initially anisotropic materials. Instead of σ_α for isotropic materials, $\sigma_\alpha{}^* = \sigma_{0°}°(\sigma_\alpha/\sigma_\alpha°)$ for equal ratio hardening materials and $\sigma_\alpha{}^{**} = \sigma_\alpha - \sigma_\alpha°$ for equal difference hardening materials, are considered as the yield stress magnitudes. By assuming the same relation about α as Eq. (3) for the influence of the Bauschinger effect, these factors in the same form as Eq. (4) are given as follows. For equal ratio hardening materials,

$$K_\alpha{}^* = (\sigma_\alpha{}^* - \sigma_\alpha{}^{*\prime})/\sigma_{0°} \tag{5}$$

and for equal difference hardening materials,

$$K_\alpha{}^{**} = (\sigma_\alpha{}^{**} - \sigma_\alpha{}^{**\prime})/\sigma_{0°}. \tag{6}$$

The relations between the Bauschinger effect, the excessive hardening factor and the phenomenon of excessive hardening are given in Table 1 for the materials used in these experiments.

TABLE 1

Relation Between Ranks for Bauschinger Effect, Excessive Hardening Factor and Phenomenon of Excessive Hardening

Material	Bauschinger Effect	K_α , $K_\alpha{}^*$, $K_\alpha{}^{**}$	Phenomenon of Excessive Hardening
Aluminum-Stabilized Steel	2(medium)	1(large)	1(easy)
Low Carbon Steel	3(medium)	2(large)	2(medium)
Aluminum	1(small)	4(small)	3(medium)
70/30 brass	4(large)	3(medium)	4(hard)

Formulation of Deformation Behavior for Prestrained Materials—The variation of the excessive hardening factor K_α [defined by Eq. (4)] with α is well approximated by Eq. (7) on the basis of comparison experiments,

$$K_\alpha \propto (1 - \cos^2\alpha). \tag{7}$$

By combining this relation with Eqs. (3) and (4), the following equation is obtained for the magnitude of subsequent stress vector in the normal stress plane,

$$\sigma_\alpha = A + B\cos\alpha + C\cos^2\alpha. \tag{8}$$

Assuming rotational symmetry for the yield surface after straining, α will be replaced by ω the angle between subsequent stress vector and prestrain stress vector in the stress space shown in Fig. 9. Finally, the yield surface for prestrained, initially-isotropic materials is represented by Eq. (9),

$$\sigma_\omega = A + B\cos\omega + C\cos^2\omega. \tag{9}$$

σ_ω in Eq. (9) is equivalent to σ in Eq. (1). Thus, the effective stress at a given strain is not a constant but a function of stress state. The coefficients A, B and C are constants with respect to ω but are functions of the subsequent strain. These three coefficients may be determined by three tests under arbitrary stress

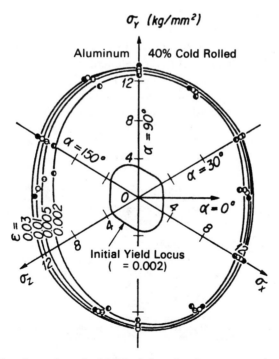

Fig. 32.   Calculated results of yield loci for aluminum prestrained by cold rolling. Substituting directions: $\alpha = 30°$, $90°$ and $150°$.

References pp. 108–109.

states. It is recommended that the uniaxial tests be applied because of their simplicity.

For prestrained, initially-anisotropic materials, the excessive hardening factors K_α^* and K_α^{**} are approximated by the same function of α as in Eq. (7). Similar to Eq. (8), the yield loci in the normal stress plane are represented by Eqs. (10) and (11) as follows: for equal ratio hardening materials,

$$\sigma_\alpha = \sigma_\alpha{}^\circ(A^* + B^* \cos\alpha + C^* \cos^2\alpha), \tag{10}$$

and for equal difference hardening materials,

$$\sigma_\alpha = \sigma_\alpha{}^\circ + (A^{**} + B^{**} \cos\alpha + C^{**} \cos^2\alpha), \tag{11}$$

where the coefficients A^*, B^* and C^* are dimensionless while A^{**}, B^{**} and C^{**} are stresses. For the yield surface in the stress space, replacing α with ω, as in Eq. (9), leads to

$$\sigma_{\omega\psi} = \sigma_{\omega\psi}^\circ (A^* + B^* \cos\omega + C^* \cos^2\omega) \tag{12}$$

$$\sigma_{\omega\psi} = \sigma_{\omega\psi}^\circ + (A^{**} + B^{**} \cos\omega + C^{**} \cos^2\omega) \tag{13}$$

for equal ratio- and equal difference-hardening materials, respectively. In these equations, $\sigma_{\omega\psi}$ and $\sigma_{\omega\psi}^\circ$ are the magnitudes of the stress vectors after and before straining, respectively, in the direction denoted by the subscript, and ω and ψ represent, respectively, the angles the subsequent stress vector makes with the prestrain stress vector, and with an arbitrarily fixed direction, such as σ_X-axis.

Calculated results of yield loci are compared with experiments in Figs. 30–32. For the results shown in Figs. 30 and 32, the information from the uniaxial tests at $\alpha = 30°$, $90°$ and $150°$ was used in the calculation. For Fig. 31, the information at $\alpha = 0°$, $180°$ and $300°$ shown in Fig. 18 was used.

Fig. 33 shows that the uniaxial stress-strain curve at $\alpha = 60°$ calculated from the experimental data in the other three directions, is in good agreement with the experimental curve in the same direction [4].

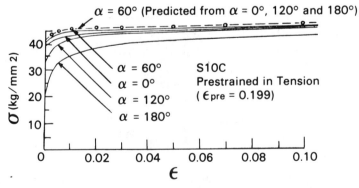

Fig. 33.   Uniaxial stress-strain curves for steel (S10C) prestrained in tension.

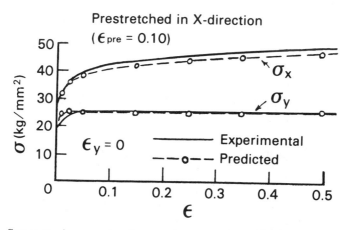

Fig. 34.   Stress-strain curve in plane strain compression in X-direction for steel (S10C) prestretched in X-direction.

The stress state under a given strain state is obtained from the yield locus by applying the normality rule for the strain vector. The calculated compressive stress-strain relations for plane strain are compared with experiments in Fig. 34. It is seen that the formulation described above is satisfactory.

Limit of Application of Formula—The formulas described in the preceding sections should not be used when the yield loci for a prestrained, initially isotropic material become significantly asymmetric with respect to the prestrain stress vector. An example of such an asymmetry is depicted in Fig. 35 where the yield loci of prerolled brass are shown.* Note that as the reduction in thickness increases, the extent of asymmetry becomes greater.

Another remark to be made here is that these formulas are still applicable to situations where there is no unloading occurring during the change of strain path from prestrain to subsequent strain. This is based on the results of a previous investigation [12] in which two types of experimental conditions—one with unloading and the other without—were employed to study the effect of unloading on subsequent yield behavior. It was found that for a given strain path, the difference in stresses measured under the two experimental conditions becomes negligible when the subsequent strain is beyond a few percent.

CONCLUSIONS

The deformation behavior of sheet metal under conditions of combined stress has been explored and is presented in a systematic fashion. However, the results obtained need to be extended: only four materials were studied; the strain range

---

* *The occurrence of asymmetrical distortion may be explained by the development of crystallographic texture due to the cold rolling [18].*

References pp. 108–109.

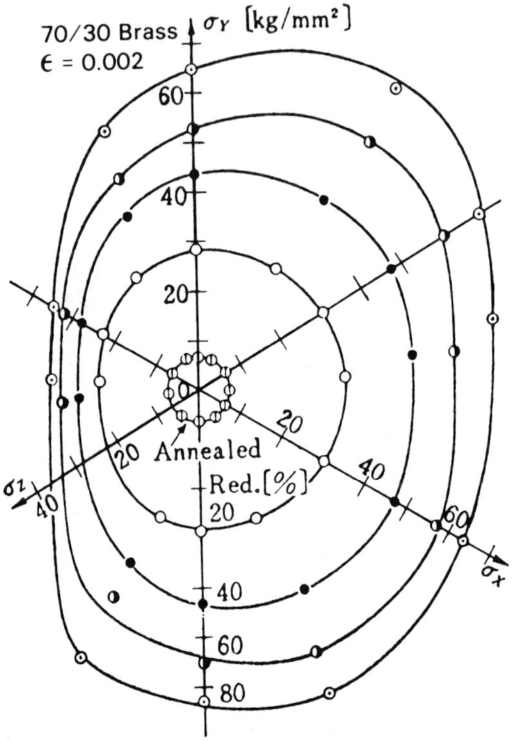

Fig. 35. Asymmetrical shape developed in yield locus for 70/30 brass prestrained by cold rolling to high reductions.

studied extended only up to 0.1; only proportional loading paths were investigated; the effect of strain rate is unknown. Nevertheless, these results should be most useful in analyzing forming problems at the present time.

ACKNOWLEDGMENTS

The author expresses his appreciation to many colleagues in his laboratory, especially to Dr. M. Nakamura and Dr. H. Shirai, for their helpful cooperation in the investigation and also wishes to thank Mrs. M. Miwa for her assistance in preparation of the manuscript. This research was supported in part by the Funds of Scientific Research, granted by the Ministry of Education, Japan.

REFERENCES

[1] K. Yoshida, Reports, Inst. of Physical and Chemical Research, Tokyo, *45* (1969), 157.
[2] Y. Tozawa, J. Japan Soc. Tech. Plasticity, *16* (1975), 740.
[3] R. Hill, The Mathematical Theory of Plasticity, Oxford (1950).
[4] Y. Tozawa, Bull. Japan Inst. Metals, *15* (1976), 607.
[5] A. N. Bramley and P. B Mellor, Int. J. mech. Sci., *8* (1966), 101.

[6] A. N. Bramley and P. B. Mellor, Int. J. mech. Sci., *10* (1968), 211.

[7] R. Pearce, Int. J. mech. Sci., *10* (1968), 995.

[8] J. Woodthorpe and R. Pearce, Int. J. mech. Sci., *12* (1970), 341.

[9] R. M. S. B. Horta, W. T. Roberts and D. V. Wilson, Int. J. mech. Sci., *12* (1970), 231.

[10] G. S. Kular and M. J. Hillier, Int. J. mech. Sci., *14* (1972), 631.

[11] H. Naziri and R. Pearce, J. Inst. Metals, *98* (1970), 71.

[12] Y. Tozawa, M. Nakamura and I. Shinkai, Proc. Int. Conf. Sci. Tech. Iron and Steel, Suppl. Trans. ISIJ, *11* (1971), 936.

[13] Y. Tozawa and H. Shirai, J. Japan Soc. Tech. Plasticity, *16* (1975), 550.

[14] H. Shirai and Y. Tozawa, J. Japan Soc. Tech. Plasticity, *16* (1975), 645.

[15] H. Shirai and Y. Tozawa, J. Japan Soc. Tech. Plasticity, *16* (1975), 1041.

[16] H. Shirai and Y. Tozawa, J. Japan Soc. Tech. Plasticity, *18* (1977), 100.

[17] H. Shirai and Y. Tozawa, J. Japan Soc. Tech. Plasticity, *18* (1977), 774.

[18] Y. Tozawa, M. Kojima, T. Kato and K. Kojima, Preprint 1977 Japan Spring Cong. Plastic Working (1977), 335.

DISCUSSION

D. Lee (General Electric Company)

I assume that you are trying to develop a hardening rule for anisotropic material. That is one of your aims. Now it appears to me that there might be some inconsistency in the way that you are treating experimental data. You said that you assumed the effective stress and effective strain defined by isotropic theory and you used them to interpret experimental data for both isotropic and anisotropic materials. I am wondering if this is justifiable for anisotropic materials?

Tozawa

With regard to the definition of effective stress and strain, it is hard to determine effective stress; but the effective strain is one kind of definition of volume strain. So, in my experiments I used the effective strain defined for isotropic material. However, I compared the difference between the results using the effective strain and the maximum geometric strain, and the difference was not large. So, we used effective strain.

S. S. Hecker (Los Alamos Scientific Laboratory)

First, could you describe how you physically conduct the biaxial compression tests. Another question is: Have you attempted to define the yield locus at smaller plastic strains than 0.002?

.Tozawa

To answer your first question requires slides. I would like to answer the second question. It is very difficult to obtain these results and since the major objective of my research is application to forming, we have used proof stress defined by 0.002 strain.

Hecker

I have two comments to that question. One of them is that I have done some work at very small plastic strains. For instance your 0.002 strain is 2000 micro strain, and I have done some work at 5 micro strains, defining a yield locus. In my results, instead of getting what you call excessive hardening at 90°, I actually find a maximum in softening at 90°, rather than hardening. So there is a possibility that the trend that you are showing reverses depending on whether you define yielding by small plastic strain offset or by very large plastic strain offset.

And on your comment of large plastic strains for the forming problem, I think that maybe one of the things that makes forming analysis very difficult is that although you have very large plastic strains involved (of the order of 20 to 50%), it is the yield stress at the moment of plastic deformation (that is the very small offset yield surface) that determines which way the plastic strain vector points. And the direction of the plastic strain vector is going to govern the onset of instability. So even though you have very large plastic strains, you are concerned about the very small plastic strain levels around the yield surface.

Tozawa

I agree with you, but in my opinion there is very little information concerning the large plastic strain region. There are many data concerning the very small strain range, but not many in the large plastic strain range. The object of my paper is to contribute at that point.

SHEET NECKING—I. VALIDITY OF PLANE STRESS ASSUMPTIONS OF THE LONG-WAVELENGTH APPROXIMATION*

J. W. HUTCHINSON

Harvard University, Cambridge, Massachusetts

K. W. NEALE

Université de Sherbrooke, Sherbrooke, Quebec, Canada

A. NEEDLEMAN

Brown University, Providence, Rhode Island

ABSTRACT

Two special solutions are used to illustrate the errors involved in the analysis of sheet necking from invoking plane stress assumptions. The first is an exact solution for the growth-rate of a small amplitude, sinusoidal thickness variation in a sheet of material characterized by $\dot{\epsilon} = \alpha \sigma^n$ in simple tension. The plane stress assumptions become accurate when the ratio of variation wavelength to average thickness exceeds four and otherwise lead to overestimates of the actual growth-rate. When this same ratio is approximately unity the relative size of the thickness variation decays with increasing deformation—an effect not predicted by a plane stress analysis. The second special solution is obtained using a perturbation expansion with the nonlinear long-wavelength solution (i.e., the plane stress solution) as the lowest order contribution. In this way explicit corrections to the plane stress solution are obtained. Selected comparisons with fully nonlinear finite element calculations are made.

INTRODUCTION

The list of variables and issues thought to be important in the analysis of necking failures in thin sheet metals has almost reached perplexing proportions.

---

* *The material in this three-part paper was presented orally in Session II under the title "Constitutive Relations for Sheet Metal" and Session IV under the title "Sheet Necking: Influence of Constitutive Theory and Strain-Rate Dependence."*

Included are not only strain hardening and plastic anisotropy characteristics but also strain-rate sensitivity, choice of plastic constitutive law, bifurcation analysis vs. imperfection growth analysis, and microscopic fracturing and void growth. In the three parts of this paper an attempt will be made to give a unified examination of the role of a number of these in the analysis of sheet metal necking.

Part I deals with the adequacy of the plane stress assumptions often invoked in the analysis of sheet necking. We will refer to a solution based on these assumptions as the long-wavelength approximation, and we obtain some indication of the range of validity of this approximation by way of two special solutions and some fully nonlinear finite element calculations.

Choice of constitutive law will be central to Parts II and III with time-independent material behavior discussed in II and rate-dependent behavior featured in III. In II a bifurcation analysis of localized necking will be given side by side with a long-wavelength analysis of the growth of initial nonuniformities of the type introduced by Marciniak and Kuczyński [1]. Differences in forming limit diagrams computed using a standard flow (incremental) theory of plasticity on the one hand and a deformation theory on the other will be noted. These differences are large when both principal in-plane strains are positive. There is reason to doubt whether a standard flow theory can be used to analyze localized necking in sheet metals. This is clearly an important issue since large scale numerical programs are currently being developed to cope with complications associated with particular sheet forming operations. It is essential that limitations of competing constitutive laws be well understood.

The effect of strain-rate dependence on constitutive behavior is incorporated in III in a simple and straightforward manner. The influence of small amounts of rate-sensitivity on the forming limit diagrams is emphasized.

PLANE STRESS ASSUMPTIONS OF THE LONG-WAVELENGTH APPROXIMATION

In Part I attention will be restricted to in-plane plane strain deformations of a sheet subject to a tensile load per unit length P acting in the x_1-direction as shown in Fig. 1. The surface of the sheet at any instant of time t and position x_1 is symmetrically disposed about its midplane at $x_3 = 0$ according to $x_3 = \pm h(x_1, t)/2$. Under in-plane plane strain all quantities are independent of x_2 and no straining in the x_2-direction is permitted. In the plane stress approximation only the stress components σ_{11} and σ_{22} are assumed to be nonzero and these are taken to be uniform over each section $x_1 = $ const. For each value of x_1 equilibrium requires

$$P = \sigma_{11} h \qquad (1)$$

Nonzero strain-rates are $\dot{\epsilon}_{11}$ and $\dot{\epsilon}_{33}$ and these are also uniform over each section. These, together with (1), the constitutive law and $\dot{\epsilon}_{22} = 0$, can be used to eliminate σ_{22} and to formulate equations governing the evolution of the thickness variation.

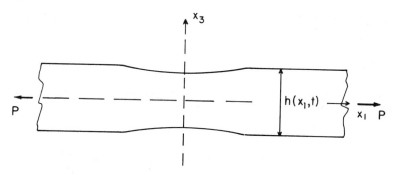

Fig. 1.   Sheet geometry.

In what follows, we give a detailed assessment of the consequences of the plane stress assumptions using two special solutions: (1) a linearized solution for the growth-rate of thickness variations of arbitrary wavelength, and (2) a perturbation solution valid for moderately-long-wavelength variations. For this purpose we consider an incompressible, strain-rate dependent material characterized in simple tension by the pure power law

$$\dot{\epsilon} = \alpha \sigma^n \qquad (2)$$

where σ is the true stress and $\dot{\epsilon}$ is the natural strain-rate. Revealing results can be obtained using analytic methods for this relatively simple class of materials since the prior deformation history in any problem appears only through the current geometry. Such results are applicable to super plastic behavior and to materials undergoing power law creep. While these results can not be expected to apply directly to problems involving more complex material behavior, they should serve to illustrate some of the most important points at issue in the validity of the plane stress assumptions.

With s_{ij} as the deviator of the true stress σ_{ij} and σ_e as the effective stress, the Eulerian strain-rate under a general stress state is taken to be

$$\dot{\epsilon}_{ij} = \tfrac{3}{2}\alpha \sigma_e^{n-1} s_{ij}, \qquad \sigma_e = (3 s_{ij} s_{ij}/2)^{1/2} \qquad (3)$$

which generalizes (2).

In the plane stress approximation the strain-rates at each position x_1 can readily be found in terms of the current thickness variation $h(x, t)$ as

$$\dot{\epsilon}_{11} = -\dot{\epsilon}_{33} = (\sqrt{3}/2)\alpha[\sqrt{3}P/(2h)]^n \qquad (4)$$

The corresponding (convected) rate of change of the thickness of the material element currently at x_1 is given by

$$\dot{h} \equiv \frac{\partial h}{\partial t} + v_1 \frac{\partial h}{\partial x_1} = \dot{\epsilon}_{33} h = -\dot{\epsilon}_{11} h \qquad (5)$$

where $v_1(x_1)$ is the velocity in the x_1-direction.

References p. 126.

LINEARIZED ANALYSIS FOR SMALL NONUNIFORMITIES OF ARBITRARY WAVELENGTH

Consider a sheet of material (3) under in-plane plane strain conditions and subject to a load per unit length P. Let $h_0(t)$ denote the evolving thickness of a perfect sheet with no variation in thickness. Equations (4) and (5) involve no approximation for the perfect sheet so that

$$-\dot{h}_0/h_0 = \dot{\epsilon}_{11} = -\dot{\epsilon}_{33} = \dot{\epsilon}_0 \equiv (\sqrt{3}/2)\alpha[\sqrt{3}P/(2h_0)]^n \qquad (6)$$

where $\dot{\epsilon}_0$ is the strain-rate in the perfect sheet. We also consider a nonuniform sheet characterized at the current instant by a symmetric thickness variation

$$h = h_0[1 - \xi \cos(2\pi x_1/l)] \qquad (7)$$

For $|\xi| \ll 1$, an exact linearized solution for the strain-rate and rate of change of the nonuniformity can be obtained along lines similar to those given for the analogous problem for a circular bar in [2, 3]. Velocity fields (v_1, v_3) can be expressed in terms of a velocity potential $\Phi(x_1, x_3)$ by

$$v_1 = \Phi,_{x_3}, \qquad v_3 = -\Phi,_{x_1} \qquad (8)$$

The solution for Φ is simpler than that for the circular bar given in detail in [2], and therefore we will proceed directly to give the end result. Let ζ be the first quadrant root of

$$\zeta^2 = \left(1 - \frac{2}{n}\right) + i\sqrt{1 - \left(1 - \frac{2}{n}\right)^2} \qquad (i \equiv \sqrt{-1}) \qquad (9)$$

and let

$$q = \pi h_0/l \qquad (10)$$

With terms of order ξ^2 neglected, the solution for Φ is

$$\Phi = \dot{\epsilon}_0 x_1 x_3 + \xi \frac{\dot{\epsilon}_0 l h_0}{\pi} \text{Re}\left[c \sin\left(\frac{2\zeta q x_3}{h_0}\right)\right] \sin\left(\frac{2\pi x_1}{l}\right) \qquad (11)$$

where Re denotes the real part and c is a complex constant satisfying

$$\left.\begin{array}{c} \text{Re}[c(1 - \zeta^2)\sin(\zeta q)] = 1 \\ \text{Re}[c\zeta(1 - \zeta^2 - 4/n)\cos(\zeta q)] = 0 \end{array}\right\} \qquad (12)$$

Let $\Delta h(x_1, t) = h(x_1, t) - h_0(t)$ be the nonuniformity in thickness and let

$$a = \Delta h(x_1, t)/h_0(t) \qquad (13)$$

measure the size of the nonuniformity relative to the evolving thickness of the perfect sheet. At the current instant, therefore,

$$a = -\xi \cos(2\pi x_1/l) \qquad (14)$$

From the above linearized solution the convected rate of a can be found to be

$$\dot{a} = -\xi n \dot{\epsilon}_0 G(n, q) \cos(2\pi x_1/l) \qquad (15)$$

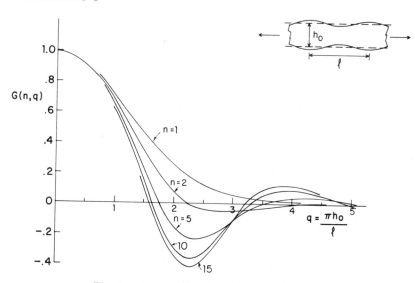

Fig. 2.   Values of $G(n, q)$ in Equation (16).

or, from (14),

$$\dot{a} = n\dot{\epsilon}_0 a G(n, q) \tag{16}$$

where

$$G(n, q) = (4/n)\,\text{Re}[c\,\sin(\zeta q)] \tag{17}$$

Curves of G vs. q for fixed values of n have been calculated numerically from (17) and (12) and are shown in Fig. 2. In the long-wavelength limit (i.e., $q \to 0$) it is seen that $G = 1$ so that

$$\dot{a} = n\dot{\epsilon}_0 a \tag{18}$$

This simple result can be obtained directly from an analysis using the plane stress assumptions of the previous section. Henceforth we will refer to the solution based on the plane stress assumptions as the long-wavelength approximation. From Fig. 2 it can be noted that (18) is reasonably accurate for values of l/h_0 greater than about 4. For nonuniformities shorter than this the long-wavelength result (18) *overestimates* the rate of growth of the nonuniformity. More surprising is the prediction that, for wavelength ratios in the approximate range $1 < l/h_0 < 2$, G is negative and the relative size of the nonuniformity actually decays, except for $n = 1$.*

A similar phenomenon has been observed by Appleby and Richmond [4] in a

---

* The analogous equation for round bars in [3] for the growth of the relative size of the nonuniformity of cross-sectional area also takes the form (16). For the round bar, however, G is never negative and thus decay of the nonuniformity as measured by its relative size does not occur for any wavelengths. In this case the long-wavelength analysis is accurate for nonuniformity wavelengths greater than about three diameters. For shorter wavelengths it also overestimates the growth-rate.

References p. 126.

numerical study of the growth of sinusoidal geometric imperfections of various wavelengths in sheets of time-independent elastic-plastic materials under similar in-plane plane strain conditions. Using the absolute size of the nonuniformity, Δh, as their measure, they find that Δh grows for $l/h_0 > 2$ and that Δh diminishes for $l/h_0 < 2$. Furthermore, they find that the "most stability", i.e., fastest rate of decay in Δh, occurred for l/h_0 around unity. This leads them to make the intriguing suggestion that, if controlled favorable nonuniformities are deliberately imposed on a sheet, its ductility might be enhanced. The present equation (16) for $\dot{a}$ can readily be converted to an equation for $\Delta \dot{h}$ using $\dot{a} = \Delta \dot{h}/h_0 + a\dot{\epsilon}_0$, which follows from (13) and (6). Thus (16) transforms to

$$\Delta \dot{h} = (n - 1)F(n, q)\dot{\epsilon}_0 \Delta h \qquad (19)$$

where $F = (nG - 1)/(n - 1)$. In this equation F has been defined such that in the long-wavelength limit, $q \to 0$, $F = 1$. The calculations of Appleby and Richmond were carried out for a material with a strain hardening exponent between $n = 4$ and $n = 5$. Using the curve for $n = 5$ in Fig. 2, we first note that $F = 0$ at a value of $q \cong 1.5$ which corresponds very closely to the transition value $l/h_0 \cong 2$ for the sign of Δh found by Appleby and Richmond. Secondly, the most negative value of F occurs when $dF/dq = dG/dq = 0$ at about $q = 2.5$ for $n = 5$. This value, too, is not far from the value $l/h_0 \cong 1$ found by Appleby and Richmond for wavelength giving the maximum rate of decay. Thus we can note that some quantitative details as well as qualitative ones, for the simple time-dependent material (3), apply to more complicated time-independent materials.

LOWEST ORDER CORRECTIONS TO THE NONLINEAR LONG-WAVELENGTH APPROXIMATION

In this section a perturbation method is used to show that the long-wavelength solution obtained by invoking the plane stress assumptions is an exact solution in an appropriate limiting sense. In addition, the lowest order corrections to this solution are produced. In particular, deviations of the stresses and strain-rates from the values of the long-wavelength solution will be given. We continue to consider a sheet of power law material characterized by (3) under in-plane plane strain conditions. Now, however, it is not assumed that the variation in thickness is necessarily small and thus full nonlinear behavior is treated.

The starting point in the analysis is the introduction of the perturbation parameter β. Consider a family of similarly shaped, symmetric thickness variations $h(\beta x_1)$ which differ from one another only in scaling in the x_1-direction depending on β. One such variation is depicted in Fig. 3. As $\beta \to 0$ the characteristic wavelength of the variation becomes large compared to the average thickness of the sheet, and it is in this limiting sense that the long-wavelength solution becomes increasingly accurate, as will be shown. For convenience let $z \equiv x_3$ and introduce a scaled variable

$$X = \beta x_1$$

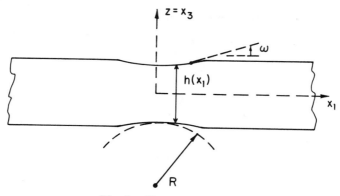

Fig. 3. Sheet geometry.

With

$$\sigma(X) = P/h(X) \tag{20}$$

and

$$\dot{\epsilon}(X) = (\sqrt{3}/2)\alpha[\sqrt{3}\sigma(X)/2]^n \tag{21}$$

the nonzero stresses and strain-rates of the previous nonlinear long-wavelength solution are given by

$$\left. \begin{array}{l} \sigma_{11} = 2\sigma_{22} = 2\sigma_e/\sqrt{3} = \sigma(X) \\ \dot{\epsilon}_{11} = -\dot{\epsilon}_{33} = \dot{\epsilon}(X) \end{array} \right\} \tag{22}$$

Field equations for the fully nonlinear plane strain problem are given below using X and z as independent variables. Equilibrium requires

$$\left. \begin{array}{l} \beta\sigma_{11,X} + \sigma_{13,z} = 0 \\ \beta\sigma_{13,X} + \sigma_{33,z} = 0 \end{array} \right\} \tag{23}$$

Velocities are related to the velocity potential by

$$v_1 = \Phi_{,z} \quad \text{and} \quad v_3 = -\beta\Phi_{,X} \tag{24}$$

Strain-rates are given by

$$\dot{\epsilon}_{11} = \beta v_{1,X}, \quad \dot{\epsilon}_{33} = v_{3,z}, \quad 2\dot{\epsilon}_{13} = v_{1,z} + \beta v_{3,X} \tag{25}$$

Strain-rates and stresses are related through (3). Under the present in-plane plane strain conditions $\sigma_{22} = (\sigma_{11} + \sigma_{33})/2$ so that

$$s_{11} = -s_{33} = (\sigma_{11} - \sigma_{33})/2 \tag{26}$$

and

$$\sigma_e = [3(\sigma_{11} - \sigma_{33})^2/4 + 3\sigma_{13}^2]^{1/2} \tag{27}$$

References p. 126.

Traction-free conditions along $y = h(X)/2$ are

$$
\left.
\begin{array}{c}
- \sigma_{11} \sin \omega + \sigma_{13} \cos \omega = 0 \\
- \sigma_{13} \sin \omega + \sigma_{33} \cos \omega = 0
\end{array}
\right\}
\tag{28}
$$

where

$$
\tan \omega = \beta h'(X)/2
$$

Here and throughout the remainder of Part I

$$
(\ )' = \frac{d(\ )}{dX}
$$

For future use, expansions of the boundary conditions in powers of β are

$$
- \sigma_{11}\beta h'/2 + \sigma_{13}(1 - \beta^2 h'^2/8) + O(\beta^3) = 0 \tag{29}
$$

$$
- \sigma_{13}\beta h'/2 + \sigma_{33}(1 - \beta^2 h'^2/8) + O(\beta^3) = 0 \tag{30}
$$

These equations are supplemented by the condition that all fields must be symmetric about $z = 0$. Overall equilibrium with the applied force per unit length P requires that for every X

$$
\int_{-h/2}^{h/2} \sigma_{11}\, dz = P \tag{31}
$$

We expand all quantities in a regular perturbation expansion in powers of β, anticipating that the zeroth order terms in the expansion are given by the long-wavelength solution (22). The expansion is of the form

$$
\left.
\begin{array}{c}
\sigma_{11} = \sigma(X) + \beta \sigma_{11}^{(1)} + \beta^2 \sigma_{11}^{(2)} + \cdots \\[4pt]
\sigma_{33} = \beta \sigma_{33}^{(1)} + \beta^2 \sigma_{33}^{(2)} + \cdots \\[4pt]
\sigma_{13} = \beta \sigma_{13}^{(1)} + \beta^2 \sigma_{13}^{(2)} + \cdots
\end{array}
\right\}
\tag{32}
$$

$$
\left.
\begin{array}{c}
\dot{\epsilon}_{11} = -\dot{\epsilon}_{33} = \dot{\epsilon}(X) + \beta \dot{\epsilon}_{11}^{(1)} + \beta^2 \dot{\epsilon}_{11}^{(2)} + \cdots \\[4pt]
\dot{\epsilon}_{13} = \dot{\epsilon}_{13}^{(0)} + \beta \dot{\epsilon}_{13}^{(1)} + \beta^2 \dot{\epsilon}_{13}^{(2)} + \cdots
\end{array}
\right\}
\tag{33}
$$

$$
\left.
\begin{array}{c}
\Phi = \beta^{-1} z \int^{X} \dot{\epsilon}(\tilde{X})d\tilde{X} + \Phi^{(0)} + \beta\,\Phi^{(1)} + \beta^2\,\Phi^{(2)} + \cdots \\[4pt]
v_1 = \beta^{-1} \int^{X} \dot{\epsilon}(\tilde{X})d\tilde{X} + v_1^{(0)} + \beta v_1^{(1)} + \beta^2 v_1^{(2)} + \cdots \\[4pt]
v_3 = -z\dot{\epsilon}(X) + \beta v_3^{(1)} + \beta^2 v_3^{(2)} + \cdots
\end{array}
\right\}
\tag{34}
$$

The terms in the expansions are assumed to be functions of X and z. The term $\dot{\epsilon}_{13}^{(0)}$ is zero but the development which follows is clearer if it is carried along for the moment.

From (24) and (25) one finds

$$v_1^{(0)} = \Phi_{,z}^{(0)}, \quad v_1^{(1)} = \Phi_{,z}^{(1)}, \quad v_2^{(2)} = \Phi_{,z}^{(2)}, \quad v_3^{(1)} = -\Phi_{,X}^{(0)}$$

$$v_3^{(2)} = -\Phi_{,X}^{(1)}, \quad \dot{\epsilon}_{11}^{(1)} = \Phi_{,Xz}^{(0)}, \quad \dot{\epsilon}_{11}^{(2)} = \Phi_{,Xz}^{(1)}, \quad 2\dot{\epsilon}_{13}^{(0)} = \Phi_{,zz}^{(0)}$$

$$2\dot{\epsilon}_{13}^{(1)} = \Phi_{,zz}^{(1)} - z\dot{\epsilon}', \quad 2\dot{\epsilon}_{13}^{(2)} = \Phi_{,zz}^{(2)} - \Phi_{,XX}^{(0)} \tag{35}$$

Substitution of (32) for σ_{11} into (31) gives

$$\int_{-h/2}^{h/2} \sigma_{11}^{(k)} dz = 0, \quad k \geq 1 \tag{36}$$

Expansion of the effective stress in powers of β gives

$$\sigma_e = (\sqrt{3}\sigma/2)\{1 + \beta(\sigma_{11}^{(1)} - \sigma_{33}^{(1)})/\sigma \tag{37}$$

$$+ \beta^2[\sigma(\sigma_{11}^{(2)} - \sigma_{33}^{(2)}) + 2\sigma_{13}^{(1)^2}]/\sigma^2 + O(\beta^3)\}$$

Next, using (32) and (33) in (3) gives the long-wavelength result (22) with $\dot{\epsilon}_{13}^{(0)} = 0$ plus

$$\dot{\epsilon}_{11}^{(1)} = n\dot{\epsilon}(\sigma_{11}^{(1)} - \sigma_{33}^{(1)})/\sigma, \quad \dot{\epsilon}_{13}^{(1)} = 2\dot{\epsilon}\sigma_{13}^{(1)}/\sigma$$

$$\dot{\epsilon}_{11}^{(2)} = n\dot{\epsilon}(\sigma_{11}^{(2)} - \sigma_{33}^{(2)})/\sigma + n(n-1)\dot{\epsilon}(\sigma_{11}^{(1)} - \sigma_{33}^{(1)})^2/(2\sigma^2)$$

$$+ 2(n-1)\dot{\epsilon}\sigma_{12}^{(1)^2}/\sigma^2 \tag{38}$$

$$\dot{\epsilon}_{13}^{(2)} = 2\dot{\epsilon}\sigma_{13}^{(2)}/\sigma + 2(n-1)\dot{\epsilon}\sigma_{13}^{(1)}(\sigma_{11}^{(1)} - \sigma_{33}^{(1)})/\sigma^2$$

We now proceed to solve for the various terms in the expansions. First note that $\dot{\epsilon}_{13}^{(0)} = 0$, by ($35_4$), implies that

$$\Phi^{(0)} = z f_1(X) + f_2(X) \tag{39}$$

where f_1 and f_2 are to be determined. Since Φ must be odd in z, $f_2 = 0$. Substitution of (32) into the equilibrium equations (23) and the boundary conditions (29) and (30), retaining the lowest order nonvanishing terms in β, gives

$$\sigma_{13,z}^{(1)} = \sigma h'/h, \quad \sigma_{33,z}^{(1)} = 0 \tag{40}$$

with

$$\sigma_{13}^{(1)} = \sigma h'/2, \quad \sigma_{33}^{(1)} = 0 \quad \text{on} \quad z = h/2 \tag{41}$$

Together, with (38), these imply

$$\sigma_{33}^{(1)} = 0, \quad \sigma_{13}^{(1)} = \sigma z h'/h, \quad \dot{\epsilon}_{13}^{(1)} = 2\dot{\epsilon}z h'/h \tag{42}$$

Equations (38_1), (35_4) and (42) can be used to find

$$\Phi^{(1)} = (4-n)\dot{\epsilon}h'z^3/(6h) + f_3(X)z \tag{43}$$

where f_3 is to be determined. From (35_3) and (39) it follows that $\dot{\epsilon}_{11}^{(1)} = f_1'$; thus,

References p. 126.

by (38_1) and (42),

$$\sigma_{11}^{(1)} = \sigma f_1'/(n\dot\epsilon)$$

But imposition of (36) implies $f_1' = 0$ and thus $\sigma_{11}^{(1)} = 0$. Without loss in generality we may take $f_1 = 0$.

Repetition of the above procedure for the next higher order terms in each of the equations leads to

$$\left.\begin{array}{ll} \sigma_{13}^{(2)} = 0, & \sigma_{33}^{(2)} = \sigma h[h'' - 4(h^{-2}h')'z^2]/8 \\[2mm] \Phi^{(2)} = f_4(X)z, & \dot\epsilon_{13}^{(2)} = 0, \quad \dot\epsilon_{11}^{(2)} = f_3' + (4-n)(\dot\epsilon h'/h)'z^2/2 \end{array}\right\} \quad (44)$$

Furthermore, (44_2) together with (38_2) and (36) give

$$f_3' = (n\dot\epsilon/24)[(5-n)h'^2 - (1+4/n)hh''] \quad (45)$$

and

$$\sigma_{11}^{(2)} = (\sigma/24)[1 - 12(z/h)^2][(5-n)h'^2 + 2(1-2/n)hh''] \quad (46)$$

This completes the solution for the stresses and strain-rates up to and including terms of order β^2.

We now examine in some detail the distribution of stress and strain-rate across the minimum section of the neck. Take the minimum section to lie at $X = x_1 = 0$, as depicted in Fig. 3, so that $h'(0) = 0$. With h, h'', σ, etc., now denoting values at the minimum section, the above results specialize to

$$\left.\begin{array}{l} \sigma_{11} = \sigma\{1 + (\beta^2/(12n))(n-2)hh''[1 - 12(z/h)^2]\} \\[2mm] \sigma_{33} = \sigma(\beta^2/8)hh''[1 - 4(z/h)^2] \\[2mm] \sigma_e = (\sqrt3\sigma/2)\{1 - (\beta^2/(24n))hh''[n + 4 + 12(n-4)(z/h)^2]\} \\[2mm] \dot\epsilon_{11} = \dot\epsilon\{1 - (\beta^2/24)hh''[n + 4 + 12(n-4)(z/h)^2]\} \end{array}\right\} \quad (47)$$

to order β^2 with $\sigma_{13} = \dot\epsilon_{13} = 0$. For $n = 4$, σ_e and $\dot\epsilon_{11}$ are uniform across the minimum section of the neck to this order. Furthermore, for $n = 4$, or at $z = 0$ for any n, σ_e and $\dot\epsilon_{11}$ are diminished below their respective long-wavelength values by an amount proportional to β^2hh''. Let R be the radius of curvature of the surface of the sheet at the neck minimum which is given by

$$\frac{1}{R} = \frac{1}{2}\frac{d^2h}{dx_1^2} = \frac{1}{2}\beta^2h'' \quad (48)$$

Noting that the combination β^2hh'' equals $2h/R$, we can therefore write to order h/R

$$\sigma_e = \frac{\sqrt3}{2}\sigma\left\{1 - \frac{h}{R}\left[\frac{n+4}{12n} + \frac{(n-4)}{n}\left(\frac{z}{h}\right)^2\right]\right\} \quad (49)$$

$$\dot\epsilon_{11} = \dot\epsilon\left\{1 - \frac{h}{R}\left[\frac{n+4}{12} + (n-4)\left(\frac{z}{h}\right)^2\right]\right\} \quad (50)$$

The hydrostatic tension distribution across the neck minimum is given by

$$\frac{1}{3}\sigma_{pp} = \frac{1}{2}\sigma\left\{1 + \frac{h}{R}\left[\frac{5n-4}{12n} - \frac{(3n-4)}{n}\left(\frac{z}{h}\right)^2\right]\right\} \tag{51}$$

To draw comparison with Bridgman's [5] formulas for the stress distribution at a plane strain neck in a sheet of time-independent plastic material, we use (47) and (48) to determine the ratios

$$\frac{\sigma_{11}}{(2\sigma_e/\sqrt{3})} = 1 + \frac{1}{4}\frac{h}{R}\left[1 - 4\left(\frac{z}{h}\right)^2\right]$$

$$\frac{\sigma_{33}}{(2\sigma_e/\sqrt{3})} = \frac{1}{4}\frac{h}{R}\left[1 - 4\left(\frac{z}{h}\right)^2\right]$$

$$\tag{52}$$

These ratios, which are independent of n, are identical to the analogous expressions obtained from Bridgman's formulas when his expressions are expanded out up to terms of order h/R. Bridgman's starting point is the assumption $\dot{\epsilon}_{11}$ and σ_e are uniform across the neck. For the present time-dependent power law material this is precisely true only when $n = 4$, to order h/R. For $n < 4$ the effective stress and the strain-rate are largest at the surface of the sheet, while for $n > 4$ they are largest at the midplane. The hydrostatic tension is maximum at the midplane if $n > 4/3$.

COMPARISONS WITH NUMERICAL SOLUTIONS

Let $\dot{\epsilon}_A$ and $\dot{\epsilon}_B$ be the strain-rates as predicted by the long-wavelength analysis at any two material cross-sections with h_A and h_B as the current thicknesses. From (4) any two such strain-rates are related by

$$\dot{\epsilon}_A h_A{}^n = \dot{\epsilon}_B h_B{}^n \tag{53}$$

Since

$$\dot{h}_A = -\dot{\epsilon}_A h_A, \qquad \dot{h}_B = -\dot{\epsilon}_B h_B \tag{54}$$

the respective true strains can be written in terms of the current thickness and the initial thicknesses, $h_A{}^0$ and $h_B{}^0$, as

$$\epsilon_A = -\ln(h_A/h_A{}^0), \qquad \epsilon_B = -\ln(h_B/h_B{}^0) \tag{55}$$

Equations (53) and (55) can be combined and integrated to give

$$1 - e^{-n\epsilon_A} = (h_B{}^0/h_A{}^0)^n(1 - e^{-n\epsilon_B}) \tag{56}$$

which provides a relation between the strains at any two cross-sections independent of the loading history.

To make a direct comparison with the long-wavelength result (56) and some full finite element calculations, we consider a sheet whose initial thickness variation is given by (7). The finite element procedure has been specially tailored to

References p. 126.

deal with the class of materials (3). Incompressible plane strain elements [6] have been designed for the purpose and a Newton-Raphson procedure is used to solve the nonlinear equations at each time step. Periodicity and symmetry in the x_1-direction permit the use of a finite element grid over just one half-wavelength of the current geometry, which is updated each time step. A zero shear stress σ_{13} and a uniform velocity v_1 over each end of the half-wavelength sector can be imposed as boundary conditions. From the homogeneity of the constitutive law (3), it can be shown that the simple property noted in conjunction with (56) generalizes such that the ratio of any strain component at one point, say ϵ_{11}^A, to any component at another point, say ϵ_{11}^B, is independent of the load history $P(t)$. Thus the results in Figs. 4 and 5 for ϵ_A/ϵ_B vs. ϵ_B, comparing the long-wavelength result (56) and the finite element results, do not depend on load history when presented in this manner.

The initial amplitude in (7) for the calculated results in Fig. 4 is $\xi = .01$ and $n = 5$. For the long-wavelength curve ϵ_A is the strain at the minimum section ($x = 0$) and ϵ_B is the strain at the thickest section ($x = l/2$). Thus, from (7), the

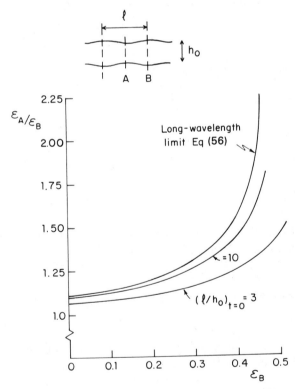

Fig. 4.   Strains at the thinnest and thickest sections of the sheet for two initial wavelength to thickness ratios compared with long-wavelength prediction. (Initial amplitude $\xi = .01$; $n = 5$).

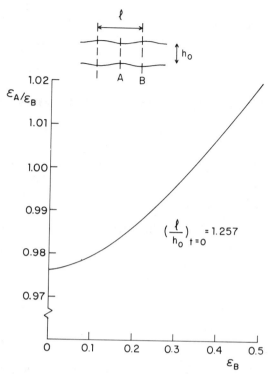

Fig. 5.   Strains at the thinnest and thickest sections of the sheet for an initial wavelength to thickness ratio chosen to give an initial decrease in the relative size of the nonuniformity. (Initial amplitude $\xi = .01$; $n = 5$).

initial ratio of thicknesses at A and B is

$$h_A{}^0/h_B{}^0 = (1 - \xi)/(1 + \xi) = .980$$

The maximum strain attained at B according to the long-wavelength approximation is given by (56) with $\epsilon_A \to \infty$, i.e.,

$$\epsilon_B{}^* = -(1/n)\ln[1 - (h_A{}^0/h_B{}^0)^n] = .470 \tag{57}$$

Finite element results in Fig. 4 are shown for initial wavelength to thickness ratios of

$$(l/h_0)_{t=0} = 3 \quad \text{and} \quad 10$$

For these curves ϵ_A and ϵ_B are the effective strains at the minimum and maximum sections defined in terms of the respective current and initial thicknesses according to (55). (The variation of the strain-rates across the thickness direction at these sections is nearly constant for $n = 5$ as will be seen below.) From the exact results of the linearized analysis in Fig. 2 it can be seen that $G \cong .6$ for $l/h_0 = 3$ which, from (16), corresponds to an initial growth-rate which is sixty percent of the long-wavelength growth-rate. For $l/h_0 = 10$, $G \approx .92$ so that the initial growth-rate is much closer to the long-wavelength prediction. For an initial ratio

$l/h_0 = 3$ the long-wavelength estimate for the limit strain ϵ_B* falls below the actual value by more than 25 percent.

For the example in Fig. 5, $\xi = .01$, $n = 5$ with the initial ratio $l/h_0 = \pi/2.5 = 1.257$. From Fig. 2, $G \cong -.23$ and thus we expect the relative size of the nonuniformity to decrease initially. This is reflected in the finite element results in Fig. 5 by the fact that the ratio ϵ_A/ϵ_B is less than unity until ϵ_B reaches about .35. Once the current wavelength to thickness ratio l/h reaches about 1.8 the linearized analysis of Fig. 2 indicates that the relative size of the nonuniformity will begin to grow. Note, however, that even at $\epsilon_B = .5$ exceedingly little nonuniformity has developed compared to responses in Fig. 4.

Distributions of $\dot{\epsilon}_{11}/\dot{\epsilon}$ across the minimum section of the neck are shown for several values of n in Fig. 6, where $\dot{\epsilon}$ is the long-wavelength value (21). The dashed line curves are computed from the long-wavelength approximation with the lowest order h/R correction (50), while the solid line curves are from the finite element calculations. For each of the n-values in these figures the current thickness was taken to be the sinusoidal variation (7) with $l/h_0 = 3$. In Fig. 6(a),

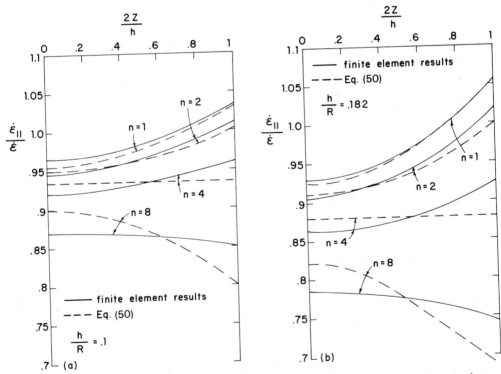

Fig. 6. Strain-rate distribution across the minimum section of the sheet—a comparison of the finite element results with the long-wavelength result including the lowest order correction given by equation (50). $\dot{\epsilon}$ is the long-wavelength result from equation (21). In (a), $l/h_0 = 3$ and $\xi = .048$ corresponding to $h/R = .1$; in (b), $l/h_0 = 3$ and $\xi = .0912$ corresponding to $h/R = .182$.

$\xi = .048$ corresponding to

$$h/R = 2\pi^2\xi(1 - \xi)(h_0/l)^2 = .1 \tag{58}$$

In Fig. 6(b), $\xi = .0912$ corresponding to $h/R = .182$. The larger is n, the larger is the order h/R correction term in (50). For $n = 8$ this correction term appears to be too large for the perturbation expansion, to order h/R, to retain accuracy in the present examples. For $n = 4$ the numerical results do show that $\dot{\epsilon}_{11}$ is approximately uniform across the neck. Even for $n = 8$, the strain-rate at the midsurface is reasonably approximated by (50).

DISCUSSION

The linearized analysis described in this paper reveals that, regardless of how *small* the amplitude of the thickness variation is, the growth-rate of the nonuniformity will be less than the prediction of the long-wavelength approximation if its wavelength is less than about four times the thickness. The perturbation analysis, on the other hand, complements the linearized analysis by showing that, regardless of how *large* the amplitude is, the long-wavelength approximation is valid if the wavelength of the variation is sufficiently large. At the neck minimum, h/R measures the extent to which departure from the long-wavelength approximation can be expected. If it is assumed that the current wavelength of the nonuniformity is sufficiently large so that the long-wavelength approximation is currently valid, it is a simple matter to calculate $(h/R)^{\cdot}$ at the neck minimum for the power law material (3). The result is

$$(h/R)^{\cdot} = (n - 2)\dot{\epsilon}h/R \tag{59}$$

in terms of the instantaneous quantities at the neck. Thus, if $n < 2$ and if the nonuniformity initially has a long-wavelength, then "localization" will not occur, since h/R decreases, and the long-wavelength analysis will actually become increasingly accurate. (The sheet will nevertheless shrink down to zero thickness at the minimum section with a finite limit strain at any other section of the form (57).) On the other hand, if $n > 2$, h/R increases and the stresses and strain-rates will progressively depart from the predictions of the long-wavelength analysis.

It is interesting to note that a slightly different criterion for localization based on steepening of the surface also leads to $n = 2$ as the transition point. Steepening of the neck surface depends on the convected rate of its slope, $(dh/dx_1)^{\cdot}$. Another relatively simple calculation, assuming the long-wavelength approximation applies, gives

$$(dh/dx_1)^{\cdot} = (n - 2)\dot{\epsilon}(dh/dx_1) \tag{60}$$

at an inflection point where $d^2h/dx_1^2 = 0$. Thus for $n < 2$ the magnitude of the slope at an inflection point decreases, while for $n > 2$ it steepens.

Localization in the above sense invariably leads to a breakdown of the long-wavelength approximation, as just discussed. Nevertheless, the major portion of the lifetime, in terms of straining outside the neck, may still be adequately

References p. 126.

represented by the long-wavelength approximation if the initial wavelength of the nonuniformity is sufficiently large, as is illustrated by the example of Fig. 4. This is an important consideration in our use of this approximation in Parts II and III where we will primarily be concerned with the amount of straining which can be attained in portions of the sheet away from the neck.

ACKNOWLEDGMENT

The work of J.W.H. was supported in part by the Air Force Office of Scientific Research under Grant AFOSR 77-3330, the National Science Foundation under Grant ENG76-04019, and by the Division of Applied Sciences, Harvard University. The work was conducted while K.W.N. was on leave at Harvard University. The support of the Faculty of Applied Sciences at the University of Sherbrooke is gratefully acknowledged. The work of A.N. was supported by the Materials Research Laboratory at Brown University funded by the National Science Foundation.

REFERENCES

[1] Z. Marciniak, K. Kuczyński, Int. J. Mech. Sci., 9 (1967), 609.
[2] J. W. Hutchinson and H. Obrecht, Fracture 1977, ICF4 (ed. D. M. R. Taplin), 1 (1977), 101.
[3] J. W. Hutchinson and K. W. Neale, Acta Met., 25 (1977), 839.
[4] E. J. Appleby and O. Richmond, to be published.
[5] P. W. Bridgman, Studies in Large Plastic Flow and Fracture, Harvard University Press (1964).
[6] A. Needleman and C. F. Shih, to be published.

SHEET NECKING-II. TIME-INDEPENDENT BEHAVIOR*

J. W. HUTCHINSON

Harvard University, Cambridge, Massachusetts

K. W. NEALE

Université de Sherbrooke, Sherbrooke, Quebec, Canada

ABSTRACT

Various factors affecting the prediction of limit strains in biaxially-stretched sheets are studied. Time-independent material behavior is assumed, and both the flow theory of plasticity as well as a finite-strain version of deformation theory are considered. A localization-band bifurcation analysis is first carried out. The influence of geometric imperfections is then analyzed using the long-wavelength approximation treated in Part I. We also discuss the predicted forming limit curves and comment on their relation to published experimental data. The main emphasis in this Part, however, is on comparisons between the corresponding predictions of flow theory and deformation theory.

INTRODUCTION

In this second Part, various aspects of localized necking in thin sheets are examined. We restrict attention to *time-independent* material response. A theoretical investigation of this phenomenon was apparently initiated by Hill [1], who considered rigid-plastic solids obeying the common flow theories of plasticity.

Hill's analysis, which is conducted within the framework of generalized plane stress, predicts that localized necking in biaxially stretched sheets only occurs for those principal strain states varying between uniaxial tension and plane strain. However, it is well-established experimentally that necking-type failures take place in the so-called "biaxial tension" range where, since both principal strains are positive, Hill's criterion predicts infinite ductility. Two approaches have been formulated to resolve this discrepancy between Hill's result and experiments. The first approach, due to Marciniak and Kuczyński [2] and extended by Sowerby

---

* *The material in this three-part paper was presented orally in Session II under the title "Constitutive Relations for Sheet Metal" and Session IV under the title "Sheet Necking: Influence of Constitutive Theory and Strain-Rate Dependence."*

References pp. 149–150.

and Duncan [3], postulates the existence of an initial nonhomogeneity in the form of a narrow band across the sheet. The localization process during straining is then described in terms of the growth of the nonhomogeneity relative to the remainder of the sheet. To examine this hypothesis, Azrin and Backofen [4] conducted a series of experiments on a number of metals under strain states varying from plane strain to equibiaxial tension. For most of the metals tested, they found that the experimental results contradicted the trend of the Marciniak-Kuczyński (M-K) theory. It is important to note, however, that the J_2 (von Mises) flow theory of plasticity was assumed in the analyses.

An alternative approach, recently proposed by Stören and Rice [5], incorporates a J_2 deformation theory of plasticity into a classical bifurcation analysis. The bifurcation mode corresponds to localized deformation in a narrow band, as in Hill's analysis [1]. In certain cases, the results with this theory showed much better agreement with experimental trends than those obtained with the M-K flow theory analysis.

These developments in sheet necking analysis closely parallel discussions which have taken place in the related field of plastic buckling. There, much attention was focused on the reliability of J_2 deformation theory bifurcation predictions vs. the severe imperfection-sensitivity often associated with J_2 flow theory calculations. In spite of many objections which were initially raised concerning the use of deformation theory, it was later shown that use of this theory could be rigorously justified in bifurcation analyses since it is equivalent to a flow theory which permits the development of yield-surface corners (see discussion in [6]). The fact that bifurcation calculations based on deformation theory are generally in much better accord with experimentally determined buckling loads than the corresponding flow theory predictions has strongly favored the use of deformation theory in this application. On the other hand, it has also been demonstrated that incorporation of realistic imperfections in a J_2 flow theory analysis usually produces buckling loads which agree adequately with test data. Consequently, it is still an open question as to which is the more fundamental approach to analyzing plastic buckling.

This Part of our study represents an extensive investigation into various factors which affect necking in thin sheets. The main emphasis throughout is on comparisons between the theoretical predictions of flow theory and deformation theory of plasticity. A finite-strain version of J_2 deformation theory, somewhat different than that proposed by Stören and Rice [5], is introduced and applied to both a bifurcation analysis and an assessment of imperfection-sensitivity. The imperfection-sensitivity analysis is based on the "long-wavelength" approximation treated in detail in Part I of this paper. We consider the full biaxial stretching range, i.e., with principal strain states varying between uniaxial and equibiaxial tension, and obtain corresponding results according to J_2 flow theory.

An important feature of sheet necking is seen to be the imperfection-sensitivity associated with the deformation theory as well as with the flow theory. This is in sharp contrast to plastic buckling results obtained using deformation theory, which are often relatively insensitive to imperfections. In the discussion of such results and relevant experimental data, we address the issue of deformation theory vs. flow theory in this context.

CONSTITUTIVE LAWS

In this study finite-strain versions of both J_2 flow theory and J_2 deformation theory of plasticity will be employed. Throughout this investigation the material is assumed to be incompressible and initially isotropic.

J_2 **Flow Theory**—For the flow theory analysis the constitutive law, with reference to fixed Cartesian coordinates, is taken to be (see for example [7–9])

$$\dot{\epsilon}_{ij} = \frac{3}{2E} \tilde{s}_{ij} + \frac{\alpha f}{E} s_{ij} s_{kl} \tilde{s}_{kl} \tag{1}$$

where for incipient yielding

$$\begin{aligned} \alpha = 1 \quad &\text{if} \quad \dot{J}_2 = s_{ij}\tilde{s}_{ij} \geq 0 \\ \alpha = 0 \quad &\text{if} \quad \dot{J}_2 < 0 \end{aligned} \tag{2}$$

Here, $\dot{\epsilon}_{ij}$ is the Eulerian strain-rate tensor, E is Young's modulus, $\tilde{s}_{ij}$ denotes the Jaumann rate of change of the Cauchy stress deviator $s_{ij} = \sigma_{ij} - \frac{1}{3}\delta_{ij}\sigma_{kk}$, and f is a function of $J_2(= \frac{1}{2}s_{ij}s_{ij})$ which can be chosen to make (1) coincide with any monotonic proportional loading history. If the uniaxial tension curve is used to determine f, the inverted form of (1) for plastic loading becomes

$$\tilde{\sigma}_{ij} = \frac{2E}{3}[\dot{\epsilon}_{ij} - Q s_{ij} s_{kl} \dot{\epsilon}_{kl}] + \delta_{ij}\dot{p} \tag{3}$$

where p is the hydrostatic pressure, and

$$Q = \frac{3}{2\sigma_e^2}\left(1 - \frac{E_t}{E}\right) \tag{4}$$

Here, the tangent modulus E_t is regarded as a function of the effective stress $\sigma_e(= \sqrt{3J_2})$ and in simple tension corresponds to the slope of the true stress-natural (logarithmic) strain curve.

In the localization-band bifurcation analysis, the assumption of approximate plane stress will be adopted. The state of uniform stress considered prior to bifurcation is such that the only nonvanishing stress components are $\sigma_{11} \equiv \sigma_1$ and $\sigma_{22} \equiv \sigma_2$. In this case, $\dot{p}$ can be eliminated from (3) by means of the relation $\tilde{\sigma}_{33} = 0$ and the incompressibility condition $\dot{\epsilon}_{kk} = 0$. The constitutive law (3) for plastic loading then reduces to

$$\begin{aligned} \tilde{\sigma}_1 &= \hat{L}_{11}\dot{\epsilon}_1 + \hat{L}_{12}\dot{\epsilon}_2 \\ \tilde{\sigma}_2 &= \hat{L}_{12}\dot{\epsilon}_1 + \hat{L}_{22}\dot{\epsilon}_2 \\ \tilde{\sigma}_{12} &= 2\hat{L}_s\dot{\epsilon}_{12} \end{aligned} \tag{5}$$

where the flow theory instantaneous moduli are given by

$$\hat{L}_{11} = \frac{4}{3} E - (E - E_t)\left(\frac{\sigma_1}{\sigma_e}\right)^2$$

$$\hat{L}_{22} = \frac{4}{3} E - (E - E_t)\left(\frac{\sigma_2}{\sigma_e}\right)^2$$

$$\hat{L}_{12} = \frac{2}{3} E - (E - E_t)\frac{\sigma_1 \sigma_2}{(\sigma_e)^2} \tag{6}$$

$$\hat{L}_s = \frac{E}{3}$$

J_2 **Deformation Theory**—To construct a finite-strain version of the deformation theory constitutive law, we make use of Hill's theory [10] for finitely deformed isotropic elastic solids. Because of isotropy, the principal directions of Cauchy stress must coincide with the axes of the Eulerian strain ellipsoid. Furthermore, the state of strain in a material element is completely specified by the three principal stretches ($\lambda_1, \lambda_2, \lambda_3$) relative to some reference configuration, together with the principal directions of strain. "Principal axes techniques" introduced by Hill [10] can then be conveniently applied to determine the deformation theory instantaneous moduli analogous to (6). Such principal axes methods will also prove to be convenient for the direct finite deformation calculation performed in a subsequent Section on Long-wavelength Analysis.

The most appropriate measure of strain in our formulation seems to be the logarithmic strain tensor which, by definition, is coaxial with the Lagrangian strain ellipsoid and has principal values

$$\epsilon_i = \ln \lambda_i \tag{7}$$

In [10, 11] Hill has presented convincing arguments in favor of this measure. For our purposes we simply note that since $\dot{\epsilon}_i = \dot{\lambda}_i/\lambda_i$ with (7) the incompressibility constraint $\lambda_1 \lambda_2 \lambda_3 = 1$ becomes exactly $\epsilon_{ii} = 0$ as well as $\dot{\epsilon}_{ii} = 0$.

In view of the above remarks, the well-known small strain J_2 deformation theory can immediately be extended to finite strains as follows

$$\epsilon_i = \mu s_i \tag{8}$$

where μ is assumed to be a function of the effective stress $\sigma_e = (3 s_i s_i /2)^{1/2}$ or the effective strain $\epsilon_e = (2 \epsilon_i \epsilon_i/3)^{1/2}$, obtainable from the uniaxial tension curve. Since σ_e and ϵ_e are respectively equal to the true stress and true strain in uniaxial tension,

$$\mu = \frac{3}{2}\frac{\epsilon_e}{\sigma_e} = \frac{3}{2E_s} \tag{9}$$

in which E_s denotes the secant modulus. (Note that $\bar{\epsilon}$ rather than ϵ_e will be used to denote effective strain in the flow theory analysis. This is defined as $\int d\bar{\epsilon}$ where $d\bar{\epsilon} = (2 d\epsilon_{ij} d\epsilon_{ij}/3)^{1/2}$. For monotonic proportional straining paths $\epsilon_e = \bar{\epsilon}$.)

In terms of the principal components of Cauchy stress, the constitutive law (8) becomes

$$\sigma_i = \tfrac{2}{3} E_s \epsilon_i - p \tag{10}$$

For a true stress-natural strain curve of the form

$$\sigma_e = K \epsilon_e^N \quad \text{(deformation theory)}$$
$$\sigma_e = K \bar{\epsilon}^N \quad \text{(flow theory)} \tag{11}$$

equation (10) can also be expressed as follows (c.f. [12])

$$\sigma_i = \lambda_i \frac{\partial W}{\partial \lambda_i} - p \quad \text{(no sum on } i) \tag{12}$$

where the strain energy density function W is given by

$$W = \frac{K}{N+1} \epsilon_e^{N+1} \tag{13}$$

Although the "total" form (10) of the deformation theory will be employed in our long-wavelength analysis, a rate form of (10) is required for the localization-band bifurcation analysis. From (9) and the conventional definitions of effective stress and strain, the rate form of (10) becomes

$$\dot{s}_i = \dot{\sigma}_i + \dot{p}$$
$$= \tfrac{2}{3} E_s \dot{\epsilon}_i - s_i (E_s - E_t) \frac{s_k \dot{\epsilon}_k}{(\sigma_e)^2} \tag{14}$$

When this is specialized to the plane stress case considered in the derivation of (5), the deformation theory instantaneous moduli corresponding to (6_{1-3}) are

$$\hat{L}_{11} = \frac{4}{3} E_s - (E_s - E_t) \left(\frac{\sigma_1}{\sigma_e} \right)^2$$

$$\hat{L}_{22} = \frac{4}{3} E_s - (E_s - E_t) \left(\frac{\sigma_2}{\sigma_e} \right)^2 \tag{15}$$

$$\hat{L}_{12} = \frac{2}{3} E_s - (E_s - E_t) \frac{\sigma_1 \sigma_2}{(\sigma_e)^2}$$

These relations also follow from a substitution of the secant modulus E_s for E in (6). Hill's formula [10] for the instantaneous shear modulus gives

$$2\hat{L}_s = \frac{\lambda_1^2 + \lambda_2^2}{\lambda_1^2 - \lambda_2^2} (\sigma_1 - \sigma_2) = \frac{2}{3} E_s \ln \left(\frac{\lambda_1}{\lambda_2} \right) \frac{\lambda_1^2 + \lambda_2^2}{\lambda_1^2 - \lambda_2^2} \tag{16}$$

A finite-strain version of J_2 deformation theory has recently been proposed by Stören and Rice [5] which, as discussed by these authors, has path independence only when the strains are small or when the principal axes of strain are fixed relative to the material. If the principal axes are fixed relative to the material the

References pp. 149–150.

present deformation theory and that of Stören and Rice coincide in relating stress to strain. For proportional loading, which in the present study corresponds to monotonically increasing ϵ_1 and ϵ_2 in fixed ratio, the two deformation theories and the flow theory (3) all coincide. Only the present deformation theory is independent of the loading path for arbitrary histories. According to the theory of [5] the instantaneous shear modulus, instead of (16), is $2\hat{L}_s = 2E_s/3$, whereas the remaining moduli are given still by (15). For equibiaxial stretching ($\lambda_1 = \lambda_2$) the instantaneous shear modulus $\hat{L}_s$ from (16) does coincide with the Stören-Rice value. However for the strain levels and ratios of interest here, spanning equibiaxial and uniaxial, the value of $\hat{L}_s$ from (16) exceeds the value $E_s/3$ used by Stören and Rice. Since indirect evidence in the range of small strains suggests that deformation theory tends to underestimate the instantaneous moduli of an actual metal when they differ substantially from those of simple flow theory, it can be argued perhaps that the present "true" deformation theory is sufficiently conservative in its estimate of $\hat{L}_s$ in the large strain range.

If the proportional straining path

$$\frac{\epsilon_2}{\epsilon_1} = \rho = \text{const} \tag{17}$$

is imposed on the sheet, then from (1) or (8)

$$\frac{\sigma_1}{\sigma_e} = \frac{2+\rho}{[3(1+\rho+\rho^2)]^{1/2}}$$
$$\frac{\sigma_2}{\sigma_e} = \frac{1+2\rho}{[3(1+\rho+\rho^2)]^{1/2}} \tag{18}$$

which can be substituted in the moduli expressions (6) or (15). Furthermore, for power-law hardening of the type (11) in the plastic range

$$E_t = NK\epsilon_e^{N-1}, \qquad E_s = K\epsilon_e^{N-1} \tag{19}$$

where

$$\epsilon_e = \bar{\epsilon} = \frac{2(1+\rho+\rho^2)^{1/2}}{\sqrt{3}}\epsilon_1 \tag{20}$$

LOCALIZATION-BAND BIFURCATION ANALYSIS

The localization-band bifurcation analysis is carried out within the context of plane stress. The analysis is similar to those performed by Hill and Hutchinson [13] and Stören and Rice [5] in that conditions are determined for which the bifurcation mode corresponds to localized plastic deformation in a narrow band while the deformation remains homogeneous elsewhere.

We consider a thin flat sheet which is currently of uniform thickness and subjected to the homogeneous stress field (Fig. 1)

$$\sigma_{11} = \sigma_1, \qquad \sigma_{22} = \sigma_2 \tag{21}$$
$$\text{all other} \quad \sigma_{ij} = 0$$

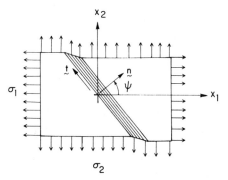

Fig. 1. Localization-band geometry.

Furthermore, we assume proportional loading up to the current state with the σ_2/σ_1 ratio being given by (18).

The velocity field describing the bifurcation mode is constrained to vary only across the band, i.e., [5]

$$v_i = F_i(n_1 x_1 + n_2 x_2) \qquad i = 1, 2 \tag{22}$$

where v_i denotes the differences between the velocity components inside and outside the band, and $n_1 = \cos \psi$ and $n_2 = \sin \psi$ are the components of the unit normal to the band. The velocity gradients corresponding to (22) are

$$v_{i,j} = F_i' n_j \equiv g_i n_j \qquad i, j = 1, 2 \tag{23}$$

and the strain-rate field is given by

$$\dot{\epsilon}_{ij} = \tfrac{1}{2}(v_{i,j} + v_{j,i}) \qquad i, j = 1, 2$$
$$\dot{\epsilon}_{33} = -(\dot{\epsilon}_{11} + \dot{\epsilon}_{22}) \tag{24}$$

Equilibrium across the band at the instant of bifurcation requires that the nominal traction-rates $\dot{T}_i$ on the band boundaries be continuous. Thus, with t_{ij} denoting the difference or jump in nominal stress-rates across the boundaries of the band

$$\Delta \dot{T}_j = n_i t_{ij} = 0 \tag{25}$$

From the following relation between nominal stress rates and Jaumann rates of Cauchy stress (for incompressible materials)

$$\dot{t}_{ij} = \overset{\triangledown}{\sigma}_{ij} + \sigma_{ik} v_{j,k} - (\sigma_{ik} \dot{\epsilon}_{jk} + \sigma_{jk} \dot{\epsilon}_{ik}) \tag{26}$$

and the constitutive law (5), the condition (25) for the above bifurcation mode becomes

$$\{n_1^2(\hat{L}_{11} - \sigma_1) + n_2^2[\hat{L}_s + \tfrac{1}{2}(\sigma_2 - \sigma_1)]\} g_1$$
$$+ n_1 n_2 \{\hat{L}_{12} + \hat{L}_s - \tfrac{1}{2}(\sigma_1 + \sigma_2)\} g_2 = 0$$
$$n_1 n_2 \{\hat{L}_{12} + \hat{L}_s - \tfrac{1}{2}(\sigma_1 + \sigma_2)\} g_1$$
$$+ \{n_1^2[\hat{L}_s + \tfrac{1}{2}(\sigma_1 - \sigma_2)] + n_2^2(\hat{L}_{22} - \sigma_2)\} g_2 = 0 \tag{27}$$

References pp. 149–150.

Bifurcation is thus possible for those states where a non-trivial solution to (27) exists, i.e., when the determinant of the coefficient matrix of the g_i vanishes:

$$\left\{ \left(\frac{n_1}{n_2}\right)^2 (\hat{L}_{11} - \sigma_1) + \hat{L}_s + \tfrac{1}{2}(\sigma_2 - \sigma_1) \right\} \left\{ \left(\frac{n_1}{n_2}\right)^2 [\hat{L}_s + \tfrac{1}{2}(\sigma_1 - \sigma_2)] + \hat{L}_{22} - \sigma_2 \right\}$$
$$- \left\{ \left(\frac{n_1}{n_2}\right) [\hat{L}_{12} + \hat{L}_s - \tfrac{1}{2}(\sigma_1 + \sigma_2)] \right\}^2 = 0 \qquad (28)$$

In general there is a bifurcation state for each band orientation ψ. However, the pertinent critical state corresponds to that n_1/n_2 ratio in (28) which minimizes the bifurcation stress. This stress will be denoted by σ_1^* and its associated value is $\sigma_2^* = (1 + 2\rho)\sigma_1^*/(2 + \rho)$.

In the flow theory analysis the appropriate moduli are (6) together with (18)–(20) for the simple power law stress-strain curve (11). Substitution of these relations in (28) leads to an expression of the form

$$f_1(\rho, \beta) + \frac{\sigma_e}{E} f_2(\rho, \beta, \bar{\epsilon}, N) + \left(\frac{\sigma_e}{E}\right)^2 f_3(\rho, \beta, \bar{\epsilon}, N) = 0 \qquad (29)$$

where $\beta = n_1/n_2$. By implicitly minimizing σ_e with respect to β in (29) and neglecting terms of the order $\sigma_e/E \ll 1$, we obtain the critical orientation first given by Hill [1]

$$\psi^* = \tan^{-1}(\sqrt{-\rho}) \qquad (30)$$

Moreover, the term $f_1(\rho, \beta)$ vanishes for this value of ψ, so that the bifurcation condition reduces to $f_2 = 0$ when terms of the order σ_e/E are neglected compared to unity. This furnishes Hill's well-known result [1] for the critical strain ϵ_1^* at localization

$$\epsilon_1^* = \frac{N}{1 + \rho} \qquad (31)$$

In Hill's analysis the material is considered to be rigid-plastic and the above represents the only localization-band solution available. Furthermore, in view of (30), bifurcation is predicted only for values of $\rho \le 0$. In contrast, a detailed numerical solution of (29), which is based on elastic-plastic material behavior, reveals that bifurcation occurs for each band orientation ψ and for the full range of ρ. However, as suggested by (29) and the subsequent development leading to (31), the bifurcation stresses become of the order of the elastic modulus E when ψ deviates from Hill's angle (30). The corresponding strains are then unrealistically large.

In the deformation theory analysis the appropriate moduli to be substituted in the bifurcation equation (28) are (15) and (16). In view of (11) and (17)–(20), we can eliminate stress quantities in (28) and, for prescribed values of ρ, N and ψ, treat ϵ_e or ϵ_1 as the eigenvalue. The bifurcation strains obtained in this case do not exhibit the strong sensitivity to variations in ψ characterized by the flow theory results. (Plots illustrating this will be displayed later.)

In general, simple expressions such as (30) and (31) cannot be obtained with the present deformation theory, so numerical solutions of (28) are required to determine the value ψ^* which minimizes the bifurcation strain ϵ_1^*. However, in the so-called "biaxial tension range" ($\rho \geq 0$), $\psi^* = 0$ is the minimizing angle and

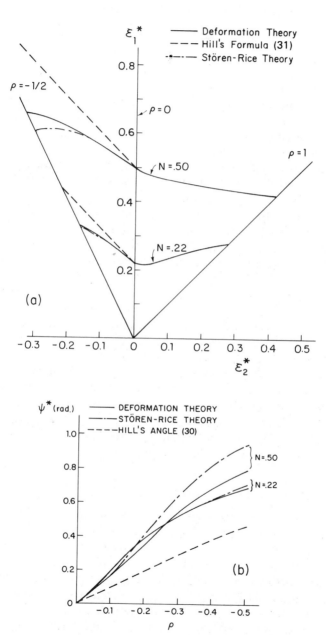

Fig. 2. (a) Forming limit curves from bifurcation analysis for various plasticity theories. (b) Localization-band angle ψ^* in negative-ρ range.

References pp. 149–150.

the critical condition (28) reduces to $n_2 = g_2 = 0$ and $\sigma_1{}^* = \hat{L}_{11}$, i.e.,

$$\epsilon_1{}^* = \frac{3\rho^2 + N(2 + \rho)^2}{2(2 + \rho)(1 + \rho + \rho^2)} \tag{32}$$

The principal axes of strain do not rotate for the associated bifurcation mode and the shear modulus $\hat{L}_s$ does not enter into (32) so that, as expected, this result is identical to that obtained by Stören and Rice [5]. In the negative ρ range, however, principal axes do not remain fixed at bifurcation and the present deformation theory leads to results which differ somewhat from those furnished by the Stören-Rice deformation theory.

In Fig. 2(a) the theoretical localization-band bifurcation strains are plotted as "forming limit curves", i.e., curves which illustrate the dependence of the critical strain $\epsilon_1{}^*$ on imposed strain ratio. Here, the solid and dashed curves refer to the deformation theory and flow theory predictions, respectively. Results are shown for a strain hardening exponent $N = .22$, typical of certain steels and aluminum alloys [4, 14], and also $N = .50$ which is representative of brass [15].

In the biaxial tension range ($\rho \geq 0$) only deformation theory predicts bifurcation, as mentioned earlier, and the corresponding necking strains are given by (32). However, Hill's flow theory formula (31) does predict localized necking in the negative ρ range. In this range, the forming limit curves of Fig. 2(a) obtained from (28) with the present version of deformation theory are observed to lie below Hill's curve, yet above the curves furnished by the Stören-Rice deformation theory [5]. The discrepancy between the two deformation theories increases with increasing strain-hardening exponent N. Nevertheless, the discrepancy even for $N = 0.50$ is not very large, and thus the choice of which deformation theory to use may not be critical in this application.

The critical angle ψ^* minimizing the bifurcation strains $\epsilon_1{}^*$ are plotted in Fig. 2(b) for strain ratios varying from plane strain ($\rho = 0$) to uniaxial tension ($\rho = -\frac{1}{2}$). According to flow theory this angle, given by (30), only depends on ρ. With deformation theory the critical orientation also depends on N, and the curves for ψ^* are above Hill's curve. The present version of deformation theory gives values of ψ^* which are between (30) and the predictions of Stören and Rice [5], and fairly close to the latter results for smaller N.

LONG-WAVELENGTH (M-K) ANALYSIS

In order to assess the effects of geometric nonuniformities on localized necking behavior, the "*long-wavelength*" approximation discussed in Part I will be applied. Consistent with the bifurcation analysis of the preceding section, our approach is within the framework of generalized plane stress. The present analysis is thus along the lines of that introduced by Marciniak and Kuczyński [2] for this problem; however, it is not restricted to the biaxial tension range ($\rho \geq 0$, $\psi = 0$) and both flow theory and deformation theory are employed.

We consider a sheet having a nonuniformity in the form of a groove or band which is initially inclined at an angle $\bar{\psi}$ (Fig. 3). The thickness along the minimum section in the groove is denoted by $h(t)$, with an initial value $h(0)$. In applying

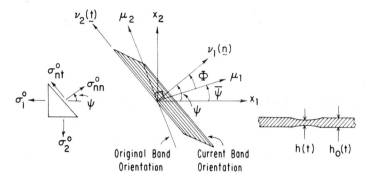

Fig. 3.   Conventions for long-wavelength analysis.

the long-wavelength approximation it is tacitly assumed that the band width-to-thickness ratio is large. The region outside the band, referred to as the "uniform" section, has thickness $h_o(t)$ with an initial value $h_o(0)$. The initial geometric nonuniformity is defined as follows

$$\xi = \frac{h_o(0) - h(0)}{h_o(0)} \geq 0 \tag{33}$$

Throughout the analysis a subscript or superscript o will denote quantities associated with behavior in the uniform region of the sheet whereas the absence of this symbol will refer to quantities along the minimum section of the groove.

In addition to the fixed $x_1 - x_2$ reference axes, two other systems of Cartesian axes are used: the set $\mu_1 - \mu_2$ aligned with the initial inclination of the groove $\bar\psi$, and axes $\nu_1 - \nu_2$ coincident with the current orientation of the groove ψ.

As in the localization-band bifurcation analysis, we take the loading imposed on the edges of the sheet to be such that

$$\frac{\epsilon_2{}^o}{\epsilon_1{}^o} = \rho = \text{const,} \qquad \frac{\sigma_2{}^o}{\sigma_1{}^o} = \frac{1 + 2\rho}{2 + \rho} \tag{34}$$

The $x_1 - x_2$ axes thus represent the principal directions of stress and strain for the uniform sections of the sheet. From the basic definition (7) of logarithmic strain, we obtain the following relation for the current groove inclination

$$\tan \psi = \frac{\lambda_1{}^o}{\lambda_2{}^o} \tan \bar\psi = \exp[(1 - \rho)\epsilon_1{}^o] \tan \bar\psi \tag{35}$$

The rotation of the band during loading is thus

$$\Phi = \psi - \bar\psi \tag{36}$$

According to the long-wavelength simplification the state of stress over each cross-section is considered to be uniform. That is, we take the stress and strain components to be quantities which are averaged through the thickness. With n and t denoting the normal and tangential directions to the current groove inclination ($n \equiv \nu_1$, $t \equiv \nu_2$), the equilibrium conditions across the band become

$$\sigma_{nn}^o h_o = \sigma_{nn} h, \quad \sigma_{nt}^o h_o = \sigma_{nt} h \quad \text{or} \quad \frac{\sigma_{nt}}{\sigma_{nn}} = \frac{\sigma_{nt}^o}{\sigma_{nn}^o} \tag{37}$$

In the uniform section, the normal and shear stress components with respect to the $n - t$ axes are given by

$$\sigma_{nn}^o = \sigma_1{}^o \cos^2\psi + \sigma_2{}^o \sin^2\psi$$

$$\sigma_{tt}^o = \sigma_1{}^o \sin^2\psi + \sigma_2{}^o \cos^2\psi \tag{38}$$

$$\sigma_{nt}^o = (-\sigma_1{}^o + \sigma_2{}^o) \sin\psi \cos\psi$$

An effective stress-effective strain law of the type (11) together with the first of (37) and the definitions

$$\epsilon_3 = \ln\frac{h}{h(0)}, \qquad \epsilon_2{}^o = \ln\frac{h_o}{h_o(0)} \tag{39}$$

leads to the following expression (c.f. [14]) for deformation theory

$$\frac{\sigma_{nn}^o/\sigma_e{}^o}{\sigma_{nn}/\sigma_e} = (1 - \xi)\left(\frac{\epsilon_e}{\epsilon_e{}^o}\right)^N \exp(\epsilon_3 - \epsilon_3{}^o) \tag{40}$$

where ξ is the initial geometric nonuniformity introduced in (33). The same expression holds for flow theory with $\epsilon_e/\epsilon_e{}^o$ replaced by $\bar{\epsilon}/\bar{\epsilon}^o$.

Equation (34) and the basic definition of effective stress yields

$$\frac{\sigma_{nn}^o}{\sigma_e{}^o} = \frac{(\rho + 2) \cos^2\psi + (2\rho + 1) \sin^2\psi}{\sqrt{3}(1 + \rho + \rho^2)^{1/2}} \tag{41}$$

Furthermore, the incompressibility condition $\epsilon_1{}^o + \epsilon_2{}^o + \epsilon_3{}^o = 0$ and (20) gives

$$\epsilon_3{}^o = -\frac{\sqrt{3}(1 + \rho)}{2(1 + \rho + \rho^2)^{1/2}} \epsilon_e{}^o \equiv -C\epsilon_e{}^o \tag{42}$$

Note that, since proportional loading occurs in the uniform section, $\epsilon_e{}^o \equiv \bar{\epsilon}^o$ in this case.

The purpose of the subsequent analysis is to express the remaining quantities in (40), which correspond to the deformation along the groove, in terms of $\bar{\epsilon}$ (or ϵ_e). The resulting relationships will of course depend on whether flow theory or deformation theory is applied. We seek equations relating $\bar{\epsilon}$ to $\bar{\epsilon}^o$ (or ϵ_e to $\epsilon_e{}^o$). From these relationships we can calculate the development of the groove and the limit strains in the uniform section of the sheet.

J_2 Flow Theory Analysis—In the flow theory analysis the material is assumed to be rigid-plastic. The constitutive law (1) then becomes

$$d\epsilon_{ij} = \frac{3}{2}\frac{d\bar{\epsilon}}{\sigma_e} s_{ij} \tag{43}$$

Using this expression together with the defining equation for σ_e, one can eliminate σ_{tt} to find

$$\frac{d\epsilon_{tt}}{d\bar{\epsilon}} = \pm \left[1 - \frac{3}{4}\left(\frac{\sigma_{nn}}{\sigma_e}\right)^2 - 3\left(\frac{\sigma_{nt}}{\sigma_e}\right)^2 \right]^{1/2} \tag{44}$$

where the appropriate sign depends on the orientation of the band.
From the compatibility condition

$$d\epsilon_{tt} = d\epsilon_{tt}^o \tag{45}$$

the relation

$$d\epsilon_{tt}^o = d\epsilon_1{}^o \sin^2\psi + d\epsilon_2{}^o \cos^2\psi \tag{46}$$

and equations (34), (37), (38) and (44), it follows after a lengthy but straightforward calculation that

$$\frac{\sigma_{nn}}{\sigma_e} = \frac{1}{AH}\left[1 - B\left(\frac{d\bar{\epsilon}^o}{d\bar{\epsilon}}\right)^2 \right]^{1/2} \tag{47}$$

where for the range $-\frac{1}{2} \le \rho \le 1$ considered here the root $\sigma_{nn} > 0$ has been chosen. Here, in a form similar to that in [14],

$$A = \frac{\sqrt{3}}{2}$$

$$B = \frac{3(\sin^2\psi + \rho \cos^2\psi)^2}{4(1 + \rho + \rho^2)} \tag{48}$$

$$H = \left\{ 1 + \left[\frac{2(\rho - 1)\sin\psi\cos\psi}{(\rho + 2)\cos^2\psi + (2\rho + 1)\sin^2\psi} \right]^2 \right\}^{1/2}$$

Substitution of (41), (42) and (47) into the equilibrium equation (40) together with (48) then gives

$$(1 - B - G)^{1/2} H \left[1 - B\left(\frac{d\bar{\epsilon}^o}{d\bar{\epsilon}}\right)^2 \right]^{-1/2} = (1 - \xi)\left(\frac{\bar{\epsilon}}{\bar{\epsilon}^o}\right)^N \exp(C\bar{\epsilon}^o + \epsilon_3) \tag{49}$$

where

$$G = \frac{(\rho - 1)^2 \sin^2\psi \cos^2\psi}{(1 + \rho + \rho^2)} \tag{50}$$

From (43) and (45)–(47), the following expression is obtained

$$\frac{d\epsilon_3}{d\bar{\epsilon}} = -\frac{A}{H}\left[1 - B\left(\frac{d\bar{\epsilon}^o}{d\bar{\epsilon}}\right)^2 \right]^{1/2} - D\frac{d\bar{\epsilon}^o}{d\bar{\epsilon}} \tag{51}$$

with

$$D = \frac{\sqrt{3}(\sin^2\psi + \rho \cos^2\psi)}{4(1 + \rho + \rho^2)^{1/2}} \tag{52}$$

A straightforward incremental solution of (49) and (51) determines the groove deformation $\bar{\epsilon}$ in terms of the prescribed uniform deformation $\bar{\epsilon}^o$. In the numerical solution, the strain ratio ρ and initial groove angle $\bar{\psi}$ are initially specified. The current groove orientation ψ is updated at each increment using (35). For the case $\bar{\psi} = \psi = 0$, equations (49) and (51) reduce to those given by Marciniak *et al* [14].

J_2 **Deformation Theory Analysis**—In the deformation theory analysis the total form (10) of the constitutive law is employed together with the principal axes techniques referred to previously. Alternatively, an incremental formulation along lines just given for flow theory could be developed, but we prefer to take full advantage of the total form of the deformation theory.

We first describe the deformation in the groove with respect to the $\mu_1 - \mu_2$ axes (Fig. 3) as follows

$$\mu_1 = \delta\bar{\mu}_1 + \gamma\bar{\mu}_2$$
$$\mu_2 = \eta\bar{\mu}_1 + \phi\bar{\mu}_2 \tag{53}$$

Here, and in the following, a bar denotes the initial coordinates of a material point whereas an un-barred coordinate refers to its current position. For the uniform deformation (34),

$$x_1 = \lambda_1{}^o\bar{x}_1, \qquad x_2 = \lambda_2{}^o\bar{x}_2 \tag{54}$$

which, when referred to the $\mu_1 - \mu_2$ axes, becomes

$$\mu_1 = (\lambda_1{}^o \cos^2\bar{\psi} + \lambda_2{}^o \sin^2\bar{\psi})\bar{\mu}_1 - (\lambda_1{}^o - \lambda_2{}^o) \sin\bar{\psi} \cos\bar{\psi}\bar{\mu}_2$$
$$\mu_2 = -(\lambda_1{}^o - \lambda_2{}^o) \sin\bar{\psi} \cos\bar{\psi}\bar{\mu}_1 + (\lambda_1{}^o \sin^2\bar{\psi} + \lambda_2{}^o \cos^2\bar{\psi})\bar{\mu}_2 \tag{55}$$

Compatibility of deformation on $\bar{\mu}_1 = 0$ then leads to the following "matching" conditions

$$\phi = \lambda_1{}^o \sin^2\bar{\psi} + \lambda_2{}^o \cos^2\bar{\psi}$$
$$\gamma = -(\lambda_1{}^o - \lambda_2{}^o) \sin\bar{\psi} \cos\bar{\psi} \tag{56}$$

(From (35), (36) and (56) it follows that $\gamma/\phi = -\tan\Phi$ as expected. Furthermore it can be shown, by describing the groove and uniform deformations in terms of the current $\nu_1 - \nu_2$ reference axes, that (56) is equivalent to (45) where the Eulerian strain rates $\dot{\epsilon}_{tt}$ are matched along the current groove orientation.)

For the uniform deformation the principal directions of stress and strain are the $x_1 - x_2$ axes. However, from (53) the deformation-gradient matrix $\mathbf{A}(A_{ij} = \partial\mu_i/\partial\bar{\mu}_j)$ for the groove behavior is

$$\mathbf{A} = \begin{bmatrix} \delta & \gamma \\ \eta & \phi \end{bmatrix} \tag{57}$$

Since the Eulerian strain ellipsoid is coaxial with the principal axes of the left Cauchy-Green matrix $\mathbf{AA}^T$, the orientation $\hat{\theta}$ of these axes relative to the $\mu_1 -$

μ_2 reference system is given by

$$\tan 2\hat{\theta} = \frac{2(\delta\eta + \gamma\phi)}{(\delta^2 + \gamma^2) - (\eta^2 + \phi^2)} \qquad (58)$$

The orientation with respect to the current $\nu_1 - \nu_2$ axes is, from (36) or Fig. 3

$$\theta = \hat{\theta} - \Phi \qquad (59)$$

Now, the principal stretches λ_1, λ_2 for the groove deformation are related to the principal values e_1, e_2 of Green's strain matrix e as follows

$$e_1 = \tfrac{1}{2}(\lambda_1{}^2 - 1), \qquad e_2 = \tfrac{1}{2}(\lambda_2{}^2 - 1) \qquad (60)$$

where

$$\mathbf{e} = \tfrac{1}{2}(\mathbf{A}^T\mathbf{A} - \mathbf{I}) \qquad (61)$$

and $\mathbf{I}$ denotes the unit matrix. In view of (57) and these relations, the principal stretches are

$$\lambda_{1,2} = \frac{1}{\sqrt{2}}[(\delta^2 + \eta^2 + \gamma^2 + \phi^2) \pm R]^{1/2} \qquad (62)$$

with

$$R = \{[(\delta^2 + \eta^2) - (\gamma^2 + \phi^2)]^2 + 4(\delta\gamma + \eta\phi)^2\}^{1/2}$$

From the constitutive law (10), the incompressibility constraint $\lambda_1\lambda_2\lambda_3 = 1$ and the plane stress approximation $\sigma_3 = 0$ the principal values of Cauchy stress become

$$\sigma_1 = \frac{2}{3}\frac{\sigma_e}{\epsilon_e}\ln(\lambda_1{}^2\lambda_2)$$

$$\sigma_2 = \frac{2}{3}\frac{\sigma_e}{\epsilon_e}\ln(\lambda_1\lambda_2{}^2) \qquad (63)$$

As mentioned previously, the principal directions of this stress tensor must coincide with those of the Eulerian strain tensor for an isotropic material. A straightforward transformation of (63) therefore provides the required stress components with respect to the current groove axes:

$$\sigma_{nn} = \tfrac{1}{2}[(\sigma_1 + \sigma_2) + (\sigma_1 - \sigma_2)\cos 2\theta]$$

$$\sigma_{nt} = \tfrac{1}{2}(\sigma_1 - \sigma_2)\sin 2\theta \qquad (64)$$

$$\sigma_{tt} = \tfrac{1}{2}[(\sigma_1 + \sigma_2) - (\sigma_1 - \sigma_2)\cos 2\theta]$$

From the above expressions and the third of (37), we obtain the relation

$$\sin 2\theta \ln\left(\frac{\lambda_1}{\lambda_2}\right) = 3MY \qquad (65)$$

in which

$$M = \frac{\sigma_{nt}^0}{\sigma_{nn}^0} = \frac{(\rho - 1) \sin \psi \cos \psi}{(\rho + 2) \cos^2 \psi + (2\rho + 1) \sin^2 \psi} \tag{66}$$

and

$$Y = \ln(\lambda_1 \lambda_2) + \tfrac{1}{3} \cos 2\theta \ln \left(\frac{\lambda_1}{\lambda_2} \right) \tag{67}$$

Furthermore, the equilibrium equation (40) becomes

$$\frac{(\rho + 2) \cos^2 \psi + (2\rho + 1) \sin^2 \psi}{\sqrt{3}(1 + \rho + \rho^2)^{1/2}} = \frac{(1 - \xi)}{\lambda_1 \lambda_2} \left(\frac{\epsilon_e}{\epsilon_e^0} \right)^N \exp(C\epsilon_e^0) \frac{Y}{\epsilon_e} \tag{68}$$

where, according to the basic definition of effective strain

$$\epsilon_e = \frac{2}{\sqrt{3}} [(\ln \lambda_1)^2 + \ln \lambda_1 \ln \lambda_2 + (\ln \lambda_2)^2]^{1/2} \tag{69}$$

For a prescribed strain ratio ρ, initial groove inclination $\bar{\psi}$, and uniform deformation $\bar{\epsilon}^0$, the corresponding groove deformation described by (53) is determined from (65) and (68) together with the matching conditions (56) for ϕ and γ. The unknown parameters δ and η enter implicitly into (65) and (68) through the expressions (58) for the principal directions of stress, (62) for the principal stretches, (69) for the effective strain and (35)–(36) for the band rotation Φ. A Newton-Raphson technique applied to these equations provides the numerical solution of ϵ_e for given ϵ_e^0. For convenience, we determined the derivatives required in this method numerically.

For the case $\bar{\psi} = 0$, the above analysis implies that $\Phi = \gamma = \eta = \theta = 0$. Furthermore,

$$\lambda_1 = \delta, \qquad \lambda_2 = \phi = \lambda_2^0 \tag{70}$$

Equation (65) is thus identically satisfied and (68) reduces to

$$(1 - B)^{1/2} \left[1 - B \left(\frac{\epsilon_e^0}{\epsilon_e} \right)^2 \right]^{-1/2} = (1 - \xi) \left(\frac{\epsilon_e}{\epsilon_e^0} \right)^N \exp(C\epsilon_e^0 + \epsilon_3) \tag{71}$$

which is similar in form to the result (49) obtained in the flow theory analysis for $\psi = 0$. Here, the expression for ϵ_3 becomes [c.f. (51) for $\psi = 0$]

$$\frac{\epsilon_3}{\epsilon_e} = -A \left[1 - B \left(\frac{\epsilon_e^0}{\epsilon_e} \right)^2 \right]^{1/2} - D \frac{\epsilon_e^0}{\epsilon_e} \tag{72}$$

where A, B, C and D are now constants given by (48), (42) and (52) with $\psi = 0$.

PLANE STRAIN CASE ($\rho = 0$)

Before proceeding to results for a wider range of ρ-values, we shall first consider separately the plane strain case ($\rho = 0$). Here, $\bar{\psi} = \psi = 0$ and the deformation and flow theory predictions coincide.

The critical bifurcation strain obtained from the localization-band formulas (31) or (32) is

$$\epsilon_1{}^* \equiv \epsilon_1{}^{0^*} = N \tag{73}$$

Furthermore, the parameters in the foregoing long-wavelength (M-K) analysis take on the following values: $B = D = G = 0$, $H = 1$ and $A = C = \sqrt{3}/2$. Equations (51) or (72) give $\epsilon_3 = -\sqrt{3}\epsilon_e/2$ as expected, and the expressions (49) and (71) reduce to

$$(\epsilon_1{}^0)^N \exp(-\epsilon_1{}^0) = (1 - \xi)\epsilon_1{}^N \exp(-\epsilon_1) \tag{74}$$

This result is identical to the relation obtained in [16] for the corresponding problem of an axisymmetric bar under uniaxial tension with ϵ_1 identified as the axial strain.

A direct numerical solution of the transcendental equation (74) furnishes the typical results shown in Fig. 4, where curves of $\epsilon_1/\epsilon_1{}^0$ are plotted against $\epsilon_1{}^0/N$ for an initial geometric nonuniformity $\xi = .005$. The solid dots on these curves indicate the maximum value of $\epsilon_1{}^0$, which, from (74), is given by

$$\frac{\epsilon_1{}^{0^*}}{N} \exp\left[-\left(\frac{\epsilon_1{}^{0^*}}{N} - 1\right)\right] = (1 - \xi)^{1/N} \tag{75}$$

This value is attained when the strain in the groove satisfies $\epsilon_1 = N$ and the corresponding x_1-load component applied to the sheet reaches a maximum. Furthermore, the classical result (73) is retrieved from (75) when the imperfection $\xi = 0$. The dashed portion of each curve in Fig. 4 is also obtained from (74), but is no longer strictly valid since elastic unloading should occur in the uniform

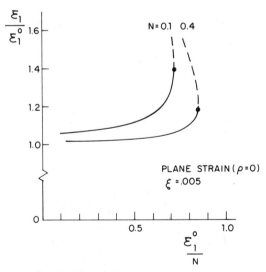

Fig. 4.   Development of ratio of strain in neck ϵ_1 to strain outside neck $\epsilon_1{}^0$ as a function of $\epsilon_1{}^0/N$ for plane strain.

region but has not been taken into account. The important point, however, is that ϵ_1^{0*} given by (75) does correctly give the maximum value of the strain attained in the uniform part of the sheet in the long-wavelength approximation. We therefore refer to this limiting value of uniform strain as the "limit" strain or critical strain for localized necking as it represents the state where the deformation becomes concentrated in the groove while the remainder of the sheet begins to unload. This criterion, which in general is equivalent to $d\bar{\epsilon}^0/d\bar{\epsilon} = 0$ (or $d\epsilon_e^0/d\epsilon_e = 0$) has been discussed in [14, 16]. It will also serve as our definition of critical strain for the other values of strain ratio ρ.

In Fig. 5 the critical strains (75) are plotted against the initial imperfection ξ for strain-hardening exponent values $N = 0.1$ and 0.4. The dashed curves shown here correspond to the asymptotic result from (75) for small ξ

$$\frac{\epsilon_1^{0*}}{N} \simeq 1 - \sqrt{\frac{2\xi}{N}} \tag{76}$$

It is clear from this expression and Fig. 5 that a small amount of geometric nonuniformity ξ substantially reduces the critical strain for necking. This type of behavior has received considerable attention in connection with the elastic and elastic-plastic buckling of structures, and is characteristic of structures which are referred to as *imperfection-sensitive*. In fact, the $\sqrt{\xi}$-type dependence indicated in (76) typifies some of the most imperfection-sensitive shell structures [17].

We emphasize again that flow theory and deformation theory give identical results for this case, so imperfection-sensitivity here is not the anomalous im-

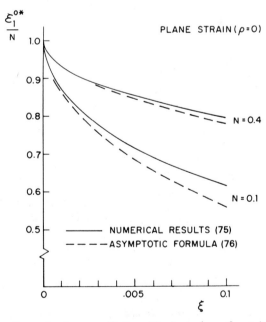

Fig. 5.   Imperfection-sensitivity of limit strain ϵ_1^{0*} outside neck.

perfection-sensitivity associated with simple flow theory when the bifurcation predictions of the flow theory are greatly in excess of those for deformation theory, as will be seen to be the case when $\rho > 0$.

RESULTS AND DISCUSSION FOR FULL RANGE OF ρ

The long-wavelength analysis was applied to generate numerical results for strain ratios varying from $\rho = -\frac{1}{2}$ to 1, corresponding to uniaxial and equibiaxial tension, respectively. For $\rho < 0$, various initial band orientations $\bar{\psi}$ were considered in the long-wavelength analysis to determine the angle which produced the minimum limit strain. For $\rho \geq 0$, the critical angle is such that the band is aligned in the x_2-direction, i.e., normal to the direction of maximum principal strain ϵ_1^0.

The effect of the initial geometric nonuniformity ξ on the limit strain ϵ_1^{o*} is depicted in Fig. 6 for $\rho = -\frac{1}{2}, 0, 1$ and $N = .22$. Again, solid curves represent deformation theory results and dashed lines refer to flow theory. As discussed in the previous Section, the plane strain ($\rho = 0$) results here coincide for both theories and exhibit the type of imperfection-sensitivity characterized by (76). The necking strains for uniaxial tension ($\rho = -\frac{1}{2}$) and equibiaxial tension ($\rho = 1$) are also seen to be reduced considerably in the presence of a small geometric nonuniformity. For the uniaxial tension case, the predicted flow theory limit strains are somewhat higher and slightly more imperfection-sensitive than the corresponding deformation theory necking strains.

The discrepancy between the two theories is more drastic for the equibiaxial tension case. Here, there is an enormous imperfection-sensitivity associated with flow theory as the bifurcation analysis in this case predicts an infinite necking strain for $\xi = 0$. Moreover, these flow theory limit strains continue to diminish

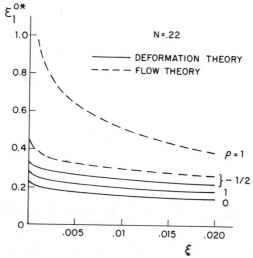

Fig. 6. Comparisons of imperfection-sensitivity of limit strain outside neck for deformation theory and flow theory for three states.

substantially as ξ increases beyond the value .01. On the other hand, the deformation theory critical strains for $\rho = 1$ do not show this severe sensitivity to the assumed value of ξ since a finite bifurcation strain ($\xi = 0$) exists according to this theory. In contrast to the flow theory results, these limit strains are rather insensitive to variations in ξ within the range .01–.02.

Figs. 7(a) and 7(b) illustrate the manner in which the limit strain ϵ_1^{o*} varies with the initial groove orientation $\bar{\psi}$ and current groove inclination ψ^* at necking. These results correspond to a strip under uniaxial tension ($\rho = -\frac{1}{2}$) with $N = .22$ and initial imperfections $\xi = .01, .001$. In these figures the numerical values $\bar{\psi}_D^*$ and $\bar{\psi}_F^*$ denote the initial orientations, calculated from (35), that would align with the current critical angles ψ_D^* and ψ_F^* predicted by the bifurcation analyses. Here, subscripts D and F refer to deformation and flow theory, respectively.

The results shown in these figures indicate that the flow theory limit strains are very sensitive to variations in $\bar{\psi}$ and ψ^*, whereas the deformation theory necking strains are relatively insensitive. It can also be seen from these curves that, as the initial imperfection decreases, the angles $\bar{\psi}$ and ψ which minimize ϵ_1^{o*} approach their corresponding values at bifurcation and, even for imperfec-

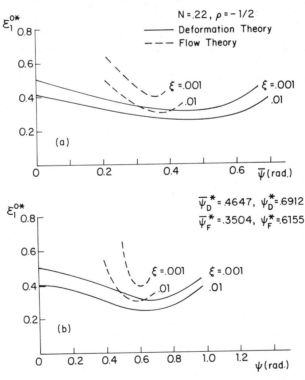

Fig. 7.   Limit strain outside neck ϵ_1^* as a function of necking band orientation for two plasticity theories: (a) in terms of initial orientation angle $\bar{\psi}$, (b) in terms of final orientation angle ψ^*. (Numerical values of ψ^* give corresponding values from bifurcation analysis.)

tions as large as $\xi = .01$ the critical values of $\bar{\psi}$ and ψ from the bifurcation analyses give reasonable estimates for the minimum limit strain.

Forming limit curves, similar to those depicted in Fig. 2, are plotted in Fig. 8 for an imperfection level $\xi = .01$ and for strain hardening constants $N = .22$, .50. In the negative ρ range, the shapes of the forming limit curves strongly resemble the related bifurcation limit curves of Fig. 2(a). A similar strong resemblance is also evident in the positive ρ range for the deformation theory results. Consequently, the effect of an initial imperfection in these cases is to essentially just shift the forming limit curves downwards.

The flow theory results in Fig. 8 for $\rho \geq 0$ are representative of the original M-K [14] analysis, where an initial nonuniformity was postulated to describe localized necking. In this biaxial tension range, the flow theory curve for $N = .22$ rises much more steeply with increasing ρ than the corresponding deformation theory curve. Furthermore, for $N = .50$ the flow theory curve also rises fairly steeply, whereas the deformation theory forming limit curve falls slightly as ρ increases.

Although it is difficult at this stage to make quantitative comparisons between theory and experiment, we shall nevertheless comment on the trends predicted in Fig. 8 and related test data. It must be emphasized, however, that many factors such as strain-rate sensitivity, anisotropy and nonhomogeneous straining may be pertinent in an experiment and that such effects have not been accounted for in

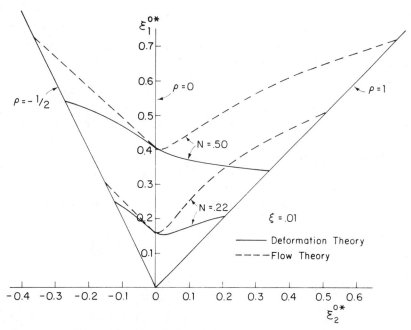

Fig. 8. Comparison of forming limit curves from two plasticity theories at two strain hardening levels for an imperfection $\xi = .01$.

the present analysis. Furthermore, the testing technique undoubtedly has an important effect on the experimentally-determined forming limit curve [18].

First we note that, in the negative ρ range, it is generally accepted that Hill's predicted curve (31) describes the trend of experiments for materials having lower N-values ($\sim$.25) [19–21]. In this case the flow theory and deformation theory predictions of Figs. 5 and 8 do not differ substantially. However, for materials with higher N-values ($\sim$.50) the discrepancy between the two plasticity theories becomes more pronounced. It is in this range that Painter and Pearce [21] have reported a poor agreement between Hill's curve and test results. In fact, the shape of their experimental curve for brass is very similar to the deformation theory curve in Fig. 8.

Another case where the results of both plasticity theories differ is in the prediction of the band angle at necking (Fig. 7) for a strip under uniaxial tension ($\rho = -\frac{1}{2}$). We simply observe here that, in view of the relative insensitivity of limit strain to variations in groove orientation with deformation theory, this theory would tend to better accommodate any experimental results for which the groove angle may be sufficiently different from the bifurcation prediction. However, available experimental evidence in this respect appears at present to be inconclusive.

In the biaxial tension range ($\rho \geq 0$), forming limit curves for reasonably isotropic metals with lower N-values ($<$.25) have been reported in [4, 18–22]. The experimental trend for these materials is such that the limit strain $\epsilon_1^{0^*}$ increases as ρ varies from plane strain to equibiaxial tension. This is in accordance with both theoretical curves in Fig. 8 for $N = .22$. However, the rate of increase observed experimentally tends to be less than that described by the flow theory curve and more representative of the deformation theory predictions. The reported forming limit curves for higher N-values ($>$.50) lend further support to the predictions of deformation theory. In this case the results of Azrin and Backofen [4] for brass and stainless steel, as well as those of Painter and Pearce [21] for brass, show a limit curve which is virtually horizontal or decreases slightly with increasing biaxiality. Only the deformation theory predictions in Figs. 5 and 8 for $N = .50$ indicate this tendency. Even by incorporating other characteristic values of ξ in the analysis, it is doubtful that flow theory can adequately predict this trend.

Throughout our analysis, the quantity ξ defined by (33) is referred to as a geometric nonuniformity. However, as discussed in [14], a material nonhomogeneity can easily be accommodated in the analysis. For example, if the constant K in (11) has values K^o and K in the uniform section and groove, respectively, then in (40) and the subsequent analysis we need only replace ξ by a new parameter

$$\bar{\xi} = 1 - \frac{K}{K^o} (1 - \xi) \tag{77}$$

which incorporates *both* material inhomogeneity and geometric nonuniformity. It has been suggested [23] that, whereas realistic levels of geometric imperfection do not appear sufficient to bring the flow theory trends in line with experiments, it might be possible to justify using higher values of $\bar{\xi}$ by appealing to material nonhomogeneities and thereby gain better agreement. For the purpose of discussion of this point suppose that $\bar{\xi} = .01$ and $\xi = 0$, corresponding to a geometrically perfect sheet with a one percent variation in the material parameter K. From (11), this implies a one percent difference in *stress* at any given strain level. It is unlikely that pre-existing nonhomogeneous work hardening levels alone could account for substantially larger stress differences than this at the higher strain levels associated with necking. On the other hand, pre-existing variations in structural metallurgical properties might justify use of substantially larger values of $\bar{\xi}$.

In summary, we are led to tentatively conclude that it is unlikely that a flow theory, based on a smooth yield surface, can be used to adequately predict localized necking in the range $\rho > 0$. This has disturbing implications for the numerical analysis of sheet metal forming. The study in the present paper was restricted to cases in which the overall straining was proportional (i.e., $\epsilon_2^0 = \rho\epsilon_1^0$) so that, in the uniform sections of the sheet, deformation theory and flow theory give identical predictions. In extensions to forming histories which are distinctly nonproportional, it is clear that a deformation theory will not adequately characterize behavior in the uniform sections and consequently cannot be reliably used under such circumstances to analyze necking. It appears to us doubtful whether either the deformation theory or the flow theory used here would be an adequate constitutive law for incorporation in a numerical program for the *combined* analyses of overall stretching and localized necking under general overall histories.

ACKNOWLEDGMENT

The work of J.W.H. was supported in part by the Air Force Office of Scientific Research under Grant AFOSR 77-3330, the National Science Foundation under Grant ENG76-04019, and by the Division of Applied Sciences, Harvard University. The work was conducted while K.W.N. was on leave at Harvard University. The support of the Faculty of Applied Sciences at the University of Sherbrooke is gratefully acknowledged.

REFERENCES

[1] R. Hill, J. Mech. Phys. Solids, *1* (1952), 19.
[2] Z. Marciniak and K. Kuczyński, Int. J. Mech. Sci., *9* (1967), 609.
[3] R. Sowerby and J. L. Duncan, Int. J. Mech. Sci., *13* (1971), 217.
[4] M. Azrin and W. A. Backofen, Metall. Trans., *1* (1970), 2857.
[5] S. Stören and J. R. Rice, J. Mech. Phys. Solids, *23* (1975), 421.
[6] J. W. Hutchinson, Adv. Appl. Mech., *14* (1974), 67.
[7] R. Hill, J. Mech. Phys. Solids, *6* (1958), 236.
[8] J. W. Hutchinson, Numerical Solution of Nonlinear Structural Problems (ed. R. F. Hartung) ASME (1973), 17.
[9] R. M. McMeeking and J. R. Rice, Int. J. Solids Struct., *11* (1975), 601.
[10] R. Hill, Proc. R. Soc. Lond. A, *314* (1970), 457.
[11] R. Hill, J. Mech. Phys. Solids, *16* (1968), 229, 315.

[12] R. W. Ogden, Proc. R. Soc. Lond. A, 326 (1972), 565.
[13] R. Hill and J. W. Hutchinson, J. Mech. Phys. Solids, 23 (1975), 239.
[14] Z. Marciniak, K. Kuczyński and T. Pokora, Int. J. Mech. Sci., 15 (1973), 789.
[15] A. K. Ghosh, Metall. Trans., 5 (1974), 1607.
[16] J. W. Hutchinson and K. W. Neale, Acta Met., 25 (1977), 839.
[17] J. W. Hutchinson and W. T. Koiter, Appl. Mech. Rev., 23 (1970), 1353.
[18] A. K. Ghosh and S. S. Hecker, Metall. Trans., 5 (1974), 2161.
[19] S. S. Hecker, Sheet Metal Ind., (1975), 671.
[20] S. S. Hecker, J. Eng. Mater. Technol. (Trans. ASME, H), 97 (1975), 66.
[21] M. J. Painter and R. Pearce, J. Phys. D: Appl. Phys., 7 (1974), 992.
[22] R. Venter, W. Johnson and M. C. de Malherbe, Int. J. Mech. Sci., 13 (1971), 299.
[23] A. Needleman, J. Mech. Phys. Solids, 24 (1976), 339.

DISCUSSION

S. Nemat-Nasser (*Northwestern University*)

One thing I was wondering if you'd like to make a comment on. The approach is essentially macroscopic which is certainly what one expects to have eventually, but it seems to me that microstructure must play quite a significant role in the whole process, and I did not see you emphasizing that.

Hutchinson

Hopefully, the role of the microstructure is reflected by the macroscopic law. More complex constitutive laws may be used than what I have discussed here. As an example, recently Needleman and one of his graduate students at Brown University have used a phenomenological characterization of the effect of void growth on yield behavior. So that is a microstructural effect which they have incorporated in the yield surface behavior. I think there will be some discussions on this tomorrow when Jim Rice talks. Now I might also add that when I said you could not use flow theory with a smooth yield surface, I did not mean to rule out that by doing something such as this one indeed might get around the problems associated with a smooth yield surface.

A. K. Ghosh (*Rockwell International*)

I have the following comment. You tended to reject the flow theory on the basis of the fact that it predicts too high a limit strain on the positive ($\epsilon_2 > 0$) side. Now professor Marciniak has made measurements of fracture strain for a number of materials, and I have done some of that type of work as well. It seems that near the balanced biaxial end of the forming limit diagram you do have fracture influencing formability, and even though flow theory predicts a high limit, flow will be terminated by fracture. So, I feel that the fact that the fracture limit comes down into the balanced biaxial end may be the controlling factor. Therefore, there is no arbitrary reason to throw out flow theory.

That's one point, and the second point—I didn't quite understand one aspect with respect to the calculations. You have shown the m effect (parameter in Hill's new yield criterion): for biaxial strength less than uniaxial the FLD goes up on the positive side, and it comes down for biaxial strength greater than uniaxial strength. I thought the experimental evidence was to the contrary. I may be wrong.

Hutchinson

Based on both Miyauchi's paper and that of Mellor, my understanding was that, for example with Mellor's material, he was seeing relatively low r values which suggests that if you used Hill's old formula the biaxial yield stress was low and, therefore, you get more strengthening in that biaxial region. But the experiments were showing the opposite. Well, that's consistent with the trends I show for this new 3-parameter yield surface. It essentially decouples their interdependence but the predominant effect is the effect of m, that is to say the ratio of σ_b over σ_u. Now, I asked Dr. Miyauchi just after lunch and he also verified that, generally speaking, when his X-parameter decreases you get more strain in the biaxial region, which is consistent.

Ghosh

I am thinking of steel versus brass. Steel shows a rising forming limit which goes with its $X > 1$, brass with a line not rising, independent of strain state, which has an X less than one.

Hutchinson

Well, maybe Mellor wants to comment on this.

P. B. Mellor *(University of Bradford, U.K.)*

I'd just like to question as well that on the biaxial tension quadrant, we have in fact Hill's new yield criterion for aluminum and for brass. If we put that into Marciniak-Kuczyński theory, we find that this brings down the limit strain dramatically at balanced biaxial tension, by an amount far more than I would have ever assumed. We find no cause to call for a fracture theory. One knows if there are plenty of voids then there is going to be fracture. So we have measured the density changes and we find these to be very, very small. Well, the point I am making is to incorporate the m value in the M-K theory brings the strains right down.

Hutchinson

And you are talking about $m < 2$?

Mellor

Surely yes.

Hutchinson

So that's completely consistent and you have helped to answer the first of Ghosh's questions too. I have no doubt that using Hill's new yield criterion in conjunction with the M-K analysis will also lead to the same trends. I am sure of it. And you can see from Mellor's comments that the trend is in the right direction. With respect to your first remark, I am sure there are materials in biaxial tension which may indeed fracture before significant strain localization occurs. I don't question it. But, I suspect that that's not the general case. And so, if you have to justify the Marciniak analysis based on flow theory and if you always have to invoke a fracture strain, then you start to worry about the theory. Can I say just a little bit more? It shouldn't be a surprise that flow theory overestimates things. You know, in plastic buckling, where we are not nearly as far into the plastic range, it has been known for 20 years, that if you use flow theory, certainly in a bifurcation analysis, you are going to overestimate the buckling load. All the engineering formulas for plates and shells based on bifurcation analysis use deformation theory moduli. Now it is also true, I believe, that most people think that if you do incorporate imperfection in a buckling calculation, with flow theory, you certainly do get much more realistic predictions. This is still somewhat of an open question. The reason, I think, there is so much more problem in the sheet metal necking is because straining is so much further into the plastic range. The discrepancies are therefore much greater. Again, I emphasize that just from a mathematical point of view I think you have strong reason to question using flow theory with a smooth yield surface, just because there is such enormous imperfection sensitivity which doesn't seem to be present in the data.

Ghosh

Yes, but when you see experimental fracture strain much lower than the predicted strain, you know the sample has fractured much earlier than. . . .

Hutchinson

Agreed. But are you suggesting that's the general case?

Ghosh

Well, in the material that you dealt with, certainly brass and A-K steel in the balanced biaxial end.

Hutchinson

That's not what Mellor said.

P. G. Hodge, Jr. *(University of Minnesota)*

I guess I can quibble about a term here which I am sure you will agree with. You made the comparison of flow and deformation theory being essentially the same under proportional loading. In the case of a test being carried out when you have homogenous stress state, proportional loading and proportional stressing are the same thing. In any application, however, you might have proportional external loading, but because you have a non-homogenous stress state you may have highly non-proportional stressing.

Hutchinson

I agree.

J. L. Duncan *(McMaster University, Canada)*

I'd just like to comment on one point. Given an experimentally determined forming limit curve and a Marciniak analysis, it's possible really to choose a shape for the yield surface that will give you a fit.

Hutchinson

I think you are probably right on that.

Duncan

Except for some degree of difficulty and that is, when the forming limit strain under plane strain conditions exceeds the strain hardening index value, the N value in your notation. I wonder whether you can comment on this point, because I think in all of your diagrams the forming limit strain that you measured was equal or less than the strain hardening index, N, at the plane strain axis.

Hutchinson

No, that's not really true. It's true at the higher N values on the right hand side of that diagram.

Duncan

I am talking about the plane strain axis, where $\epsilon_2 = 0$.

Hutchinson

Well, then I should say that there is no issue in plane strain, because the deformation theory and the flow theory coincide in all calculations. But, I think the answer to your question is, if it's perfect and if you don't introduce an imperfection, then the necking does occur at N. If you introduce an imperfection, the limit strain is reduced fairly substantially as Neale will discuss tomorrow.

SESSION III

ROLE OF FRICTION

Session Chairman
J. J. JONAS

McGill University
Canada

FRICTION AND LUBRICATION IN
SHEET METAL FORMING

W. R. D. WILSON

University of Massachusetts, Amherst, Massachusetts

ABSTRACT

The type of lubrication regime which occurs in a metal forming operation has a strong influence on the frictional conditions as well as on other important factors such as product surface finish and tooling wear rates. The various types of regimes which can occur are reviewed, together with what is known about their incidence in sheet metal forming operations. In the light of this information it can be seen that the commonly used methods of characterizing friction in analytical models of metal forming processes will often lead to erroneous results. Suggestions for improved methods of characterizing friction are made.

INTRODUCTION

Friction and lubrication are of vital importance in most metal forming operations. Effective lubrication systems result in low friction levels which reduce the loads imposed on tooling and workpieces. This can eliminate problems with tooling or workpiece failures or permit a reduction in the number of steps required to form a part. Lower force levels also reduce tooling deflection and can improve the dimensional accuracy of the product.

The presence of an effective lubricant film also reduces the amount of direct surface contact between the workpiece and tooling. This increases tooling life by reducing wear and improves product quality by reducing the incidence of scoring and other surface defects. On the other hand, an overly thick lubricant film may result in a roughened workpiece surface due to lack of constraint by the tooling.

Thermal problems are often eliminated by effective lubrication which not only reduces the amount of frictional heating but may also provide an insulating film between the workpiece and the tooling. This may allow the use of higher speeds or reduce thermally induced residual stresses or metallurgical damage in the product.

Despite the importance of lubrication in metal forming, there is relatively little interaction between researchers in the two fields. Thus it is common to find

References p. 174.

sophisticated treatments of plasticity combined with naive assumptions about frictional conditions in a single model of a forming process. The predictions of such a model can often be extremely misleading. For example, a model which allows for the effects of tooling geometry on the mode of deformation, but neglects the effects on friction and lubrication may suggest tooling geometries which are far from optimum.

At present, understanding of the basic mechanisms of metal forming lubrication is advancing at a rapid rate. In particular it is possible to explain many observed phenomena using mechanical models of the lubrication processes involved. Recent advances in this area are described in a paper by Wilson [1]. However, most of this activity has been concentrated on "bulk" deformation processes such as extrusion, rolling and forging and relatively little on sheet metal forming.

The purpose of the present paper is to explain some of the fundamentals of the theory of lubrication and how this can be applied to sheet metal forming operations. In particular the commonly used methods of characterizing frictional conditions will be discussed and improved methods will be suggested.

LUBRICATION REGIMES

The most important concept in lubrication is the idea that different types of lubrication regimes can occur. Failure to recognize this fact leads to most problems in modeling lubrication processes. The most important factor in deciding which lubrication regime occurs is the thickness of the lubricant film separating the surfaces. The various types of lubrication regimes which can occur in sheet metal forming operations are described below and illustrated in Fig. 1.

Thick Film Regime—In the thick film lubrication regime the mean film thickness of the lubricant is substantially larger than the roughnesses of the surfaces involved as shown in Fig. 1(a). Quantitatively the mean film thickness might be greater than ten times the sum of the arithmetic average roughnesses of the surfaces. Under these circumstances the effect of surface roughness on the lubrication process is negligible and the lubricant can be modeled as a continuum between smooth surfaces.

For the case of incompressible isoviscous liquid lubricants in the thick film regime the relationship between the local film thickness h and pressure p can be expressed by the well known Reynolds' equation [2]

$$\frac{h^3}{12\mu}\frac{dp}{dx} = \frac{(U + V)}{2}h + C_1 \tag{1}$$

where x is the distance along the film, μ is the lubricant viscosity, U and V are, respectively, the workpiece and tooling surface velocities and C_1 is a constant. Steady two-dimensional flow has been assumed. The friction stress τ_f on the surfaces is due to viscous shear in the lubricant and is given approximately by

$$\tau_f = \mu\frac{(U - V)}{h} \tag{2}$$

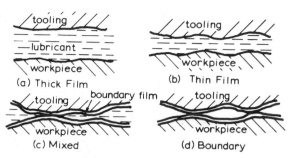

Fig. 1.   Lubrication regimes.

Most of the analyses of metal forming lubrication in the literature involve solving equations (1) and (2) (or their close relatives) together with additional relationships between the parameters involved which are imposed by considerations of metal flow. Such analyses, which allow the prediction of the local lubricant film thickness and friction stress as a function of the lubricant and workpiece properties, and system geometry and speeds, are reviewed in [1].

The analyses of thick film lubrication in metal forming have been facilitated by the discovery that the lubricant film can be broken up into a series of zones. In each zone it is possible to make some simplifying assumptions about the lubrication system. For example, consider the lubricant film between the ironing ring and the workpiece in the ironing operation shown in Fig. 2. The film can be divided into three zones; an inlet zone; a work zone; and an outlet zone.

The amount of lubricant which will be entrained is decided in the inlet zone. The pressure in this region is controlled by the lubricant hydrodynamics, and must build up from the ambient value far upstream to a maximum value equal approximately to the yield strength of the material at the boundary between the inlet and work zones. In the inlet zone the surfaces are effectively rigid [3]. Thus the surface velocities are known ($U = U_1$, $V = 0$) and the shape of the film is decided by the tooling. Integration of the Reynolds' equation (1) and elimination of the various arbitrary constants using the above pressure boundary conditions permits the calculation of the film thickness h_1 at the boundary between the inlet and work zones. For a conical entry of semi-angle θ as shown, and an isoviscous lubricant, h_1 is given by [4]

$$h_1 = \frac{3\mu U_1}{\sigma \tan \theta} \tag{3}$$

where U_1 is the inlet velocity and σ is the flow strength of the workpiece. Equation (3) indicates that the entrained film thickness will increase with increase in speed or viscosity and decrease with increase in workpiece strength or tooling semi-angle.

In the work zone, the pressure in the lubricant film is largely controlled by the plasticity of the workpiece. This means that the pressure gradients in the work zone are very small compared with those in the inlet zone. Thus in the work zone the term on the left of equation (1) is negligible and after evaluating the

References p. 174.

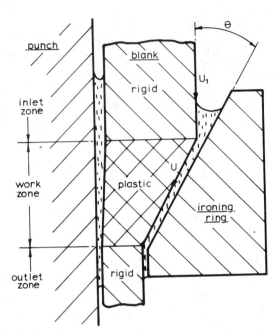

Fig. 2.   Ironing.

constant C_1 from the conditions at the boundary between the inlet and work zones the equation may be rewritten

$$h = \frac{h_1 U_1}{U} \tag{4}$$

Equation (4) shows that as the workpiece velocity U increases, the film thickness h will decrease. If the variation of workpiece surface velocity with position is known, then the local film thickness can be calculated from equation (4) and equation (2) can be used to calculate the local value of the friction stress τ_f. Given τ_f, plasticity will dictate the local value of the pressure p in the work zone.

In the outlet zone, the conditions are somewhat like those in the inlet zone. The pressure is largely controlled by the lubricant hydrodynamics but the elasticity of the tooling and workpiece may also be important. However, in any case the outlet zone plays a relatively minor part in the process. Its main function is to allow the lubricant pressure to fall back to ambient. This means that the film thickness in the outlet is slightly less than that at the outlet boundary of the work zone.

In practice, the method outlined above must be modified to allow for the variation of the viscosity of real liquid lubricants (such as oils) with pressure and temperature. The viscosity of a typical mineral oil lubricant will double with each increase in pressure of 35 MPa (5000 psi) and half with each increase in temperature of 15°C (27°F). Thus, under the severe conditions of pressure and temperature imposed on the lubricant in a metal forming operation, its viscosity may be orders of magnitude different from the ambient value.

A convenient model for the effects of pressure and temperature on viscosity expresses the viscosity μ at pressure p and temperature T as

$$\mu = \mu_s e^{\gamma p - \alpha(T - T_s)} \tag{5}$$

where μ_s is the viscosity at zero pressure and the surface temperature T_s, and γ and α are the pressure and temperature coefficients of viscosity respectively.

The effect of pressure induced viscosity changes is to increase the film thickness between workpiece and tooling. Thus, if only pressure induced viscosity changes are considered ($\alpha = 0$) equation (3) becomes [5]

$$h_1 = \frac{3\mu_s \gamma U_1}{\tan\,\theta(1 - e^{-\gamma\sigma})} \tag{6}$$

This equation yields values of h_1 which are often much larger than those given by equation (3). Furthermore in the range of practical interest the term $e^{-\gamma\sigma}$ is very small. Since this is the only term that contains the workpiece flow strength σ, the film thickness is insensitive to the strength of the workpiece material.

Increases in temperature tend to decrease the lubricant viscosity and entrained film thickness. Temperature increases can be broken into two components. The first of these is associated with increases in the temperature of the workpiece and tooling surfaces. This is relatively easy to account for by decreasing the value of μ_s in equation (6). The second kind of increase in temperature occurs within the film due to the heating from viscous shear. This results in substantial variations in viscosity across the film and invalidates the usual Reynolds' equation (1). A simple method of handling this problem is given by Wilson and Mahdavian [6]. If their method is applied to the ironing problem it can be shown that, in the range $0.2 < \gamma\sigma < 12$, h_1 is given by

$$h_1 = \frac{3\mu_s \gamma U_1 C}{\tan\,\theta(1 - e^{-\gamma\sigma})} \tag{7}$$

where C is the thermal correction factor given by

$$C = \left[1 + 0.56(\gamma\sigma)^{0.38}\left(\frac{\mu_s \alpha U_1^{\,2}}{4k}\right)^{0.8}\right]^{-1} \tag{8}$$

and k is the thermal conductivity of the lubricant.

Fig. 3 shows how the film thickness predicted by equations (7) and (8) varies with speed for given workpiece and lubricant properties, and process geometry. It can be seen that at low speeds the film increases in proportion to speed in agreement with the isothermal theory. At higher speeds the thermal theory falls further and further below the isothermal theory. The film reaches a maximum thickness at about 0.5 m/s and thereafter decreases with further increase in speed. The maximum in the film thickness/speed curve is characteristic of thermal theory

References p. 174.

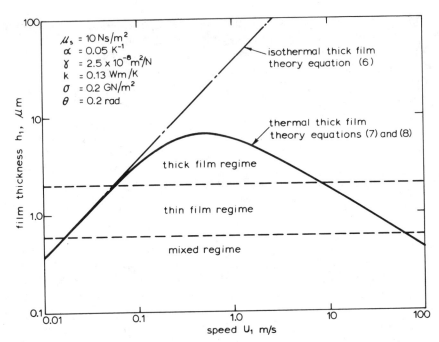

Fig. 3. Film thickness in hypothetical liquid-lubricated ironing operation.

and occurs when

$$\frac{\mu_s \alpha U_1{}^2}{4k} \simeq 10 \tag{9}$$

If it is assumed that the combined workpiece and tooling roughness is 0.2 μm (8 μin) AA, then the film thickness ranges corresponding to the thick film, thin film and mixed regimes can be sketched in as shown. It can be seen that the system analyzed will be in the thick film regime at speeds between 60 mm/s (2.3 in/sec) and 8 m/s (25 ft/sec). At speeds above or below this range the film thickness will be insufficient for the thick film regime to be maintained and the validity of the theory becomes questionable.

It should be noted that Fig. 3 shows only the effect of viscosity reduction due to temperature gradients within the lubricant film. The viscosity μ_s is assumed constant. In practice, the tooling temperature will tend to rise and μ_s will decrease with increase in speed. This will make the fall off in film thickness with speed even more severe than that indicated.

Pressure and temperature induced viscosity changes also have an important influence on the friction in the work zone. The subject is covered in a paper by Mahdavian and Wilson [8]. In general the viscosity increase due to pressure tends to increase friction, while heating due to viscous shearing in the lubricant or plastic deformation of the workpiece tends to reduce friction. Thermal effects can also have some effect on the film thickness distribution in the work zone [8], [9].

Thick film lubrication is not restricted to liquid lubricants. Soft solid lubricants such as waxes, soaps, polymers and soft metals can all engender a thick film regime, usually more effectively than liquid lubricants under slow speed conditions. A number of attempts have been made to analyze the entrainment of solid lubricants by treating them as viscous liquids with viscosities which depend on speed [10] or temperature [11]. However, this approach is incompatible with the essentially plastic behavior of solid lubricants at low speeds.

Wilson and Halliday [12] recently used the upper-bound method of plasticity to model the entrainment of a rigid plastic lubricant coating in a drawing (or ironing) process. They found that the entrained film thickness was independent of speed and proportional to the thickness of the lubricant coating applied to the workpiece. The thickness of the entrained film decreases sharply with increasing entry angle θ and with an increase in the ratio S of the workpiece to lubricant flow strength. Increasing the adhesion between the lubricant and the workpiece, and decreasing the friction between the lubricant and the die, increases the amount of lubricant entrained.

For the ideal case of perfect adhesion of the lubricant to the workpiece and zero friction between the lubricant and the tooling the ratio H of the entrained film thickness to the applied coating thickness may be expressed by the semi-empirical equation

$$H = \exp(A\theta^B) \tag{10}$$

where

$$A = -66.1\exp(0.00972S) \tag{11}$$

and

$$B = 1.05 + 8.63/S \tag{12}$$

Halliday [13] gives similar equations for other friction and adhesion conditions.

Halliday and Wilson [14] have shown that equation (10) is in excellent agreement with measurements of the thickness of wax and soap films entrained in the low speed drawing of aluminum strips through small angle dies. Some of their results are shown in Fig. 4. With larger angle dies the system is no longer in the thick film regime and the mean film thickness is approximately equal to the workpiece surface roughness.

At higher speeds ($U_1 > 1$ m/s) measurements on wax lubricated drawing made by Wistreich [15] indicate that the entrained film thickness decreases with increasing speed. This is presumably due to thermal softening of the lubricant due to plastic heating. No theory for this effect has as yet been developed.

It is usual to assume that the friction stress τ_f in a thick film regime with a solid lubricant is equal to the shear strength of the lubricant. However it should be remembered that many solid lubricants such as polymers have properties which are sensitive to both pressure and temperature and this should be allowed for in frictional models of solid lubrication. There is ample scope for further work in this area.

References p. 174.

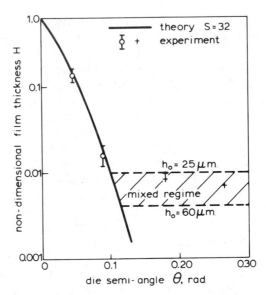

Fig. 4.   Film thickness in wax lubricated strip drawing [14].

Thin Film Lubrication—The thin film regime as shown in Fig. 1(b) occurs when the mean lubricant film thickness is of the same order as the surface roughness but the amount of contact between the surfaces is relatively small. Thus most of the load between the surface is still carried by a lubricant film which is much thicker than the molecular size of the lubricant. Under such circumstances the lubricant can be modeled as a continuum but the surface roughnesses must be taken into account.

The area of thin film lubrication with liquid lubricants is currently receiving a lot of attention. As yet no analyses have been developed for metal forming operations but the basic theory for thin film liquid lubrication has developed to such a point that such analyses should shortly be forthcoming.

The basic theory for isoviscous thin film liquid lubrication is laid out by Christensen [16]. In a system such as the ironing process shown in Fig. 2, the local pressure p and film thickness h will fluctuate with time due to the passage of the workpiece surface roughness. Thin film theory relates the mean local pressure $\bar{p}$ to the mean values of various simple functions of the film thickness. Thus for roughness lays in the direction of motion, or multi-directional lays,

$$\frac{E(h^3)}{12\mu}\frac{dp}{dx} = \left(\frac{U+V}{2}\right)h + C_2 \qquad (13)$$

while for roughness lays normal to the direction of motion

$$\frac{1}{12\mu E(h^{-3})}\frac{dp}{dx} = \left(\frac{U+V}{2}\right)\frac{E(h^{-2})}{E(h^{-3})} + C_3 \qquad (14)$$

In each case, E is the expectancy operator defined so that $E(f(h))$ is the average value of the function $f(h)$, and C_2 or C_3 is a constant. These can be evaluated from information on the mean film thickness $\bar{h}$ and the probability density functions for the surface roughness. The close affinity between equations (1), (13) and (14) is evident.

The friction stress τ_f in a thin film isoviscous regime also fluctuates with time at a given point. The mean value $\bar{\tau}_f$ is given by

$$\bar{\tau}_f = \mu(U - V)E(h^{-1}) \tag{15}$$

It should be noted that this equation does not account for any friction due to surface contact.

Although no direct analyses of liquid lubricated metal forming processes have been published it is possible to infer the effects of surface roughness on processes such as the ironing operation shown in Fig. 2 from existing analyses of bearing problems. These suggest that with roughened surfaces with lays in the direction of motion or with multidirectional lays the mean film thickness entrained will be smaller and the friction levels higher than with a smooth surface. On the other hand, surfaces with roughness lays normal to the direction of motion give thicker mean films and lower friction levels than a smooth surface.

As in the thick film regime analytical methods for solid lubricants in the thin film regime are only now being developed. Lak and Wilson [17] have used the upper-bound method of plasticity to study the transport of solid lubricant films between rough surfaces. However, there is no analytical method available to handle the entrainment of solid lubricants under conditions where surface roughness is important.

Mixed Regime—The mixed lubrication regime shown in Fig. 1(c) occurs when the film thickness is such that a significant fraction of the load between the surfaces is borne by contact between the roughness peaks. The remainder of the load is carried by the pressure in the lubricant film separating the surfaces in the roughness valleys.

In fact properly formulated lubricants will contain materials which will react chemically with the surfaces forming tightly adhering boundary lubricant films of the order of the lubricant molecular size. These films will tend to prevent direct metal to metal contact between the surfaces involved.

In the mixed regime the mechanisms at work in the roughness valleys and at the roughness peaks are quite different. The lubrication processes in the valleys are like the thin film regime described in the previous Section while the contacting peaks operate in the boundary lubrication regime which will be described in the next Section. Thus any realistic model of mixed lubrication must apply separate methods of analysis to the peaks and valleys.

These are two important modes of coupling between the two lubrication mechanisms. Firstly, the total load is shared between the relatively thick film in the valleys and the contacting peaks. Thus an increase in load support from the lubricant in the valleys will reduce the severity of the lubrication problem at the peaks. Secondly, the boundary films at the contacting peaks tend to be continually destroyed by rubbing, and regenerated by the action of fresh active material from

References p. 174.

the lubricant in the valleys. If the thin film mechanisms controlling lubricant flow in the valleys are not effective in providing fresh lubricant to the contact, then the boundary lubrication mechanism will become less effective.

Friction in the mixed regime is divided between components due to shear of the relatively thick films in the valleys and shear of the boundary films on the contacting peaks. The mean friction stress $\bar{\tau}_f$ may be calculated from

$$\bar{\tau}_f = a\tau_b + (1 - a)\bar{\tau}_t \tag{16}$$

where a is the fraction of surface area in contact, τ_b is the shear strength of the boundary film and $\bar{\tau}_t$ is the mean shear stress in the lubricant in the valleys.

Although the mixed regime is of considerable practical interest in metal forming, attempts to model it are relatively rare. Of particular note is the pioneering work of Tsao and Sargent [18], [19] who have modeled liquid lubricated strip rolling. They used thick film theory of the type described earlier to calculate the mean film thickness entrained into the interfaces between the rolls and strip. From information on the surface roughness distribution of the surfaces they then calculated the fraction of the surface area in contact and hence the friction stress. Their analyses reflect the rapid increase in friction level that accompany a transition from thick film to the boundary regime.

Boundary Regime—In the boundary regime as shown in Fig. 1(d) the load between the surfaces is supported purely by contact between the roughness peaks. If the contact between the peaks is plastic and tangential loading has no effect on the deformation, then under light loads the fraction a of the surface area in contact will be given by

$$a = \bar{p}/p_i \tag{17}$$

where $\bar{p}$ is the mean pressure between the surfaces and p_i is the indentation pressure of the softer surface (about five times its shear strength).

If both surfaces are perfectly clean then they will tend to weld or adhere perfectly at the roughness peaks in contact. The shear strength of the junctions would then be equal to the shear strength τ_s of the softer surface. Thus the mean friction stress $\bar{\tau}_f$ between the surfaces would be given by

$$\bar{\tau}_s = a\tau_s \tag{18}$$

If the fraction a of surface area in contact increases linearly with mean pressure $\bar{p}$ as indicated by equation (17) then the mean friction stress $\bar{\tau}_f$ will also increase linearly with $\bar{p}$ and the coefficient of friction μ will be constant and given by

$$\mu = \frac{\bar{\tau}_f}{\bar{p}} = \frac{a\tau_s}{\bar{p}} = \frac{\tau_s}{p_i} \simeq 0.2 \tag{19}$$

In practice the high tangential loading of the junctions associated with perfect adhesion would reduce the effective value of the indentation pressure p_i. This would result in an increase in the area of contact which would result in a coefficient of friction many times that given by equation (19). Under such circumstances sliding is accompanied by a progressive increase in friction and severe surface damage, an intolerable situation in a metal forming process.

Boundary lubricants are materials including fatty acids and their derivatives, and various organic compounds of sulfur, phosphorus and chlorine which react with metal surfaces to produce very thin tightly adherent films. These films have the capability of reducing the adhesion between the surfaces. This not only reduces friction directly by reducing the shear strength of the junctions but also, by reducing tangential loading on the junctions, tends to suppress growth of real area of contact. If the shear strength of the boundary film is τ_b then the coefficient of friction will be given by

$$\mu = \frac{a\tau_b}{\bar{p}} = \frac{\tau_b}{p_i} = \frac{\tau_b}{p_i} \qquad (20)$$

In metal forming operations the mean pressure $\bar{p}$ between tooling and workpiece may exceed the indentation pressure p_i of the softer surface (the workpiece). Under these circumstances equation (17) fails since a cannot be greater than unity. The friction stress τ_f can no longer increase linearly with $\bar{p}$ and the coefficient of friction must fall with increase in $\bar{p}$. When the mean pressure $\bar{p}$ between the surfaces is greater than the indentation pressure p_i, a approaches unity and the friction stress τ_f approaches the shear strength τ_b of the boundary film.

There is an immense body of literature on boundary lubrication [20]. In practice pure boundary lubrication probably results in friction and wear levels which are intolerable for a production sheet metal forming process. However, this does not mean that considerations of the effectiveness of a lubricant in the boundary regime are unimportant. Boundary lubrication is of obvious importance in the mixed and thin film regimes, and even in systems designed to run in the thick film regime boundary lubrication may be of vital importance during start-up or when a piece of metallic debris is entrained and surface contact is unavoidable.

LUBRICATION IN SHEET METAL FORMING

The term "sheet metal forming" covers a wide variety of processes. Some operations, such as bending, are usually unlubricated, while others such as shallow drawing do not place stringent requirements on the lubricant. On the other hand, operations such as deep drawing and ironing require careful attention to lubrication to ensure success. Since deep drawing and ironing are of considerable economic importance and present significant lubrication problems, these processes will be discussed in some detail.

An analysis of the published work on lubrication of sheet metal forming operations such as that conducted by Newnham [21] indicates that the lubricant's mechanical properties and boundary lubricity, as well as workpiece surface roughness, and deformation speed have important influences on lubrication. Thus, it seems likely that, while thick film or boundary lubrication may occur under unusual circumstances, significant regions in most processes operate in a thin film or mixed regime. Under such circumstances, it is highly desirable to be able to estimate the lubricant film thickness which is present in different parts of the interface between workpiece and tooling. This will permit selection of important lubrication parameters on a logical basis and estimation of resultant frictional conditions.

Deep Drawing—Fig. 5 shows the deep drawing of a cup shaped part. The primary areas to be lubricated are the blank/die and blank/hold-down ring interfaces (1), the blank/die-corner interface (2), the blank/punch interface (3) and the blank/ punch head interface (4). These will be discussed in detail in turn.

Lubricant will initially be squeezed from the blank/die and blank/hold-down ring interfaces as the tooling starts to close. This process has already been analyzed in another context and Moore [22] gives the following equation for the time t after the application of a load W required to reduce the film thickness from an initial value h_0 to a value h

$$t = \frac{3\pi\mu(D_2{}^4 - 2D_1 D_2{}^3 + 2D_1{}^3 D_2 - D_1{}^4)}{64W}\left(\frac{1}{h^2} - \frac{1}{h_0{}^2}\right) \tag{21}$$

where μ is the lubricant viscosity and D_1 and D_2 are the inner and outer diameters of the interface respectively.

When drawing starts the residual film at the blank/die and blank/hold-down ring will be wiped inwards by the motion of the blank. During this phase there will be some redistribution of load within the annular interface due to the tendancy for the blank to thicken as it is drawn inwards. This will tend to thin out the film near the inner diameter. However it seems likely that relatively thick films can be maintained.

For solid lubricants the conditions in the blank/die and blank/hold-down ring are relatively mild. Thus it is likely that the film thickness present at the start of the operation will be only changed slightly during drawing.

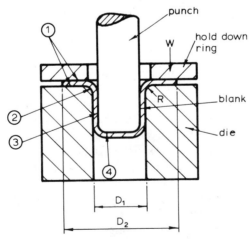

Fig. 5.   Deep drawing.

Lubrication at the blank/die radius is particularly crucial in deep drawing. It is unfortunate that little analytical work has been done on this area. However, the geometry is almost identical to the foil bearing analyzed by Blok [23]. He gives the film thickness h over the radius as

$$\frac{h}{R} = 1.405\left(\frac{\mu U}{T}\right)^{2/3} \tag{22}$$

where R is the radius of the die corner, μ is the lubricant viscosity, U the surface velocity, and T the tension in the strip per unit width. This equation could readily be modified to include thermal and pressure viscosity effects if these are significant.

At present there is no equivalent theory to deal with the entrainment of solid lubricants into the blank/die corner interface. The method developed by Wilson and Halliday [12] should be used to model this situation as a first step in understanding the lubrication of the blank/die corner interface by solid coatings.

The geometry of the entry to the blank/punch interface is even closer to the foil bearing geometry. However in this case the important tension is the circumferential tension in the cup wall and equation (22) becomes

$$\frac{h}{R} = 1.405\left(\frac{2\mu U R}{T D_1}\right)^{2/3} \tag{23}$$

where D_1 is the punch diameter.

As in the case of the blank/die corner interface there is no available analysis for the entrainment of solid lubricants into the blank/punch interface.

Lubrication at the blank/punch head appears to be a rather complex process to analyze. During the initial contact of the punch and blank lubricant will be squeezed out of the interface. As deformation progresses the blank will carry lubricant round the punch radius. In this mode there will be strong interactions between the lubrication and deformation processes since the metal flow is strongly influenced by friction, which is affected by lubricant transport, which is in turn controlled by metal flow. Analytical modeling of the lubrication process would seem to require a combination of foil bearing theory with squeezing and stretching effects such as those investigated by Wilson in his analysis of forging [24].

All the analytical approaches described above assume a thick film regime. It is obvious that each could be modified to use thin film and mixed theories as experience in using these is developed. This will allow refined predictions of film thickness and eventually friction in the various parts of the deep drawing process.

Ironing—An ironing process is shown in Fig. 2. Although lubrication conditions are usually more stringent in ironing than in deep drawing, the theoretical base for analysis of ironing lubrication is more advanced. This is largely because of the great similarity between the lubrication of the ironing ring/workpiece interface in ironing and the die/workpiece interface in wire and strip drawing, and hydrostatic extrusion. Thus, as discussed in the Section on Thick Film Regime there

References p. 174.

are available experimentally validated models for the entrainment of liquid [7] and solid lubricants [12]. In the case of liquids the models include the effect of both temperature and pressure induced viscosity changes on the lubrication process. These models can be used with some confidence to predict the lubrication conditions at the ironing ring/workpiece interface.

The lubrication processes at the workpiece/punch interface are less well understood. The inlet zone geometry appears to be controlled by the slight amount of sinking of the workpiece onto the punch which occurs. It is possible that elastic effects may be of great significance. However, given equal lubricant supplies, it is probable that much thicker films can be entrained into the workpiece/punch interface than into the ironing ring/workpiece interface since the former interface has a much greater degree of conformity than the latter.

As with deep drawing there are no available thin film or mixed lubrication models for the entrainment of lubricants in ironing. A logical first step in developing such models would be to use Christensen's theory [16] for the inlet zone problem common to ironing, drawing and extrusion.

CHARACTERIZATION OF FRICTION

In most metal forming processes friction between the workpiece and tooling has an important effect on the mode of deformation and the required forces. In analyzing such processes it is important to have some method of characterizing the frictional stress at the interface. This section describes the most commonly used methods of characterizing friction and discusses what problems they can introduce. Some more sophisticated methods are then suggested.

Commonly Used Methods—The most commonly used method of characterizing friction in any contact, lubricated or not, is the use of the constant coefficient of friction. In the present context the friction stress τ_f is assumed to be given by

$$\tau_f = \mu p \tag{24}$$

where p is the local interface pressure. The coefficient of friction is usually assumed to be a property of the workpiece and the materials, and lubricant. It is usually assumed to be independent of the process geometry and sliding speed at the interface.

The use of a constant coefficient of friction is based on research by Da Vinci, Amontons and Coulomb [25]. Almost all engineers and scientists accept the validity of this method of characterization without question, probably as a result of their early training in mechanics. In fact, the idea of a constant coefficient of friction so permeates this area of endeavor that the terms "friction" and "coefficient of friction" are often used interchangeably and frictional data are quoted in terms of coefficients of friction whether they satisfy equation (24) or not.

An alternative method of characterizing friction in metal forming operations is the use of a constant friction factor m such that the friction stress τ_f is given by

$$\tau_f = m\tau_w \tag{25}$$

where τ_w is the shear strength of the workpiece material (usually a mean value to allow for strain hardening). This model seems to be based on the idea that the interface can withstand a limited shear stress. This is the case if the surfaces are separated by a film of a plastic solid lubricant, in which case the friction factor m is the ratio of the shear strength of the lubricant to that of the workpiece.

As with the coefficient of friction, the friction factor is usually assumed to be a property of the workpiece and tooling materials and lubricant and independent of other parameters. However, the friction factor has an important advantage over the coefficient of friction as far as metal forming analyses are concerned. With the former characterization, the friction stress at a point is independent of the pressure which greatly facilitates the more sophisticated types of plasticity analyses.

Some other simple characterizations of friction have been used in the literature. However, their advantages, if any, are usually limited to a narrow range of operation of a particular process and the coefficient of friction and friction factor approaches account for the vast majority of existing analyses.

Problems with the Commonly Used Methods—It is important to understand what is expected from a frictional characterization. It is useful to have relatively simple models of forming processes which relate "dependent variables" such as forming force to "independent variables" such as material flow strength, tool geometry, and the general level of friction stress at the tooling/workpiece interface. In such analyses, the simple constant coefficient of friction, or constant friction factor characterization are usually adequate.

In real processes, the "independent variables" are often interrelated in a complex manner. In order to predict, for example, the real effect of a change in tooling geometry on forming force, the effect of tooling geometry on friction must be included. The commonly used methods of frictional characterization are usually inadequate in dealing with such effects. Some examples of this will be discussed below.

As mentioned earlier, the idea of different lubrication regimes is of vital importance in analyses of friction and lubrication. Not surprisingly, the biggest failures of the commonly used methods of frictional characterization are associated with their inability to provide appropriate models for the different regimes. Particularly serious failures occur at transitions from one regime to another.

Tooling geometry can have an important effect on the amount of lubrication entrained into the workpiece/tooling interface, which in turn can decide the lubrication regime and the level of friction. Fig. 6 shows the effect of the die semi-angle θ on the coefficient of friction μ measured in the drawing of wax lubricated aluminum strips by Wilson and Cazeault [26]. The conditions match those used by Halliday and Wilson [14] in the film thickness measurements shown in Fig. 4. The large increase in friction associated with the transition from the thick film via the thin film and mixed regimes to the boundary regime is evident. The use of a constant friction factor is totally inappropriate.

As discussed earlier, not only the system geometry but also the surface roughnesses, speed, and workpiece and lubricant properties can affect the entrained

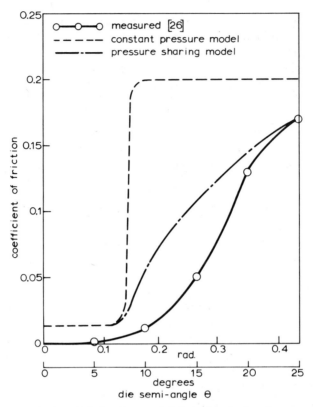

Fig. 6.   Friction in wax lubricated strip drawing.

film thickness and hence the possibility of a transition from one lubrication regime to another.

Another problem with using the constant coefficient of friction or friction factor characterization is that different lubrication regimes may exist in different parts of the workpiece/tooling interface. The formation of boundary lubricated zones in otherwise thick film or thin film lubricated extrusion [4] and forging [24] are good examples of this effect.

Improved Methods—More general methods of characterizing friction are usually models for the mixed lubrication regime, which can extend to the special cases of the thick or thin film regime in one direction, and to the boundary regime in the other direction. There are two simple possibilities for this type of model: a constant pressure model; and a pressure sharing model.

The constant pressure approach which was used by Tsao and Sargent [18], [19] uses thick film lubrication theory to calculate the mean film thickness in the interface. Using information on the roughness height distributions, the fractional area of contact a can be calculated and equation (16) (or a close relative) can be used to calculate the friction.

Fig. 6 includes a calculated friction curve based on the constant pressure method. The mean film thickness was calculated by using Wilson and Hallidays equations (10), (11) and (12). The surface roughness distributions were assumed to be Gaussian and the coefficient of friction μ was calculated from

$$\mu = a\mu_b + (1 - a)\mu_t \tag{26}$$

where the coefficient of friction μ_b in the boundary regime was assumed to be 0.2 and the coefficient of friction μ_t in the thick regime was calculated from the shear strength of the lubricant as 0.016.

It is evident that the constant pressure model predicts a much more rapid transition between regimes than actually occurs. This is probably due to its failure to allow for the beneficial effects of surface roughness on lubricant entrainment. The model also substantially overestimates friction in the thick film regime (at small die angles). This is due to overestimating μ_t which implies that the shear strength of the wax film is reduced probably by thermal softening.

The pressure sharing approach assumes that the mean film thickness entrained is equal to the surface roughness. Thick film theory is then used to calculate what pressure the lubricating film is capable of sustaining in the interface. The remainder of the required pressure is assumed to be carried by surface contact. If b is the fraction of the pressure carried by contact, the overall coefficient of friction is given by

$$\mu = b\mu_b + (1 - b)\mu_t \tag{27}$$

A friction curve calculated using the pressure sharing approach is also shown in Fig. 6. Wilson and Halliday's equations were used to calculate the pressure the lubricant film can sustain. As before μ_b and μ_t were assumed to be 0.2 and 0.016 respectively.

The pressure sharing model shows much better agreement with the experimental data than the constant pressure model. The pressure sharing model still tends to overestimate μ at small die angles. This error can be reduced by correcting for thermal softening of the lubricant (reducing μ_t). The residual error is probably associated with thin film lubrication effects.

CONCLUSIONS

Four main lubrication regimes can occur in sheet metal forming with liquid or solid lubricants: the thick-film regime; the thin film regime; the mixed regime; and the boundary regime. The characteristics of lubrication and friction are different in each regime and it is vital to recognize these differences in modeling sheet metal forming processes.

The common methods of characterizing friction in sheet metal forming operations such as the use of a constant coefficient of friction or a constant friction factor are inadequate in many cases. In particular, they fail to reflect the large changes in frictional conditions associated with transitions between lubrication regimes. These changes can be generated by a mixed model of the lubrication process. Of two such models considered, a pressure sharing model which assumes

References p. 174.

that part of the pressure in the workpiece/tooling interface is carried by the lubricant film in the roughness valleys, and part by contact of roughness peaks, shows best agreement with the experimental friction measurements of Wilson and Cazeault [26].

There is ample scope for research on lubrication and friction in sheet metal forming. In particular there is a real need for lubrication analyses of sheet metal forming operations. Initially such analyses could use existing thick film methods, but eventually good thin film and mixed lubrication analyses will be necessary to represent the real lubrication situation in many processes.

REFERENCES

[1] W. R. D. Wilson, SME Technical Paper MS77-31, 1977.
[2] O. Reynolds, Trans. Roy. Soc., *177 Pt. 1* (1886), 157.
[3] D. Dowson, B. Parsons and P. J. Lidgitt, Symposium on Elasto-Hydrodynamic Lubrication, Inst. Mech. Eng., London, 1972, 107.
[4] W. R. D. Wilson, Int. J. Mech. Sci., *13* (1971), 17.
[5] W. R. D. Wilson and J. A. Walowit, J. Lubric. Technol. (Trans. ASME, F), *93* (1971), 69.
[6] W. R. D. Wilson and S. M. Mahdavian, J. Lubric. Technol. (Trans. ASME, F), *96* (1974), 572.
[7] W. R. D. Wilson and S. M. Mahdavian, J. Lubric. Technol. (Trans. ASME, F), *98* (1976), 27.
[8] S. M. Mahdavian and W. R. D. Wilson, J. Lubric. Technol. (Trans. ASME, F), *98* (1976), 22.
[9] W. R. D. Wilson, J. Lubric. Technol. (Trans. ASME, F), *95* (1973), 541.
[10] G. H. Tattersall, J. Mech. Eng. Sci., *3* (1961), 378.
[11] E. Felder and G. Breinlinger, Proc. 4th North Amer. Metalworking Res. Conf., Battelle, Columbus (1976), 158.
[12] W. R. D. Wilson and K. Halliday, Wear, *42* (1977), 125.
[13] K. Halliday, Ph.D. Thesis, University of Massachusetts, Amherst, 1977.
[14] K. Halliday and W. R. D. Wilson, Proc. 5th North Amer. Metalworking Res. Conf., SME, Detroit (1977).
[15] J. G. Wistreich, Met. Rev., *3* (1958), 97.
[16] H. Christensen, Wear, *17* (1971), 149.
[17] S. Lak and W. R. D. Wilson, J. Lubric. Technol. (Trans. ASME, F), *99* (1977), 230.
[18] C. Tsao and L. B. Sargent, ASLE Paper No. 75-LC-3C-1 (to be published in ASLE Trans.)
[19] C. Tsao and L. B. Sargent, ASLE Paper No. 76-LC-5A-1 (to be published in ASLE Trans.)
[20] F. F. Ling, E. E. Klaus and R. S. Fein (eds.), Boundary Lubrication—A Survey of World Literature, ASME, New York, 1969.
[21] J. A. Newnham, Metal Deformation Processes Friction and Lubrication (J. A. Schey ed.), Marcel Dekker, New York, 1970.
[22] D. F. Moore, Principles and Applications of Tribology, Pergamon, Oxford, 1975, 115 and 373.
[23] H. Blok and J. J. Van Rossum, Lubric. Eng. (1953), 316.
[24] W. R. D. Wilson, J. Lubric. Technol. (Trans. ASME, F), *96* (1974), 539.
[25] F. P. Bowden and D. Tabor, Friction, Anchor Books (Doubleday), Garden City, 1973, 9.
[26] W. R. D. Wilson and P. Cazeault, Proc. 4th North Amer. Metalworking Res. Conf., Battelle, Columbus (1976), 165.

DISCUSSION

H. C. Rogers *(Drexel University)*

Just a quick question. In your various theories, does it make any difference whether you are actually talking about sliding friction where there is no deformation of either the workpieces or the tool, or whether you are having a substantial amount of deformation in the workpiece compared to the tool? Since these did not factor in, what does that cancel out?

Wilson

It is important to measure friction under conditions which reproduce not only the plastic deformation present in the real process, but also the factors controlling lubricant entrainment. One of the things that people like to do is to take a small piece of the tooling material, draw a piece of workpiece material over it and measure the coefficient of friction. Unfortunately, there is almost no relationship between the results of such screening tests and friction measurements in the real process. For example, there is no way to compare friction measurements made with a pin-on-disc system to those I described made in an actual drawing operation. The pin-on-disc system will only yield one value of the coefficient of friction, while the drawing results vary by a factor of about thirty as the die angle is changed. No matter what value of coefficient of friction the pin-on-disc system yields, it is bound to be significantly different from that found under some drawing conditions. Looking at speed effects causes even more problems, because there is no guarantee at all that there is any relationship between the speed sensitivity of the system you are using and that of the actual forming systems. The real systems behave in different ways as speed is changed. For example, in extrusion, rolling, and wire drawing, the film thickness tends to increase linearly with speed under isothermal conditions while in forging the film thickness tends to be proportional to the cube root of speed. Thus, it is very difficult to obtain results on the effect of speed from a forging test (such as the Kunogi-Male and Cockcroft ring test) and apply these in extrusion. Only by completely understanding the lubrication conditions in the real operation and simulating them in a screening test is there any hope of success.

P. A. Stine *(General Electric Company)*

What type of laboratory screening test would you propose?

Wilson

Well, I think our approach will continue to be to measure the physical properties of the lubricant such as its viscosity, and pressure-viscosity and temperature-viscosity properties, and then to use these in mathematical models of the lubrication process in the system we are interested in. There is no other way to go, since unless you actually run screening tests on the real system, you can come up with erroneous results.

J. J. Jonas *(McGill University, Canada)*

What is the pressure sensitivity of things like molybdenum disulfide or graphite?

Wilson

I don't know if there is any available information about them. The materials that we have been working on show very little sensitivity of shear strength to the imposed pressures.

K.-K. Chen *(General Motors Research Laboratories)*

I'd like to make a comment on your thin film lubrication case. You used Christensen's method in averaging the Reynolds equation. That way, you will separate the pressure and the film thickness and so you can average the thickness and the pressure gradient separately. However, we know that pressure is related to film thickness so you can not, in general, average the quantities separately.

Wilson

The area we are talking about here is at least as complicated as the area of plasticity in sheet metal forming operations. All I could do in my paper was just touch on a few key pieces of work. I just came back from a conference in Lyons (the Leeds-Lyons Symposium on Roughness Effects in Lubrication) where they spent a whole week talking about roughness effects in lubrication. There are many different methods of modeling the thin film regime. If anyone is interested in this area, the thing to do is to get hold of a copy of the proceedings of that conference.

E. H. Lee *(Stanford University)*

Although my question refers to a different kind of forming, I think it might possibly have some kind of application. One often hears the statement that in hydrostatic forming, the fact that the forming is done through a fluid that goes through the die extremely good, there is almost zero friction. Do you have any idea what the coefficient of friction is or whether it works in that case?

Wilson

In hydrostatic extrusion, the low friction levels are not due to the tendency of the high fluid pressure to push lubricant through the die. The film is so thin that the leakage effect is negligible. The low friction is due to the copious supply of lubricant available and the fact that its viscosity is elevated by the pressure in the container. Pressures in the container are of the order of 1400 MPa which might increase the lubricant viscosity by a factor of one million. The actual friction levels achieved are very low and hence difficult to measure.

M. A. Bragard *(Centre de Recherches Metallurgiques, Belgium)*

The roughness of sheet metal is made of valleys and plateaus. Do you have any information on the variation of the surfaces, the very top surface, and the plateaus?

Wilson

You want to know how it changes during draw? Well, we haven't worked on draw much. We are working very actively on this problem on a simple model system. We have been looking for some time at the roughness such as "orange-peel" that occurs. We have done some mathematical modeling to try to predict and it is a very complicated problem but we can get within a factor of two of it. This is dealing with the roughening process as an extrusion process which is much more difficult. You can go to local deformation process where the peaks and asperities may or may not be lubricated in thin film, thick film, boundary—all these are possibilities. It is quite a difficult chore.

H. Conrad (*University of Kentucky*)

Several years ago, we studied the lubrication of titanium with molten glass using the Male-Cockcroft ring test. We found minimum friction occurred at a specific ratio of viscosity to flow stress of the material. I am wondering if any of that comes out of your analysis?

Wilson

This would suggest that pressure viscosity effects are unimportant in this system, in which case theory would predict that the ratio of viscosity to flow strength would be important in deciding film thickness.

DRAWBEAD FORCES IN SHEET METAL FORMING

H. D. NINE

General Motors Research Laboratories, Warren, Michigan

ABSTRACT

During sheet metal forming on a double-action press, drawbeads on the blankholder supply a restraining force which controls the flow of metal into the die. As metal is drawn through the drawbead, the restraining force has two components, bending deformation and friction. In this study, these components were separated and analyzed. Drawing load, clamping load and strain were measured for sheet steels and aluminums which were drawn through a short section of simulated drawbead. The bending deformation component was isolated by replacing the shoulders and bead of the drawbead by rollers which reduced the friction to a negligible value. The experimental data is analyzed in terms of the materials effects which must be included in a mathematical model of drawbead forces. It was found that an isotropic hardening assumption would not adequately model the strain hardening but that a kinematic hardening assumption that takes cyclic hardening into account is necessary. It is also shown that strain rate hardening must be modeled for steel but not for aluminum. Friction is adequately represented by a coefficient of friction, μ. Experimental values of μ were obtained. Drawbead forces were calculated employing these experimental coefficients of friction and good agreement was found between the measured and calculated frictional components.

INTRODUCTION

The design of metal stamping dies is a complex process. It is also an inexact process so that considerable trial and error adjustments during tryout are often required to finish the building of a die that will produce acceptable parts. In an effort to improve the die design process, a research program is underway to understand metal flow in dies particularly by means of numerical modeling on a computer.

Complete dynamic computer modeling of the forming operation is beyond present capability. In order to reduce the problem to a workable level certain critical areas of the die have been selected for initial study. This paper describes

References p. 203.

work on one critical area, the drawbead. Experiments determine the effect of combined mechanical deformation, materials factors and friction on the drawing forces as sheet metal is drawn through drawbeads. By analyzing the experimental information, the importance of the various effects can be assessed. This provides a test of the assumptions and gives guidance on factors which should be included in a mathematical model of drawbead forces. The experiments also provide necessary materials parameters input to a mathematical model.

The only similar work found in the literature was that of Swift [1] who in 1948 developed a mathematical analysis for a simulated drawing of sheet metal over a die throat. This is similar to drawing sheet metal over the first shoulder of a drawbead groove. However, the drawbead problem is complicated by additional deformation over a bead and a second groove shoulder so that Swift's theory is not applicable to our experiments.

Drawbeads—In double-action press forming of sheet metal, the first action lowers a binder (or blankholder) onto the periphery of the sheet to hold it in position, Fig. 1a. The second action lowers the punch which draws the sheet through the binder into the die cavity and forms the part, Fig. 1b. It is frequently necessary to control the rate of metal flow into the die cavity. The force exerted by the binder on the sheet metal supplies a restraining force which controls the metal flow. For flat binder areas (right hand side of Fig. 1) the restraining force is largely friction. The force may be varied by controlling the clamping force between the upper and lower binder.

In some applications the friction does not adequately control the flow of metal through the binder. Drawbeads are then added to the binder. A drawbead consists of a semi-cylindrical bead on one binder face which fits into a groove with quarter-cylindrical shoulders on the opposing binder face (Fig. 1a inset). The restraining force of drawbeads has two components: the deformation force from bending as the metal flows through the drawbead and the friction which comes from the rubbing of the sheet metal against the drawbead surfaces.

Drawbead Forces—From a design viewpoint it is necessary to know the drawing force and the clamping force for drawbeads. The drawing force is that force required to pull the metal through the beads. The clamping force is the force necessary to hold the two opposing binder faces at a fixed separation which, of course, sets the depth of the bead between the shoulders of the groove.

Fig. 2 traces the development of the forces on a specific area element as it passes through the drawbeads and defines the problem for analysis. Assuming some small clearance is maintained no forces are exerted on the element prior to position 1. At position 1 a bending force must be applied to bend the metal to conform to the radius of drawbead shoulder A. In this type of bending the metal is strained in tension at the top surface and compression at the bottom surface, Fig. 2b (1). An unstrained position, the neutral axis, lies somewhere between the surfaces as shown.

Since the metal has been plastically deformed into a cylindrical shell of bead shoulder radius, no further bending stresses need be applied as the cylindrical

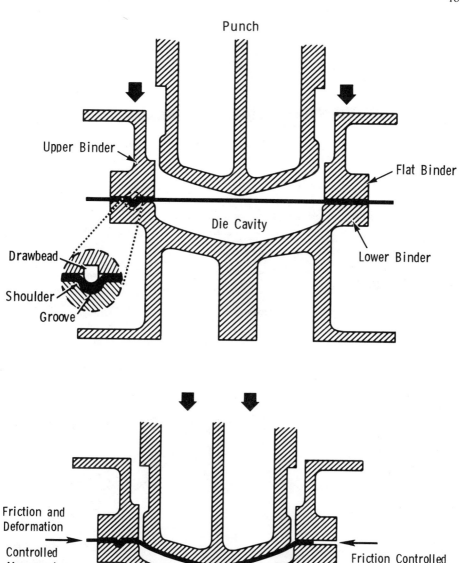

Fig. 1.   Double acting press:
  a) Action 1; the binder (blankholder) moves down to hold the sheet metal in
     position for forming.
  b) Action 2; the punch moves down into the die cavity drawing sheet metal into
     the die through the binders.

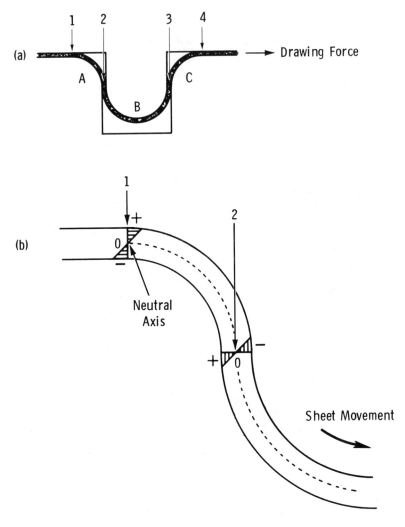

Fig. 2.
  a) Schematic of positions in the drawbead where bending, unbending and sliding
     friction take place. At positions 2 and 3 the sheet first unbends and immediately
     rebends in the opposite direction.
  b) Representation of tensile (+) and compressive (−) strain as sample first bends
     (1) and then unbends and rebends (2) in the opposite direction from the original.
     The neutral axis is indicated by 0 and the dashed line.

shell slides from 1 to 2 in Fig. 2a. However, during sliding, additional force must
be applied to overcome the friction between the sheet and the drawbead shoulder.
On reaching position 2 the metal sheet is straightened which requires additional
bending force. Now the top surface undergoes compression and the bottom
surface, tension. However, the net effect is to return each surface to near zero
strain. Thus, the force to pull the metal through the beads increases with each

bend and friction contribution. This is a net tensile force which is superimposed on the tensile and compressive stress at each subsequent bend. There is therefore a net tensile strain associated with the traverse of the sheet through the drawbeads.

Beyond position 2 the metal is bent to conform to the radius of the bead B. From position 2 to 3 the same type of bending and frictional forces develop as from positions 1 to 2. Note, however, that the metal at each surface strains in the opposite sense and the friction path is doubled. At positions 3 to 4 the bending forces and frictional forces repeat the sequence of 1 to 2. Thus throughout the path of the metal flow through the drawbeads, the drawbead forces derive from two separate contributions—deformation from bending, and friction.

For this experimental paper deformation and friction will be treated in two separate sections. As shown in Fig. 3, in the first section we will determine the deformation forces by using a simulated drawbead fixture using roller beads to eliminate friction. The results can be used to check which factors are important in a mathematical model. In the second section, the experiment will employ fixed beads. Using the prior results the contribution of friction will be isolated and a coefficient of friction will be determined. The results will then be available for comparison to calculations of combined deformation and friction in drawbeads.

DEFORMATION FORCES

The deformation forces arises from a complex combination of geometrical and materials factors. To aid in the development of a mathematical model the experiments must be designed to determine the validity of various assumptions and

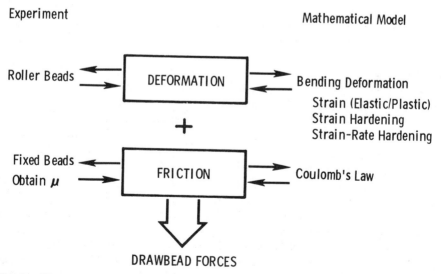

Fig. 3. The experimental and mathematical model approaches to obtaining the deformation and friction forces for drawbeads.

References p. 203.

the significance of various mechanical and materials effects. In addition experimental drawing and clamping forces must be measured to test the mathematical model.

One factor which obviously affects the magnitude of the deformation forces is magnitude of the local strain. This, in turn, is determined by the geometry of the drawbeads and the thickness of the sheet metal. The smaller the radius of a shoulder or bead, the greater the surface strain that will be imposed on the sheet metal as it is bent to that radius. Moreover, clearances between upper and lower binder and the drawbead elements are important because if they are too large or if the bead does not penetrate fully into the bead groove, the actual radii will be greater than that of the assumed reference shape and the bending strain will be less than would be calculated. Furthermore the magnitude of plastic strain as compared to elastic strain determines a plasticity assumption (i.e. rigid-ideally plastic, elastic-plastic).

Strain hardening will also affect the deformation forces. In order to develop a mathematical model, the strain hardening must be represented by a constitutive equation. In most deformation studies, strain hardening increases monotonically since the strain is increased monotonically in one direction. However, in drawbeads the deformation is complex. Both tensile and compressive strains occur simultaneously on opposite sides of the sheet and these both vary from maximum at the surface to zero at the neutral axis, Fig. 2b. In addition, from the above discussion of Fig. 2, strain is reversed four times which results in strain cycling between tension and compression rather than monotonically increasing strain. Cyclic strain hardening can be considerably different from unidirectional strain hardening [2], and formulating the proper constitutive equation for this case may be difficult.

Consideration must also be given to whether strain rates are high enough that a strain-rate hardening term needs to be added to the deformation hardening expression. Finally, strain state may influence the strain hardening. The experiments were specifically designed to determine the importance of these factors.

Experiment—A key requirement for this research was to study the deformation forces without the perturbing influence of friction. Experimentally this was accomplished by building a "frictionless" drawbead in which the shoulders of the groove and the bead of the drawbead were replaced by rollers, Fig. 4. These rollers are backed-up by sets of ball bearings. Since the surfaces of the rollers are free to rotate with the sheet metal, there is no sliding friction. The system is not, of course, completely frictionless since a small force is needed to overcome the friction of the rollers and the backing ball bearings. The coefficient of friction of ball bearings is about 0.001 to 0.002 depending on the lubrication conditions. Friction in the bearings of these rollers is, therefore, negligibly small compared to that for regular beads.

The support pieces which hold the groove shoulder rollers are designed so that the clearance between the shoulders can be adjusted by shims. This makes it possible to maintain a constant, tight clearance while testing sheet metal of different thicknesses. In addition adjustable-height rollers (Fig. 4) are provided which hold the incoming sheet metal strips in the original plane of entry as the roller bead penetrates the groove shoulders. The positioning guides are necessary to obtain a well-defined reference shape for metal movement so that a valid comparison between measured and calculated forces and strains is possible.

The drawbeads are incorporated into a fixture which was designed to fit into an MTS universal testing machine. This fixture holds the bead and groove shoulder support pieces so as to simulate the action of the binder in a press. Fig. 5 shows a schematic drawing of the apparatus. A hydraulic cylinder ram moves the bead support piece so that the bead penetrates between the groove shoulders. The position of the bead is read from a dial indicator. The back side of the groove shoulder support piece presses against a load cell built into the supporting fixture to measure the clamping force. The drawing force is measured by a load cell placed between the actuator of the MTS testing machine and a grip capable of holding strips up to 50 mm wide.

The materials investigated were aluminum-killed (A-K) steel, rimmed steel, and 2036-T4 aluminum. Table 1 lists the materials and their properties as determined in tensile tests. In the usual manner K and n were obtained from the true stress-true strain plots assuming power law strain hardening. R, the plastic anisotropy parameter, was obtained by measuring the change in shape of photogridded circles on uniformly strained tensile samples.

Sheet metal test strips were cut 400 mm in the sheet rolling direction and 50 mm wide. The strips are inserted between the bead and shoulder rollers, aligned vertically and clamped in the grips. As the bead is driven into the groove, the test strip bends around the drawbeads. The samples are then pulled through the drawbeads at 85 mm/sec for a distance of about 125 mm. This is the estimated average speed at which metal is drawn in production presses. The load vs stroke for both drawing load and clamping load are recorded on x-y recorders. Typical

TABLE 1

Mechanical Properties of Test Materials

	Initial Thickness	K^*	n^*	R^*
Rimmed Steel	0.76 mm	576 MPa	0.18	1.09
	0.86 mm	559 MPa	0.23	1.05
	0.99 mm	519 MPa	0.119	1.15
A-K Steel	0.76 mm	529 MPa	0.24	1.53
	0.86 mm	491 MPa	0.21	1.62
	0.97 mm	529 MPa	0.23	1.61
2036-T4 Aluminum	0.81 mm	643 MPa	0.26	0.69
	0.89 mm	643 MPa	0.24	0.67

* Tabulated values are averages over the plane of the sheet obtained by using the form: $X_{average} = \dfrac{X_{0°} + 2X_{45°} + X_{90°}}{4}$ where 0°, 45° and 90° refer to the strain direction relative to the sheet rolling direction of the material.

References p. 203.

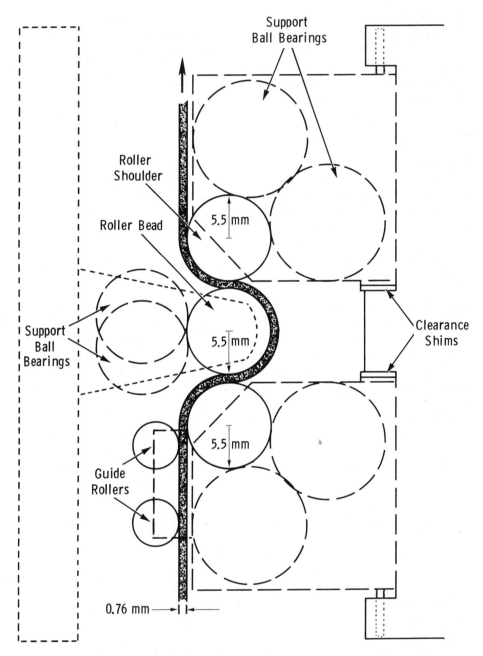

Fig. 4.
    a) Schematic drawing of roller drawbead fixture.

b) Roller drawbead fixture.

load traces are shown in Fig. 6. Load readings are made after the load reaches a steady state and load values are the average of at least three separate tests.

Strain as a function of position in the beads was measured from precision spaced lines etched on both sides of special sheet metal strips in a 0.5 mm square grid pattern. The test was stopped when the gridded section had been drawn into the roller beads and shoulders of the drawbead and then it was carefully removed from the fixture to avoid bending. The curvature of the strip prevented direct reading of the strain grid with a microscope so the gridded surfaces of the sample were replicated with Bioden R.F.A. acetyl cellulose film. After stripping, the replicas were laid flat so that the entire gridded section could be read accurately.

Strain rates were estimated from strips drawn over a range of speeds from 0.42 mm/sec to 420 mm/sec. At the higher drawing speeds, the normal x-y recorder would not follow the load and stroke so all strain rate tests were recorded on a storage digital oscilloscope and then re-recorded on an x-y recorder.

Results—To obtain information on bending deformation the drawing loads for specimens of A-K steel, rimmed steel and 2036-T4 aluminum were measured for an 11 mm diam roller bead and 11 min diam roller shoulders and are shown in Table 2. The values of the drawing load for preliminary calculations are also shown.

The surface strain distributions within the drawbead area for both sides of a 1.0 mm thick A-K steel strip are shown in Fig. 7. The magnitude of the surface strain as the specimen bends from groove shoulder to bead to shoulder is shown. The strain is determined from changes in the original grid size of the electroetched square grids on the strip. The total accumulated strain (the sum of the absolute values of all strain changes) is 46% for the top surface and 53% for the bottom surface. The net strain is, however, a tensile strain of 10% for the top surface

References p. 203.

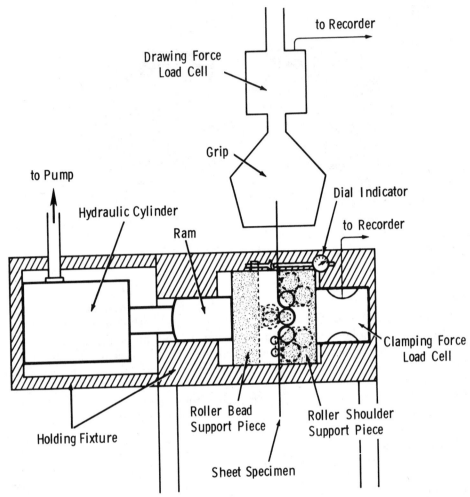

Fig. 5.   Schematic diagram of apparatus for measuring drawing and clamping forces.

and 9% for the bottom surface. The thickness of the strips was reduced 10% while the width remained unchanged. The strain distribution on each metal and thickness were similar.

Note from Fig. 7 that the unbending and rebending of the strip as it passes from the shoulder to the bead (and again from the bead to the shoulder) takes place in two to four grid spacings. This means that the strain change takes place within 1 mm to 2 mm of strip travel. At a speed of 85 mm/sec and an average surface strain change of about 16%, the surface strain rate would be about 7/sec to 14/sec. However, since the strain rate is zero at the neutral axis, the average strain rate is about 3.5/sec to 7/sec. This is about three orders of magnitude higher than a standard tensile test strain rate of 3×10^{-3}/sec. Consequently a strain-rate sensitive material (like steel) could have a significantly different deformation hardening behavior at a drawing rate of 85 mm/sec than that measured in a quasi-static tensile test.

To test the effect of strain rate, test strips of A-K steel, rimmed steel and 2036-T4 aluminum were drawn through the drawbeads at speeds from 0.42 mm/sec to 420 mm/sec. These drawing rates correspond to strain rates of about 0.035/sec to 35/sec. The drawing forces for rimmed steel and A-K steel increased by about 25% over the range of strain rates, Fig. 8. However, 2036-T4 aluminum showed no strain rate effect, i.e., draw forces measured over the same speed range as for steel showed no change. Forces for these materials are shown in Table 2.

Discussion—Experiments carried out on the "frictionless" roller drawbeads were designed to assess the validity of particular mathematical assumptions and the importance of certain materials effects relating to bending deformation. These will be discussed individually.

Plasticity Assumption—In Fig. 7 the lowest surface strain change for any bend is 4%, from 14% to 10% tensile strain. Since there is zero strain at the neutral axis, the average strain change is 2%. This is high relative to the strain at the onset of plastic flow (~0.1%). Therefore a rigid-plastic model should be adequate for calculating drawbead forces. Swift [1] used a rigid-plastic model to calculate the simulated bending of sheet metal over a die throat.

Strain Hardening Assumption—Strain hardening makes an important contribution to a calculation of the force required to deform metals and must be included in a mathematical model. Strain hardening is normally obtained from a unidirectional tensile test. Under tensile deformation strain hardening increases monotonically with strain. However, the deformation in drawbeads is cyclic (i.e., changes from tension to compression). In Fig. 7 for 1.0 mm A-K steel the total accumulated strains (sums of absolute values of the individual strain changes)

TABLE 2

Drawbead Forces for "Frictionless" Drawbeads (kN)

	0.86 mm A-K Steel Drawing Load	0.86 mm Rimmed Steel Drawing Load	0.89 mm 2036-T4 Aluminum Drawing Load
Experimental (high strain rate)	3.9	3.9	3.2
Calculated (static)	3.7	3.6	4.2
Experimental (quasi-static)	3.1	3.1	3.2

References p. 203.

are 46% and 53% on the top and bottom surfaces, respectively. This exceeds the usual strain-to-fracture (40%) for A-K steel under tensile deformation. Thus, the hardening under cyclic strain may be different from that calculated by assuming a monotonically increasing hardening proportional to the total accumulated strain. On the other hand, the measured net strain is about 10%. Tensile test samples, made from the section of drawbead test strips which had been drawn through the roller bead showed only an additional 15% residual elongation strain (or about 25% total strain-to-fracture). This is much less than usual strain-to-fracture (40%) for A-K steel in a tensile test. This suggests that the assumption that the hardening is proportional to the net strain of 10% would underestimate the strain hardening in the bead. Insight into the difference between cyclic and monotonic strain hardening can be gained by noting in Table 2 that in preliminary (time-independent) calculations based on an isotropic hardening assumption the drawing load is about 16% too high for rimmed steel and about 20% too high for A-K steel compared to the measured drawing forces for quasi-static rates. For aluminum

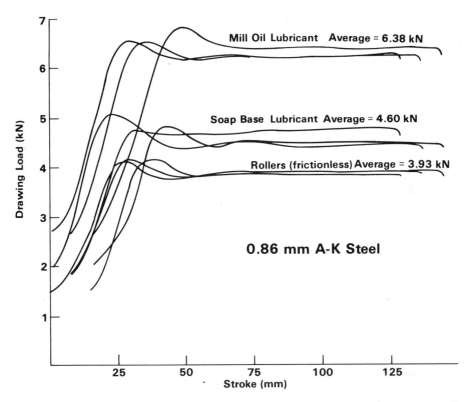

Fig. 6.   Force vs. stroke recording for the drawbead forces through 11 mm diam "frictionless" and standard drawbeads for 0.86 mm A-K steel. Mill Oil and Soap Base lubricants were measured for the standard beads.
        a) Drawing force.

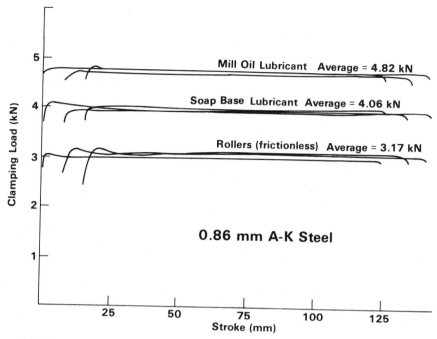

b) Clamping force.

the calculated drawing force exceeds the experimentally measured force by about 30%. Obviously, too much strain hardening is calculated from an isotropic hardening assumption. The proper approach must somehow incorporate the effects of cyclic strain hardening.

To be able to obtain an accurate constitutive equation, cyclic stress-strain curves need to be obtained for these sheet metals at strain amplitudes comparable to the cyclic strain experienced in drawbeads. Experimentally this is difficult because of buckling at high compressive strains. Further complications arise from the fact that in bending sheet metal the strain varies from a maximum at one surface to zero at the neutral axis to a maximum of opposite sign at the other surface. Thus, the cyclic stress-strain curves over a whole range of mixed strains are needed. In addition the cycling should be done under plane strain conditions to correspond to the strain state for bending. Experimental techniques for overcoming these difficulties are currently being considered.

Strain-Rate Effect—For strain-rate sensitive materials the hardening due to strain rate must also be considered. The drawbead drawing and clamping forces were measured at a drawing speed of 85 mm/sec which corresponds to an average strain rate of approximately 5/sec. From the results of strain-rate hardening tests, Fig. 8, the increase in force due to testing at 85 mm/sec as compared to 0.42 mm/ sec (normal quasi-static tensile test speed) is about 20%. Thus, the strain rate

References p. 203.

H. D. NINE

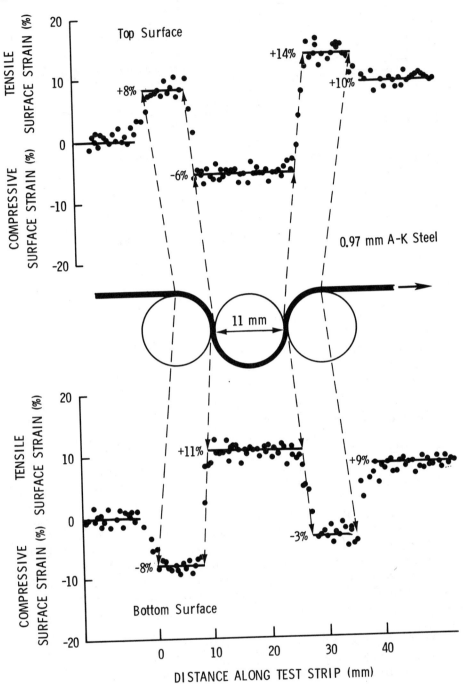

Fig. 7. Surface strain as a function of the position of sheet metal in the drawbead for both sides of the sample. Dots are individual grid readings. The solid lines are average values at the individual data points rounded off to the nearest whole percent. The strain horizontal scale matches the circumference of the 11 mm diam rollers.

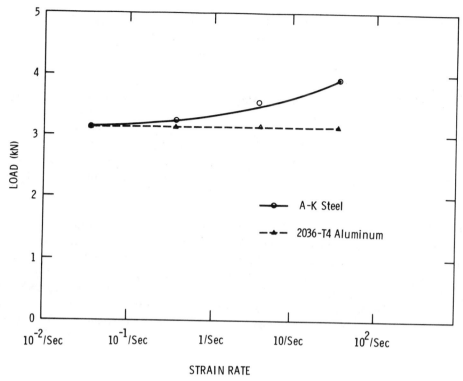

Fig. 8.   Load vs strain-rate for A-K steel and 2036-T4 aluminum during drawbead deformation.

effect on drawbead forces is significant for steels and should be considered in a mathematical model.

It may be noted that the agreement between the experimental forces measured at press rates and the preliminary calculated force for 0.86 mm A-K steel was within 5%, Table 2. Since the computation did not include strain rate effect, the agreement was better than would be expected. This is due to a canceling effect. The drawing force was measured at the relatively high strain rate of about 5/sec so that omitting the strain rate effect would make the calculated value too low. However, the cyclic strain effect was also omitted in the calculation and this would make the calculated drawing force too high. Omitting both effects tends to cancel and give apparent agreement between calculated and experimental values. The drawing force measured at a quasi-static rate, Table 2, shows that without a strain-rate effect, the experimental drawing force of steel, like aluminum, is lower than calculated.

In order to include the strain-rate effect in a mathematical model, tensile stress-strain curves must be obtained for the materials as a function of strain-rate from quasi-static rates to about 10/sec. It would be necessary to obtain stress-strain curves at enough different rates to determine what form a strain-rate modification to the basic strain hardening equation would take.

References p. 203.

Strain-State Effect—As noted above, the width of the test strips did not change during deformation through the drawbeads while the thickness was reduced 10%. This shows that the test strips were deformed in plane strain. The usual strain hardening parameters n and K are, however, obtained from uniaxial tensile tests. It has been shown by Ghosh and Laukonis [3] that n and K values are dependent on strain state. In the preliminary calculations, drawing loads are directly proportional to the input value of K while variations in n show a relatively small effect. While the effect on the drawbead forces of changes between plane-strain and uniaxial tension values of n and K would be smaller than cyclic or strain-rate effects, the accuracy of the calculation could be improved if plane strain values were available. Work has been initiated in this area but results are not yet available.

Deformation Summary—The experimental results for the deformation part of drawbead forces show that the following assumptions would be important for a mathematical model: 1) a rigid-plastic assumption would be adequate, 2) a new hardening rule should incorporate cyclic strain effect, as the commonly used isotropic hardening assumption has been shown not adequate, and 3) steels would require a strain-rate sensitivity assumption. An accurate representation for the latter two assumptions will be possible only when experimental information becomes available.

FRICTION

The second contribution to the drawbead forces, friction, develops in a complex manner.

According to the Bowden and Tabor theory on the origin of friction [4, 5], when two metal surfaces are brought into contact under load, the actual contact between the two surfaces on a microscopic scale is only at the tips of high points (asperities). The high pressure developed at individual contact points causes them to weld locally. Sliding between the two metal surfaces can then occur only by shear of the welded junctions. In addition, as the surfaces slide relative to one another, the asperities of the harder surface may plow paths through the softer material. The shearing of the welded junctions and the plowing are responsible for friction.

Lubrication, of course, considerably alters the friction. If a thick lubricant film completely separates the sliding metal surfaces, friction is a function of only the lubricant viscosity, the speed and pressure. This is known as hydrodynamic lubrication. On the other hand if a film of lubricant only a few molecules thick separates the sliding surfaces, considerable asperity contact will take place. This type of lubrication is called boundary lubrication. For boundary lubrication, bulk lubricant properties, like viscosity, are not very important. Instead, complex adsorption and physio-chemical reactions between the lubricant and the metal surfaces assume controlling importance. Direct metal to metal contact may be prevented by such surface coatings particularly if they reform rapidly.

In sheet metal forming it is not usually possible to maintain complete hydro-dynamic conditions between the sliding surfaces. Usually, mixed lubrication prevails in which there is solid contact at the tips of major asperities but there is bulk lubrication in the intervening void spaces. In this case the shear stress, τ, to slide one surface over the other can be expressed by

$$\tau = K[\tau_m k + \tau_l(1 - k)] \tag{1}$$

where K is the fraction of the apparent contact area which is in actual surface contact

k is the fraction of K that is in metal to metal contact

τ_m is the shear flow stress for the metal

and τ_l is the shear flow stress for the lubricant.

Although this discussion of friction and lubrication and the expression for the shear force are oversimplified, they will serve to establish the reasons for the approach taken here in studying friction in drawbeads.

Examining Eq. (1) it can be seen that many factors which influence the shear stress, τ, are not readily measured. For example, the fraction of the surface in actual contact and the division of that fraction between metal to metal contact and lubricant to metal contact are not known. Further, the shear flow stress of the metal at the tips of the asperities is a function of the local strain hardening, the strain rate and the temperature. Surface oxidation and other surface effects arising from chemical interactions between the lubricant and metal surfaces also change the lubrication. Calculation of the sliding shear stress under these conditions becomes an intractable problem.

To overcome these difficulties, the concept of a coefficient of friction is used. It is generally accepted that this concept goes back to Amontons, who formulated laws of friction and to Coulomb, who expanded on Amontons' work [4]. They found that, except at very high pressure where the actual contact area approaches the apparent contact area, the sliding shear stress (the friction stress) may be expressed by

$$\tau = \mu P \tag{2}$$

where μ is a constant and P is the normal stress. It follows from Eq. (2), that μ, the coefficient of friction, is defined by

$$\mu \equiv \frac{\tau}{P} = \frac{\tau A}{PA} \frac{F}{N} \tag{3}$$

where A is the apparent contact area, F is the shear force and N is the normal force. By measuring the sliding and normal forces, a coefficient of friction can be obtained which can then be used to calculate the sliding shear force for a known or measured normal force.

While μ is a very convenient simplification, it must be applied with caution. From Eqs. (1) and (3), the conditions under which τ is measured greatly affects the value of τ and consequently μ. For example, μ will be quite different in

References p. 203.

drawbeads where metal deformation is taking place than between flat plates. The deformation in drawbeads exposes new metal surface and changes the surface roughness of the sheet metal. These changes in turn influence all the variables in Eq. (1). Thus, μ is not a materials constant but depends on the system in which it occurs. It is therefore imperative that the coefficient of friction be measured under the same conditions (e.g., sheet metal drawn through drawbeads) for which μ needs to be known.

Coefficient of Friction—Obtaining a coefficient of friction for sheet metal being drawn through drawbeads presents a more complex problem than a flat plate measurement of μ because the drawing force to drawbeads contains both deformation and friction components. To obtain a coefficient of friction, that part of the drawing force due to friction must be extracted from the total.

In section I of the paper a method is described for measuring the drawbead deformation forces, i.e., the drawbead forces without friction. The friction drawing force can then be obtained by measuring the drawing force in fixed drawbeads and subtracting the measured drawing deformation force. From the measured forces in fixed and "frictionless" drawbeads, then, a coefficient of friction may be obtained by the following analysis.

In Fig. 9 if s is a parameter indicating the position of a point or a sheet metal

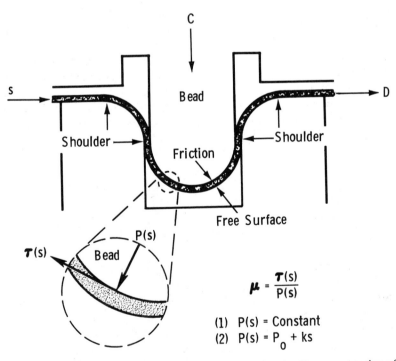

Fig. 9.   Development of pressure, $P(s)$, through drawbeads. Two assumptions for $P(s)$ are made to obtain a coefficient of friction.

strip as the strip is drawn through the drawbead, the normal stress (pressure), P, and the friction stress, τ, are functions of s. The normal stress increases following each bend and also increases because of the increasing back stress due to friction. Since for drawbeads the sheet metal thickness is significant compared to the shoulder and bead radii, bending moments must also be considered in a calculation of the normal pressure distribution. For drawbeads, therefore, $P(s)$ has a complex distribution. However, if Coulomb's law is applicable, i.e., μ does not vary with pressure, some simple assumptions can be made which will allow μ to be obtained from measured drawing and clamping forces.

If it is assumed that $P(s)$ is a constant, a simple expression for μ is obtained.

$$\mu = \frac{\tau(s)}{P(s)} = \frac{D_{d+f} - D_d}{\pi C_{d+f}}$$

where D_{d+f} = measured drawing force for fixed drawbeads
  D_d = measured drawing force for "frictionless" drawbeads
and C_{d+f} = measured clamping force for fixed drawbeads.

By analogy to a flat plate friction experiment, the friction force is $D_{d+f} - D_d$ and the normal force is C_{d+f}. However, because the bead geometry is semicylindrical and the pressure is applied to both sides of the metal strip, an effective clamping load is obtained by multiplying the measured clamping load, C_{d+f}, by π.

A second expression for μ may be obtained if it is assumed that the pressure due to friction increases linearly, $P(s) = P_0 + ks$. Expressions for the drawing and clamping forces at a point on the drawbead can be written. By integrating these expressions over the two bead shoulders and the bead, the following expression is obtained,

$$[\pi D_d + (\pi - 2)C_d]\mu^2 + \left[(\pi - 2)D_d - \left(\frac{\pi - 2}{2} \right) D_{d+f} + 2\pi C_{d+f} \right] \mu$$
$$+ [\pi(D_d - D_{d+f}) + (\pi - 2)(C_{d+f} - C_d)] = 0$$

where C_d = measured clamping force for "frictionless" drawbeads.

Details of the analysis to obtain μ for each of these methods are found in the appendix.

Experiment and Results—In order to obtain the total drawbead forces, including both deformation and friction, a set of drawbeads with fixed groove shoulders and fixed bead was built, Fig. 10, to the same dimensions as the roller bead fixture, Fig. 4. The bead and shoulders were made of production bead material, a W-2 water hardening steel in the soft condition. The bead and shoulders were replaceable in case of excessive wear or damage. The same clearance adjustments were built into the fixed beads as the roller beads.

The fixed beads were mounted in the same fixture as the "frictionless" roller beads to measure the drawbead forces. All experimental parameters such as materials, strip sizes and thicknesses and drawing speeds were kept identical to those for the "frictionless" cases and measurements were made in the same way.

References p. 203.

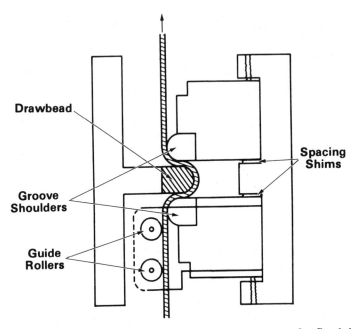

Fig. 10.   Bead and groove shoulders mounted in support pieces for fixed drawbeads.

Drawbead forces were measured for the steels with two lubricants. One lubricant was a mill oil (MO), which is normally used to prevent rusting of stored steel. In lubricity tests, it was shown to be a relatively poor lubricant. The second lubricant was a proprietary soap base lubricant (SB), which was the best lubricant found. Drawbead forces were measured for aluminum with these same two lubricants. In addition, a third lubricant, especially developed for forming aluminum, (AL), was used. Drawing force versus stroke for these lubricants on 0.86 mm A-K steel are shown in Fig. 6.

The drawing and clamping forces measured for both the "frictionless" 11 mm diam roller drawbeads and the standard 11 mm diam drawbeads which are required to obtain a coefficient of friction, μ, are shown in Tables 3 and 4. Values for various thicknesses of A-K and rimmed steel and 2036-T4 aluminum are given. Coefficients of friction derived from both constant P and linear $P(s)$ are given for each case for each lubricant.

Discussion—Individual contributions to the friction force from specific effects are not distinguished by the coefficient of friction approach taken here. Thus, the friction will be discussed in comparison to the most similar work found in the literature and in relation to some preliminary calculations.

There is reasonably good agreement between coefficients of friction obtained by the two different methods as shown in Tables 3 and 4. The results are also plotted against clamping load in Fig. 11. For the MO lubricant the constant P method gives values from 3 to 16% lower than those obtained by the linear $P(s)$

TABLE 3

Drawbead Forces and Coefficient of Friction for Test Lubricants on Steel

		Sheet Thickness (mm)					
		Rimmed Steel			A-K Steel		
Force (kN)	Drawbead	0.76	0.86	0.99	0.76	0.86	0.97
Drawing	Rollers	3.3	3.9	5.3	3.3	3.7	5.0
Clamping	Rollers	3.0	3.5	4.7	2.7	3.2	4.0
Drawing	Fixed A	5.7	6.4	8.4	5.6	6.4	8.0
Clamping	Fixed A	4.2	4.8	5.8	4.2	4.8	5.8
Coefficient of Friction (A)							
μ (Constant P)		0.181	0.163	0.168	0.176	0.162	0.164
μ (Linear $P(s)$)		0.206	0.184	0.200	0.192	0.171	0.178
Drawing	Fixed B	4.1	4.9	6.3	4.1	4.6	5.7
Clamping	Fixed B	3.4	4.2	5.2	3.4	4.0	4.9
Coefficient of Friction (B)							
μ (Constant P)		0.070	0.070	0.059	0.069	0.052	0.061
μ (Linear $P(s)$)		0.077	0.071	0.050	0.065	0.039	0.051

A = Mill Oil Lubricant B = Soap Base Lubricant

methods. The percentage of difference between coefficients of friction derived from the two methods is greater for the SB lubricant. However, the absolute differences fall in about the same range as for the MO lubricant. For the MO lubricant the friction part of the draw force is between 38% and 45% of the total draw force. However, for the SB lubricant the friction drops to from 13% to 19% of the total force. Since the coefficient of friction depends directly on the difference between the drawing forces with and without friction, the scatter in calculated coefficients of friction increases as the difference between the values of the "frictionless" and standard draw forces decreases (as it does for better lubricants). An indication that Coulomb's law applies for drawbeads is shown by the reasonably good agreement between coefficients of friction, μ, obtained by the two methods.

For the MO lubricant, μ tends to be slightly higher for the lightest load cases, Tables 3 and 4. It is a common observation that μ is higher for light loads. Also μ for A-K and rimmed steel and 2036-T4 aluminum are very similar for comparable thicknesses of stock. These materials have somewhat similar yield characteristics, Table 1.

For the MO lubricant on the steels and 2036-T4 aluminum there is only a small variation in μ (3% to 6%) for the different thicknesses of test strips. However, the clamping loads vary by about 40% (Tables 3 & 4 and Fig. 11). This is an indication that μ does not vary much with normal load and gives another indication that Coulomb's law is applicable to this drawbead analysis.

Swift [1] in measuring coefficients of friction in an apparatus which simulated bending over a die throat, noted above, found values of friction for no lubricant

References p. 203.

TABLE 4

Drawbead Forces and Coefficient of Friction for Test Lubricants
on Aluminum

Force (kN)	Drawbead	Sheet Thickness (mm) 2036-T4 Aluminum	
		0.81	0.89
Drawing	Rollers	2.6	3.2
Clamping	Rollers	2.2	2.8
Drawing	Fixed A	4.7	5.7
Clamping	Fixed A	3.6	4.8
Coefficient of Friction (A)			
μ (Constant P)		0.184	0.166
μ (Linear $P(s)$)		0.198	0.171
Drawing	Fixed B	3.0	3.9
Clamping	Fixed B	2.9	3.5
Coefficient of Friction (B)			
μ (Constant P)		0.044	0.065
μ (Linear $P(s)$)		0.025	0.062
Drawing	Fixed C	3.3	4.0
Clamping	Fixed C	3.3	3.4
Coefficient of Friction (C)			
μ (Constant P)		0.069	0.077
μ (Linear $P(s)$)		0.047	0.081

A = Mill Oil Lubricant
B = Soap Base Lubricant
C = Special Aluminum Lubricant

(dry) of about 0.12 for mild steel and 0.11 for aluminum. For a good lubricant he found a range of μ from 0.04 to 0.078 for mild steel and about 0.028 for aluminum. The agreement between μ for Swift's good lubricant and the drawbead results for the SB lubricant is excellent. Swift's values of μ for dry surfaces are somewhat lower than those found for drawbeads for the poor (MO) lubricant. However, Swift does not indicate whether he considers a "dry" specimen as one washed in an organic solvent. Residual oil on metal surfaces can be a relatively good lubricant. In tests with drawbeads, galling (i.e., locally heavy pickup of the sheet metal on the bead surface) occurs if the strips are washed in an organic solvent. Since galling was not reported by Swift, his specimens may not have been solvent washed and thus his μ might tend to be low compared to the values reported here for a poor lubricant. In addition, three bends and rebends occur during transit of the metal through drawbeads compared to one bend over a die throat. Since more deformation occurs, more new metal surface is exposed in a drawbead than in Swift's test. This would also tend to increase the magnitude of μ for a drawbead test compared to Swift's test.

Plots of the drawing force vs coefficient of friction for preliminary calculations for 0.86 mm A-K steel and 0.89 mm 2036-T4 aluminum are shown in Fig. 12. The experimental values of drawing force and μ (average between the two methods) from Tables 3 and 4 obtained from the drawbead simulator are plotted on the same graph. It may be seen that the calculated curves closely parallel the curves drawn through the experimental points for both materials.

Differences in the calculated and experimental forces at zero friction (i.e. deformation force alone) are due to the fact that the calculation does not include cyclic strain effects for aluminum and that the strain-rate effect tends to cancel the cyclic effect for steel as already discussed. In both cases, however, the fact that the experimental curve and the calculated curve are parallel as μ increases from zero shows that agreement is good for the frictional part of the calculation.

Friction Summary—Coefficients of friction were calculated from experimental drawing and clamping forces by two different methods. With this experimental μ, good agreement between experimental friction drawing forces and those calculated is shown. The good agreement between experimental and calculated drawing forces suggests that a coefficient of friction approach (Coulomb's law) would be applicable to calculation of drawbead friction forces.

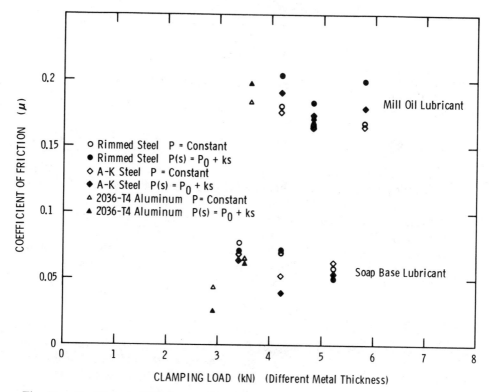

Fig. 11. Coefficient of friction vs clamping load for steel and aluminum. Clamping load is changed by drawing different thicknesses of metal strips through the drawbeads.

References p. 203.

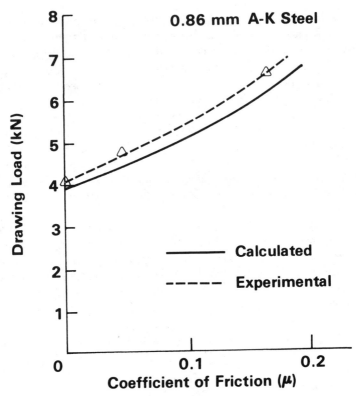

Fig. 12. Plots of force vs coefficient of friction for preliminary calculations compared to measured values, Δ, for:
a) A-K steel

CONCLUSIONS

1. Metal deformation and friction contributions to the drawing and clamping forces for drawbeads were separated. Friction contributes up to 50% of the drawing load for poor lubricants.
2. Deformation modeling for drawbeads must include cyclic strain hardening and strain-rate hardening (for steel).
3. Friction measurements for drawbeads show that Coulomb's law applies, and drawing load versus friction coefficient agrees with a computation based on Coulomb's law.

ACKNOWLEDGMENTS

The author appreciates the help of R. J. Dusman in designing the apparatus and E. G. Brewer in setting up the measurements on the MTS testing machine. Neng-Ming Wang of the Mathematics Department contributed with many helpful discussions and together with Professor B. Budiansky, Harvard University, aided substantially in developing expressions for the coefficient of friction. W. G.

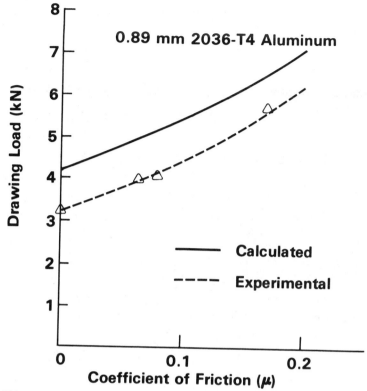

b) 2036-T4 aluminum.

Brazier of Fisher Body Division made many helpful suggestions during the course of this work.

REFERENCES

[1] M. A. Swift, Engineering, *166* (1948), 333.
[2] R. W. Landgraf, J. D. Morrow and T. Endo, J. Materials, *4* (1969), 176.
[3] A. K. Ghosh and J. V. Laukonis, Proc. 9th Biennial Congress IDDRG, ASM, Ann Arbor, (1976), 167.
[4] F. P. Bowden and D. Tabor, *The Friction and Lubrication of Solids,* Oxford Press, 1954.
[5] F. A. Bowden and D. Tabor, *The Friction and Lubrication of Solids,* Oxford Press, Part II, 1964.

APPENDIX

METHODS TO OBTAIN COEFFICIENT OF FRICTION, μ

I—Uniform Contact Pressure—The measured clamping force, C, (in Fig. A-1a) is the sum of all the components in the clamping direction of the required force.

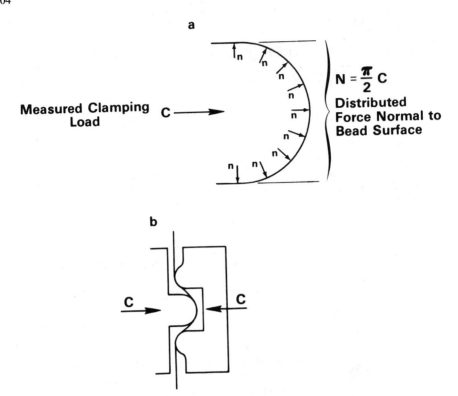

Fig. A-1.
  a) The normal force distributed over the semi-cylindrical bead surface is $\pi/2$ times the measured clamping force.
  b) The clamping force distributed on both sides of the sheet doubles the effective clamping force.

The required normal force to obtain a coefficient of friction is the distributed force normal to the semi-cylindrical bead surface, N. Assuming that the contact pressure is uniform over the bead surface, the required normal force, N, is obtained by integration over the bead surface to give:

$$N = \frac{\pi C}{2}$$

Now it is recognized that the contact pressure is not uniform but changes along the bead surface. However, as long as the contact pressure remains in the range over which μ is not sensitive to pressure (Coulomb's law), this simplifying assumption seems reasonable.

Since the sheet metal strip is loaded on both sides (on one side by the bead and on the other side by the groove shoulders), the measured clamping force, C, needs to be multiplied by two to get the total clamping force, Fig. A-1b.

Now the coefficient of friction may be obtained from the measured drawing and clamping forces from Eq. (3) in the text, as follows:

$$\mu = \frac{\tau}{P} = \frac{\dfrac{D_{d+f}}{A} - \dfrac{D_d}{A}}{2 \cdot \dfrac{\pi}{2} \cdot \dfrac{C_{d+f}}{A}} = \frac{D_{d+f} - D_d}{\pi C_{d+f}}$$

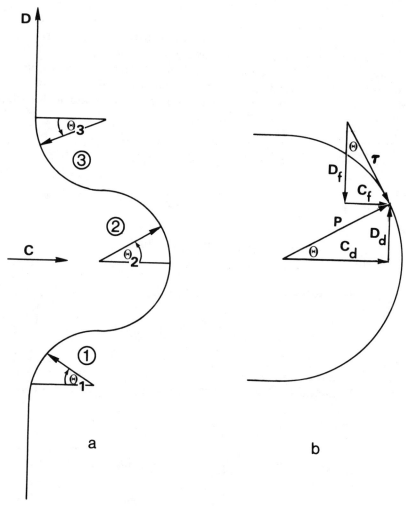

Fig. A-2.
   a) The directions of integration of the angles over the drawbead.
   b) The resolution of the stress at a point on the bead from bending deformation, P, into the drawing direction D_d and the clamping direction C_d. The friction stress, τ, is resolved likewise into D_f and C_f.

where D_{d+f} is the measured drawing force for standard drawbeads
$\quad\quad\,\, D_d$ is the measured drawing force for "frictionless" drawbeads
$\quad\quad\,\, C_{d+f}$ is the measured clamping force for the standard drawbeads
and $\quad A$ is the contact area.

II—Linearly Increasing Contact Pressure—An alternative method of obtaining a coefficient of friction is to analyze the deformation and friction forces into components in the drawing and clamping directions. These components are, of course, the experimentally measured quantities.

Fig. A-2a shows the direction of integration of θ over each shoulder, 1 and 3 and the bead 2. Fig. A-2b shows the deformation stress, P, and the friction stress, $\tau (\tau = \mu P)$, resolved into the drawing direction components, D_d and D_f respectively, and the clamping direction components, C_d and C_f, at a point on the bead. It also shows the angle chosen for θ. Then the combined forces due to deformation plus friction in the drawing direction, D_{d+f}, and in the clamping direction, C_{d+f}, are given by:

$$D_{d+f} = -\int_0^{\pi/2} PA \sin\theta_1 \, d\theta_1 - \int_{-\pi/2}^{\pi/2} PA \sin\theta_2 \, d\theta_2 + \int_0^{\pi/2} PA \sin\theta_3 \, d\theta_3$$

$$+ \mu \left[\int_0^{\pi/2} PA \cos\theta_1 \, d\theta_1 + \int_{-\pi/2}^{\pi/2} PA \cos\theta_2 \, d\theta_2 + \int_0^{\pi/2} PA \cos\theta_3 \, d\theta_3 \right]$$

$$C_{d+f} = \int_{-\pi/2}^{\pi/2} PA \cos\theta_2 \, d\theta_2 + \mu \int_{-\pi/2}^{\pi/2} PA \sin\theta_2 \, d\theta_2$$

$$= \int_0^{\pi/2} PA \cos\theta_1 \, d\theta_1 + \int_0^{\pi/2} PA \cos\theta_3 \, d\theta_3$$

$$+ \mu \left[\int_0^{\pi/2} PA \sin\theta_1 \, d\theta_1 + \int_0^{\pi/2} PA \sin\theta_3 \, d\theta_3 \right]$$

where P is the contact pressure at a point on the drawbead and $A \, d\theta$ is the contact area. The terms containing μ are the friction terms and the others are the deformation terms.

For the "frictionless" case $\mu = 0$ and $P = P_0$, then

$$D_d = -\int_0^{\pi/2} P_0A \sin\theta_1 \, d\theta_1 - \int_{-\pi/2}^{\pi/2} P_0A \sin\theta_2 \, d\theta_2 + \int_0^{\pi/2} P_0A \sin\theta_3 \, d\theta_3$$

and

$$C_d = \int_{-\pi/2}^{\pi/2} P_0A \cos\theta_2 \, d\theta_2 = \int_0^{\pi/2} P_0A \cos\theta_1 \, d\theta_1 + \int_0^{\pi/2} P_0A \cos\theta_3 \, d\theta_3$$

If P is assumed to increase linearly along the shoulders and bead, then $P = P_0 + a\theta$. Substituting into the integrals for P, integrating and simplifying, an expression for μ is obtained in terms of D_{d+f}, C_{d+f}, D_d and C_d,

$$[\pi D_d + (\pi - 2)C_d]\mu^2 + \left[(\pi - 2)D_d - \left(\frac{\pi - 2}{2}\right) D_{d+f} + 2\pi C_{d+f} \right] \mu$$
$$+ [\pi(D_d - D_{d+f}) + (\pi - 2)(C_{d+f} - C_d)] = 0$$

This expression may be solved for μ from the quadratic formula. The basic assumption of Coulomb's law, that μ does not change over the load range, is also necessary for this method of obtaining μ.

DISCUSSION

H. A. Armen (Grumman Aerospace Corporation)

In your kinematic hardening model, did you take into account the effect of accumulated stress and strain on the yield stress?

Nine

Well, at this point we really do not have a kinematic hardening model. The preliminary calculation we examined is an isotropic hardening model. There are difficulties in a kinematic model. While at this point, we know what we want to do, we do not know exactly how to do it. The problem is that for each material, the cyclic strain can be very different, depending on whether it is aluminum or whatever. So we are in the midst of trying to get together an experiment to try to do a cyclic experiment. There are a great deal of difficulties in this, many of which may have been overcome by Professor Tozawa in his earlier talk. But, with sheet metal, of course, we get tensile strains in a proper way, from a tensile sample. But, on a compressive cycle, the sheet metal buckles. Professor Tozawa has a way of glueing sheets altogether that may be appropriate. Another way: Daeyong Lee was kind enough to send us a reprint of a paper in which he describes a method using an hour glass shaped specimen for trying to get the cyclic strain. We really need a two-cycle, high-amplitude fatigue test similar to what Ron Landgraf at Ford is doing.

J. W. Justusson (General Motors Research Laboratories)

Have you ran a calculation of the membrane stresses to see how they compare with the bending stresses?

Nine

I did not mean to imply here that I ran the calculations. Actually, these are due to N.-M. Wang of General Motors Research Laboratories. But the original model, in fact, was only a membrane stress model. In that case, friction will rise in an exponential manner through the bead. This is not supported by experimental evidence. The actual distribution of the pressure in a draw bead is obtained from the way the wear occurs. We found that we got heavy wear on one side of the bead and on the other with a kind of skip in between. So there actually is a rather complex distribution of the pressure in the bead. Actually, when Dr. Wang changed the preliminary model to take into account the bending stresses, he got a pressure distribution which showed that the metal touched the first side, skipped a big area, and came down on the other side again. Thus, apparently, adding the bending stresses did work out if we can believe the agreement between Dr. Wang's analysis and our experiment. So, yes, bending is important.

G. M. Goodwin *(Chrysler Corporation)*

When you are running a test at various deflections, say by changing the thickness of your sample, did you notice (I refer to the slide involving the family of lubricants at different thicknesses of the material) the influence that the surface texture may have on friction?

Nine

Actually, that is a problem. And I guess that is why I said that these results *indicate* rather than *prove* that Coulomb's Law applied. But, that is a problem. First, when we go to different thicknesses of material in order to get the different clamping loads, it is a different lot of material. It is rolled differently. Now, I think we did not see a trend. If you recall that slide, μ is not monotonically increasing or decreasing with load. It skips around without being really very much different. We tried to control the conditions as well as we possibly could. But, of course, we are not able to control the surface roughness the way one would really like. I don't know whether there is a way of doing that.

Goodwin

Well, unless you can vary the clamping force by some other means.

Nine

Another way of varying the clamping force which we are set up to do is to use different size of rollers. This will make the deformation force higher, and make the clamping force higher. Again there is a little problem. Because then the amount of deformation changes. This may change the coefficient of friction since it may depend on the extent of deformation. The way we try to overcome the problem which Professor Wilson discussed is by modeling exactly the system which we are going to try to look at. And, if we then change the amount of deformation by changing the size of the bead radii, then we may get into trouble by having a different amount of deformation. So, there is not an obvious way. If you have, I would sure like to hear it.

Goodwin

Well, I see no obvious way. But I do know that deformation is why we went to drawbeads. The deformation makes a lot of difference compared to tests between two flat blocks.

Nine

I might say that as it turns out, as Gorton Goodwin is indicating, that we have got a very good correlation between our results and production experience. And, I think it is simply because we have modeled exactly the system that is out there in the plant. And so, we have tested lubricants with the apparatus and got very good correlation with production experience.

T. Altan (*Battelle Memorial Institute*)

I have two questions—one is: I presume at the high strain rate, 5 per second you were mentioning, you will have, of course, some heat up in the material. Would you comment on the effect of this heat up on the force? That is my first question. The second question is a kind of odd thing. Looking at the panel you have there, are there any new concepts in machine design or press design which will allow hold down force control during the process? I would imagine if you would press on one side more than the other and let the material flow, you could get into a lot of problems.

Nine

That has not reached out into the plants yet. One thing I meant to point out is that this orange tape here indicates the depression left by the drawbead. It gives you a scale as to how large the drawbeads are in relation to the rest of the panel. But those are trimmed off in the final part. In our plants the normal way the drawbead are utilized is by shimming the blankholder to a fixed clearance. Now you can, under the different sides of the die, adjust the shims a little bit so that you can keep one side up or down in relation to the other. That is part of this die try-out that I was talking about.

Altan

I was thinking about it, maybe computer control or something like this. . . .

Nine

I guess that is what this conference is about. Getting back to your question regarding temperature. We have not tried to measure temperature. There is a rapid transient as the sheet pulls through the bead at the rate of 85 mm/sec. With thin stock (as we are discussing here) there was not really noticeable heating. We did try to do some work on thicker stock; and there was extreme heating. But we did not get good results there, so we really are not prepared to discuss the experiment.

M. A. Bragard (*Centre de Recherches Metallurgiques, Belgium*)

Galling sometimes happens in the neighborhood of the drawbead. Could you interpret your results with the galling?

Nine

Yes, we did find galling. What he is referring to, is that in drawbeads or in any rather severe draw even over the lip of the die, you can get galling and the outward manifestation is that the remaining part of the sheet metal is scratched up very badly. What really is happening there is that the lubrication breaks down. You get a cold welding of the sheet material to the die interfaces or the drawbead interfaces or whatever. So actually it makes a little pile up of the sheet material. Aluminum is particularly bad that way; although iron or steel will also do that.

We, for the purpose of this study, stayed outside the region of galling because that puts in another real difficult thing to try to deal with. One thing we did note, however, was that the drawing force itself is not extremely sensitive to galling. That is, we could get galling and have a very low drawing force; and not get galling and have a very high drawing force. Of course, this is all dependent on the microscopic interactions at the asperities as Prof. Wilson was showing earlier. So there was not a good correlation there. Galling is very destructive and something we need to avoid. Therefore in screening a lubricant for example, even if it has low drawing load you would not pick that one if it galled.

W. Johnson (*University of Cambridge, U.K.*)

Just say how you derived the strain rate.

Nine

Yes. We know the spacing of our precision grid which was 0.5 mm. The pulling rate was 85 mm/sec. So we got a strain-rate but of course we measure at the surface.

Jonas

Is that the total strain divided by the time?

Nine

Yes. But because the sheet conforms to the radius of the bead so rapidly (over such a short distance) the strain-rate is very high. By measuring the strain distribution in the grid at the bending points we found that the bend occurs in a distance comparable to sheet thickness. Thus, knowing the drawing rate, we can find the bending rate and from that we can find the strain rate at the surface.

Jonas

I was going to ask Professor Wilson if he wants to just comment on the idea that Coulomb's Law might work here, for this particular case. Do you want to say anything about that?

W. R. D. Wilson (University of Massachusetts)

I think the 0.2 coefficient of friction indicates probably a good boundary regime. If you look at the 0.05 there is a lot more scatter by a factor of two or three. It might just be a model mixed regime. I'll be interested, though, in what happens when you go up by an order of magnitude in speed but that's going to be very difficult to do.

Nine

I would think that based on what you said, the reason we are getting around problems is that we model exactly the system in which we are interested. So this particular experiment in friction is not going to apply to other situations as you point out. It is specific to drawbeads.

SESSION IV

INSTABILITY PROCESSES

Session Chairman
S.S. HECKER

Los Alamos Scientific Laboratory
Los Alamos, New Mexico

SHEET METAL FORMING LIMITS

Z. MARCINIAK

Technical University of Warsaw, Poland

ABSTRACT

The maximum strain that the sheet metal can suffer in stretch-forming processes is limited by the phenomena of strain localization and fracture. Both are strictly connected with the stability loss of the workpiece. Different possible modes of instability are discussed in the paper. Also the post-stability stages of the forming process are analyzed. It is shown that the limit strain depends mostly upon the rate of strain-localization process that occurs in the unstable stage. The influences of stress ratio, material properties (strain hardening ability, strain rate sensitivity, temperature sensitivity, inhomogeneity) and the forming conditions (friction, surface pressure) on the strain concentration process and on the limit strain are analyzed.

BASIC ASSUMPTIONS

The limits of strains which a material can suffer in the sheet metal forming process are determined by various kinds of unfavorable phenomena which appear at certain stages of the forming process and disturb the proper course of the stamping operation. The following phenomena are of primary importance:

1. Strain localization on certain areas of the drawpiece leading to ductile fracture of the sheet metal in the most highly strained area.
2. Failure of the drawpiece caused by exceeding its strength in a site located outside the area of forming. For example the tearing of the bottom of a drawpiece during the deep-drawing operation.
3. Galling on the tool-material interface.
4. Wrinkling of sheet metal during stamping which is one of the forms of compression instability.
5. Fracture of tools caused by exceeding their strength.

This paper deals with the analysis of the processes of strain concentration leading to the fracture of sheet metal in the most highly strained area.

This paper tries to take an integrated view of the phenomena connected with the loss of stability and the fracture of sheet metal in stretching, taking into

References p. 233.

account their mutual interactions. In this approach it is necessary to deal with all phenomena which affect the change of yield stress and also with material softening due to the growth and coalescence of voids and microcracks, and due to the heat generated during straining, etc. This idea was put forward by Backofen [1], Rice [2], Baudelet and others.

As a result, the present analysis will be based on the assumption that the changes of yield stress σ can be expressed as a product of two functions

$$\sigma = K \cdot \sigma_\rho. \tag{1}$$

Stress σ_ρ is treated here in a classical sense as a quantity expressing the hardening of the material depending on the strain, the strain path and the temperature. On the other hand the coefficient K takes into consideration the softening of material caused by growth and coalescence of voids [1]. Its value depends, in the first place, on the sign of stress prevailing in all the earlier stages of the process. For the purposes of this paper it is not necessary to specify the form of the functions σ_ρ and K. It is sufficient to assume that σ_ρ grows monotonically together with the strain and that K decreases from 1 to 0 all the faster, the higher are the tensile stresses compared to the compressive ones. As a result, the value of the product $K \cdot \sigma_\rho$ attains the maximum at point M, Fig. 1, and then falls to 0, which means a complete loss of material cohesion. The curve $\sigma(\epsilon)$ is thus not solely a characteristic of the material but depends also on the conditions of strain. For example

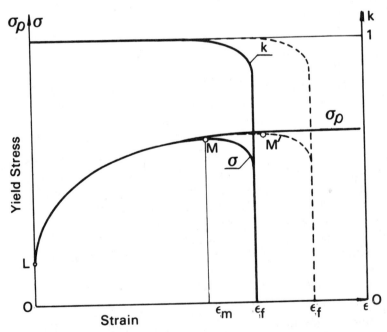

Fig 1.   Yield stress σ as a product of strain-hardening function σ_ρ and strain-softening coefficient K.

an increase of compressive stresses delays fracture and results in a change in the shape of the curve $\sigma(\epsilon)$ as is illustrated by the broken line in Fig. 1.

FORMS OF INSTABILITY IN STRETCH-FORMING PROCESSES

Let us now analyze the possible forms of instability which can take place during the process of sheet metal stretching. As an example we shall examine the process of tube expansion, Fig. 2, under internal pressure and axial tensile force. This load is so chosen as to make the axial elongation of the tube equal to zero during all stages of straining, $\epsilon_2 = 0$. We shall therefore assume that the tube is being formed under conditions of plane strain.

The course of a tube expansion process can be divided into the following stages:

1. Uniform straining stage, Fig. 2a.
   The uniform strain distribution is ensured by the intensive strain hardening of the material, which at this stage of the process is capable of compensating for both the increase in the tube diameter and the reduction of the thickness of its wall.

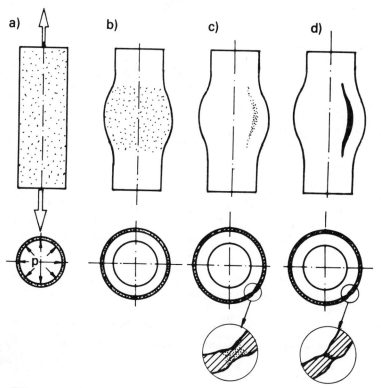

Fig 2. Tube expansion: (a) uniform straining stage; formations of (b) a bulge; (c) a localized neck; and (d) shear bands.

2. Formation of a bulge, Fig. 2b.
The loss of the tube stability is expressed here by a nonuniform distribution of strains along the length of the tube. As a result a local bulge emerges, while the remaining parts of the tube are unloaded and pass into an elastic state. A characteristic feature of this stage of the process is that the length of the bulged section of the tube depends on the diameter of the tube only and not on the thickness of its wall.

3. Formation of a localized neck, Fig. 2c.
In the course of further deformation of the bulged section of the tube a groove starts to be formed on its surface. The groove is the result of the local thinning of the tube wall (i.e., of the nonuniform distribution of strains along the perimeter of the bulge.) In contrast to the previously discussed stage, the strain concentration area i.e. the width of the localized neck depends on the thickness of the wall only and not on the diameter of the tube. The formation of such a neck attests to the onset of instability of the sheet metal.

4. Formation of shear bands, Fig. 2d.
During a further straining of the material in the middle of the neck, a band of localized shear strain occurs, setting askew the surface of the sheet. The thickness of the layer in which further strain is concentrated does not depend either on the thickness of the wall or on its diameter. The emergence of this shear band is the result of the loss of stability of the material and is connected with the loss of its work hardening ability.

5. Fracture of the material.
The growing strain in the band of localized shear strain is accompanied by an uncontrolled process of growth and coalescence of internal voids, leading to a complete loss of cohesion of the material in its most strained layer.

As can be seen from this example the area of the plastically strained zone diminishes gradually, together with the emergence of subsequent forms of instability.

The process of differentiation of strains in various parts of the tube is illustrated by Fig. 3.

In this diagram the strains at different points of the tube are plotted against the strain in the most strained part of the tube, i.e. at point C.

The particular lines show the straining history in various places of the tube, namely:

- The broken line OTD represents the process of straining at point D, which is placed outside the area of bulging (ϵ_D). The point T is the tube instability point.
- The line OSA corresponds to point A placed on the equator of the bulge outside of the groove. The point S refers to the onset of instability of the wall.
- The line OMB represents point B placed in the middle of the localized neck but outside of the shear band. The point M shows the onset of the material instability.

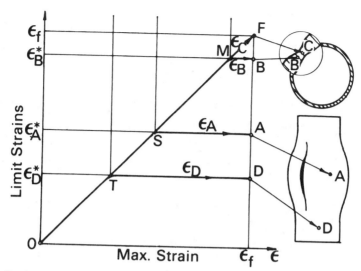

Fig. 3. Strains at various locations of the tube: D—outside the bulged section; A—inside the bulged section but away from the localized neck; B—in the neck; C—in the shear bands.

- The straight line OF corresponds to point C placed in the band of localized shear strain. The point F represents the fracture of the sheet metal.

This figure shows also the limit strains $\epsilon_D{}^*$, $\epsilon_A{}^*$, $\epsilon_B{}^*$ at various parts of the tube represented by points D, A and B. The process of gradual concentration of strains and narrowing of the plastically strained area has been presented by the simple example of a thin-walled tube. However the phenomena described can also be noticed in any other processes of sheet metal forming. They are connected with the three basic forms of stability loss during stretching, namely:

1. The stability loss of the drawpiece,
2. of the sheet metal,
3. and of the material itself.

SHEET METAL INSTABILITY

Let us concentrate our attention on the conditions of a stability loss of the sheet metal which takes the form of a localized neck.

In the above discussed example of the tube expanding process the stability loss of its wall is represented as a sudden change of the kinematic field taking place at the onset of instability. This is shown in Fig. 4 (which is a part of the previous diagram) by the broken line OSA, where the point S is the point of stability loss of the sheet metal. The sudden passage from a stable to an unstable stage of the process, at point S, can take place only in a particular case when there are no additional factors accelerating or delaying the process of strain localization.

In real metals, as a result of the action of certain factors which will be discussed

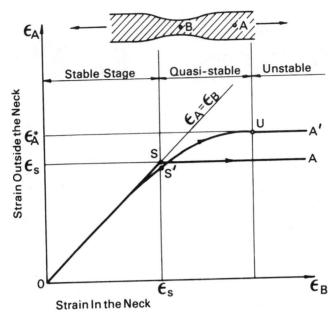

Fig. 4.   Strain outside the neck versus strain in the neck.

later, the border between the stable and unstable stage of the process is blurred and the process of localization of strains in the neck occurs gradually as shown by line $OS'UA'$ in Fig. 4. As can be seen, this is a line on which there is no particular point which can be related to the onset of instability. However, for purely conceptual reasons it is convenient to divide the process of localization of strains in the neck into three stages:

1. The stable stage, corresponding to segment OS' in Fig. 4.
2. The intermediary stage, which we shall call quasi-stable stage, corresponding to the curved segment $S'U$. In this stage, the process of strain concentration in the groove (point B) is accompanied by further ever slower straining of the material in areas located outside the groove (point A).
3. The unstable stage, in the course of which only the neck is strained, whereas the material outside the neck is unloaded and passes to an elastic state. The horizontal segment UA' corresponds to this stage.

The position of point S', which represents the conventional border between the stable and quasi-stable stage of the process depends on which criterion of sheet metal instability is assumed. Various proposals concerning the definition of the point of stability loss of sheet metal were put forward by H. W. Swift [3], P. B. Mellor [4], and others. Lately an interesting proposal which generalizes Swift's criterion of stability so as to cover inhomogeneous materials was presented by P. B. Mellor [5]. In order to analyze the quasi-stable stage of the process it is however more convenient to apply the criterion proposed by Hill [6].

The onset of instability of sheet metal is here defined as the extreme point of a product of the thickness of the sheet metal t and its yield stress σ.

According to this criterion the loss of stability of sheet metal during a proportional straining process will occur when the sum of principal strains is equal to the strain-hardening index, defined as

$$n = \frac{d\sigma}{d\epsilon} \cdot \frac{\epsilon}{\sigma} . \qquad (2)$$

When the value of n does not vary with strain, the condition of sheet metal stability represents a straight line KL on the graph ϵ_1 versus ϵ_2, Fig. 5. However, if the value of the strain-hardening index n diminishes as the strain increases, the sheet stability criterion will correspond to the curve marked by the solid line PKL in Fig. 5. In the case of plane strain in which the thickness strain ϵ_3 is zero, the loss of stability of sheet metal occurs at point P, at which the strain-hardening index n diminishes to zero.

QUASI-STABLE DEFORMATION

In accordance with Hill's criterion, the stable stage of the process of deformation ends at point S' (Fig. 4) or on the line PKL shown in Fig. 5. We shall now deal with the second, quasi-stable stage of this process which takes place immediately after this loss of stability.

At the onset of sheet instability the hardening of the material is balanced by the thinning of the sheet. Under these circumstances some additional factors, which in the stable stage of the process played a secondary role, come now to the forefront. The most important factors are the following:

- Initial inhomogeneity of sheet metal
- Change in the stress ratio

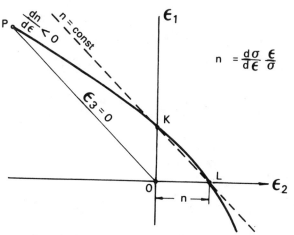

Fig. 5. Influence of strain hardening index n and its variation with strain on sheet metal stability limits.

- Normal anisotropy
- Strain-rate sensitivity
- Gradient of strains along the groove
- Surface pressure
- Friction on the surface of the tool
- Triaxial state of stresses in the neck
- Inertia forces.

Some of these factors delay, and others accelerate, the process of strain concentration which occurs in the quasi-stable stage. Their cumulative influence determines the rate of the process of strain localization in the groove as well as the limit strain $\epsilon_A{}^*$ outside the groove.

We shall now discuss the influence of some of the mentioned factors on the process of localization of strains in the quasi-stable stage.

Sheet Inhomogeneity—Sheet metal shows always a certain degree of inhomogeneity. This is both a geometrical inhomogeneity arising from the unequal thickness of the sheet metal or from the roughness of its surface [5], [8] and a physical inhomogeneity caused by an inhomogeneous distribution of impurities and voids, grain boundaries and specific orientation of crystallographic planes etc.

The initial inhomogeneity of sheet metal can be marked by an index f defined as the ratio of the minimal value of product of yield stress and sheet thickness for the weakest point of the sheet metal to its mean value.

The initial inhomogeneity of the sheet metal causes a nonuniform strain distribution on its surface, with the concentration of strains in the weakest area of the sheet beginning in the earliest stages of straining [7], [1]. In the first stable stage the process of concentration takes, however, place at a very low rate because of the intensive hardening of the sheet metal. This is shown by the line OS' in Fig. 6 which runs, at first, almost alongside the straight line OS. However, as the

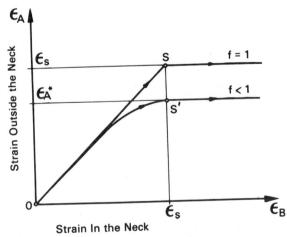

Fig. 6.   Influence of sheet metal inhomogeneity on strain localization.

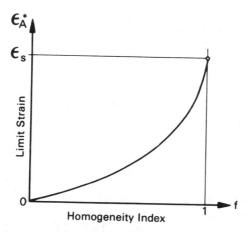

Fig. 7.   Effect of sheet metal in homogeneity on limit strain ϵ_A^* (outside the neck).

strain hardening ability diminishes the differences of strains in the emerging neck ϵ_B and outside it increase quickly. At the moment of loss of stability (point S') the limit strain outside the neck ϵ_A^* is significantly lower than that in the neck.

The drop of limit strain ϵ_A^* caused by the inhomogeneity of the material is illustrated by Fig. 7. It may be noted that the derivative $d\epsilon_A^*/df$ increases infinitely when the index f tends to one. Thus, limit strain is extremely sensitive to the inhomogeneity of sheet metal especially when the material is almost homogeneous.

Change in Stress Ratio—In the case of plane strain, the localized neck does not cause any change in the principal stress ratio. Strain ϵ_2 along the groove is equal to zero in the neck as well as outside it.

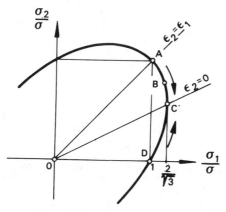

Fig. 8.   Change of stress ratio to plane strain state (from A to C or from D to C) during neck formation.

References p. 233.

However, if for example the stress ratio σ_2/σ_1 outside the neck is equal to one (in other words we have balanced biaxial tension), the process of strain concentration should move the point B in Fig. 8 which represents the state of stresses in the groove toward the point C [7]. It results in an increase in strength of the groove. This increase is connected with the Huber-Mises criterion of plasticity and does not appear in the case of the Tresca condition. Thus, the change in the stress ratio in the groove is an additional factor stabilizing the process of straining. It delays the process of strain concentration and causes an increase of limit strain $\epsilon_A{}^*$. This increase depends on the strain ratio, which is illustrated by Fig. 9. The lines KL and $JK'L'$ represent strain states at the onset of instability of the sheet metal, in and outside the groove. The line $HK'NL'$ representing limit strains is the well known forming limit curve. The distance between the forming limit curve (line $HK'NL'$), representing the end of the quasi-stable stage of the process, and the line $JK'L'$ which corresponds to the beginning of this stage, illustrates the increase in strain which occurs in the quasi-stable stage of the process. This increase is a result of the stabilizing influence of the change in the stress ratio in the neck assuming Huber-Mises criterion of plasticity.

It must be pointed out that this extra strain which occurs in the quasi-stable

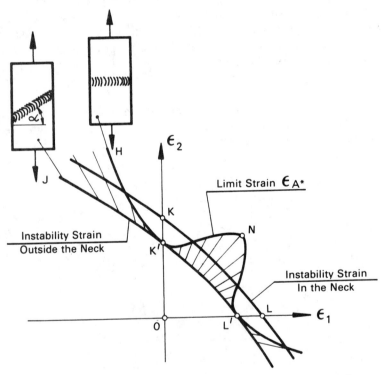

Fig. 9. Shaded region indicates the increase of limit strain due to change in stress ratio.

stage, is extremely sensitive to the shape of the yield locus, i.e. the Baushinger effect, non-isotropic strain hardening, normal anisotropy and so on.

Normal Anisotropy—The normal anisotropy of sheet metal, as expressed by Lankford's coefficient (r), affects the process of the quasi-stable stage of straining through its influence on the shape of the yield locus and the increase in the value of the σ_2/σ_1 ratio as it approaches the plane strain state.

For materials of r value greater than one, the yield locus becomes more elongated, as shown in the upper graph of Fig. 10. As a result, the forming limit curve for this anisotropic material, marked by the dashed line, crosses the isotropic curve as shown in the lower graph of Fig. 10.

Strain-Rate Sensitivity—It is obvious that this feature of the material delays the process of localization of strains, and results in an increase of limit strain $\epsilon_A{}^*$ adjacent to the groove. The process of strain localization for a strain-rate sensitive material ($m > 0$) and for a rate-insensitive material ($m = 0$) has been presented in Fig. 11.

Strain Gradient Along the Groove—In order to analyze this influence let us consider, first, an ideal case of an infinitely long groove, whose shape and

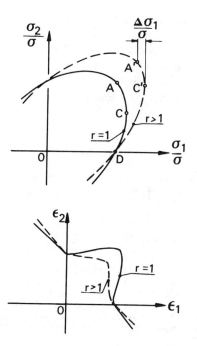

Fig. 10. Effect of the normal anisotropy coefficient r on yield locus and on forming limit curve.

References p. 233.

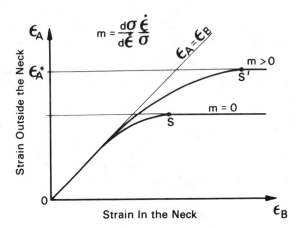

Fig. 11.   Influence of strain rate sensitivity on strain localization.

dimensions of the cross-section do not depend on the location of the section along the groove.

The grid lines marked on the surface of the sheet metal in directions parallel and perpendicular to the groove remain a rectilinear and orthogonal in all stages of the straining process.

When the strain varies along the groove the initially rectangular grid changes into a curvilinear one, as shown in Fig. 12b, and additional shear stresses τ acting on the parallel edges of the isolated strip, must be taken into account in equilibrium conditions. The gradient of strains along the groove acts, thus, in a stabilizing manner on the strain localization process, and causes a more uniform distribution of strains along the groove and also perpendicular to it.

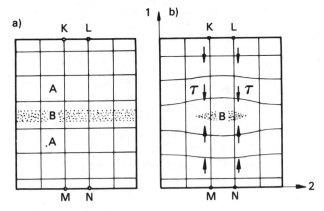

Fig. 12.   Strain gradient along the groove, B, and resulting shear stress τ acting on the strip KMNL.

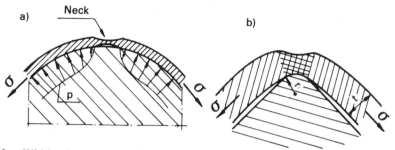

Fig. 13.   Width of tool contact is (a) much greater and (b) smaller than sheet thickness.

Tool Pressure (Fig. 13)—In the majority of stamping operations, localized necking occurs where the sheet metal is in contact with the tool and is being submitted to the pressure of the tool. Initially, this pressure effectively causes reduction of the tensile flow stress and facilitates the plastic flow of the material. As the necking process goes on, the tool pressure decreases locally. Its value may drop even to zero if a clearance should appear between the surface of the sheet metal and the surface of the tool. A reduction of tool pressure results in a growth of the strength of the neck in relation to its neighboring areas which are still subjected to the pressure of the punch. This delays the process of strain concentration in the groove and increases the value of limit strain. This effect depends on the thickness of the sheet metal in relation to the punch radius and, among other things, explains the fact that limit strain increases with sheet thickness. The tool pressure can also play the role of a factor initiating and accelerating the process of strain localization. This occurs when the pressure is localized in a small area whose width is very small compared to the thickness of the sheet metal.

Tool Friction—If the friction forces acting on both sides of the neck are directed towards the groove, like in area A in Fig. 14, the process of strain localization

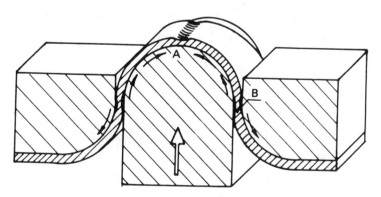

Fig. 14.   Tool friction.

References p. 233.

is delayed and the limit strain outside the neck is increased. Friction forces can also accelerate the concentration of strains and constitute an important factor initiating the neck formation. The groove which is being formed in area B (see Fig. 14), located on the border between the zones of contact to the punch and the die, can serve as an example.

Triaxial Stress State.—As a result of change in geometry in connection with the formation of the localized neck, a tensile stress normal to the surface of the sheet metal emerges in the middle of the neck. The effect of this stress is to increase the tensile flow stress in the plane of the sheet, thus effectively strengthening the material in the neck in relation to its neighboring area. Consequently, the triaxial state of stress in the middle of the neck is another factor stabilizing the process of straining. Moreover, it has a significant influence on the profile of the emerging neck.

Inertia Forces.—Let us consider the process of explosive expanding of a cylindrical tube shown in Fig. 15. The formation of localized necks in the tube is connected with the circumferential component of the inertia forces. These inertia forces acting on both sides of the emerging groove have opposite signs and are directed towards the groove. Thus, they contribute to a more uniform distribution of strains in the circumferential direction.

In conclusion it can be stated that the main factors stabilizing the process of straining after the loss of stability of the sheet metal and delaying the concentration of strains in the neck are: the strain-rate sensitivity of the material, the change in the stress ratio in the groove, the gradient of strains along the length of the groove as well as the contact pressure and the friction on the tool surface. On the other hand, the factors accelerating the localization of strains include: the

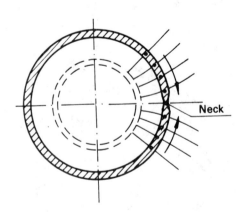

Fig. 15. Schematic illustration of the circumferential component of inertia forces (shown by arrows) resulting from neck formation in explosive tube expansion.

initial inhomogeneity of the sheet metal, the concentrated tool pressure as well as the unfavorable direction of friction forces.

The cumulative influence of the above mentioned factors determines the process of strain concentration in the neck during the quasi-stable stage as well as the value of limit strain outside the neck. It must be pointed out that depending on the conditions of forming each individual factor can exert favorable or unfavorable influences. As illustrated in Table I, for different forming operations the set of stabilizing or destabilizing factors affecting the course of the quasi-stable stage and the value of limit strain, is different from one another. Therefore, the forming limit curve depends not only on the properties of the sheet metal but also on the conditions of forming.

Table I. Stabilizing (+), neutral (0) and destabilizing (−) factors in several forming operations

		Variation in Stress Ratio	Surface Pressure	Friction	Strain Gradient	Inertia Forces
Nakazima Test		+	+	+	+	0
Explosive Bulging		+	0	0	+	+
Balanced Biaxial Stretching		+	0	0	0	0
Tensile Test		0	0	0	+	0
Bottom Stretching		0	0	−	0	0

THE MATERIAL INSTABILITY

The conditions accompanying the loss of stability of a material have been analyzed by Drucker, Iliuszyn, Rice and others. For materials which are rigid-plastic, isotropic, rate insensitive and obeying the associated rules of plastic flow, all proposed criteria of material stability merge to the condition that the instability takes place at point M corresponding to the extreme of the strain hardening curve $\sigma(\epsilon)$ (Fig. 16) for a given process of straining. Past point M there begins the unstable stage of the process. It manifests itself in the form of shear bands connected with the planes of the largest shear stresses. With the absence of additional stabilizing factors the process of localization of strains in these narrow layers can occur suddenly, as shown by the broken line OMN in Fig. 16.

However, most frequently the straining process proceeds in conditions favoring a more uniform distribution of strains. The factors stabilizing the process may include, for example:

- The material's strain-rate sensitivity
- The change of stress ratio in the layers of localized strain causing an increase in the strength of that layer

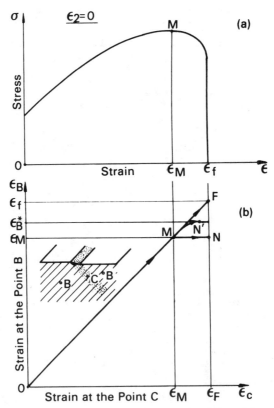

Fig. 16.   (a) Definition of material instability (point M) and instability strain ϵ_M; (b) Strain outside versus strain inside shear bands.

- The action of adjacent, less strained areas of the drawpiece remaining in a stable state which is connected with the gradient of strains along the shear band.

The joint action of these factors results in an increase of strain outside the shear bands. This process is illustrated by the line OMN' in Fig. 16 where, the curved segment MN' represents the quasi-stable stage of the process. The limit strain at point B, ϵ_B^*, exceeds ϵ_M which corresponds to the onset of instability of the material. In the extremely intensive action of stabilizing factors the line MN approaches the straight line MF and ϵ_B^* can attain the value of fracture strain ϵ_f. It can therefore be stated that the limit strain ϵ_B^* can vary within the limits marked by the instability strain and the fracture strain

$$\epsilon_M \leq \epsilon_B^* \leq \epsilon_f$$

depending on the intensity of the stabilizing factors.

However, it should be pointed out that the position of the extreme point M, as well as the shape of the descending part of the curve $\sigma(\epsilon)$, are connected with the phenomenon of growth and coalescence of voids which depend on the magnitude of the tensile stresses. For plane stress ($\sigma_3 = 0$), the stress ratio depends on the strain ratio ϵ_2/ϵ_1. Hence, points M and F representing respectively the loss of stability and the fracture strain of the material depend on ϵ_2/ϵ_1 as is shown in Fig. 17. The lines labeled by M and F form the border of the area within which the points representing the limit strain ϵ_B^*, with regard to material instability, are located. The positions of these points depend on the intensity of all other factors stabilizing the strain concentration process, and can not be determined if the exact conditions of forming are not known.

The phenomena of stability losses of the material and of the sheet metal can interact with each other. We shall explain the mechanism of this interaction by the plane strain process shown in Fig. 18. Points S and M on the strain hardening

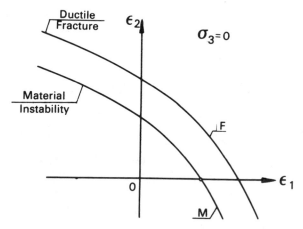

Fig. 17.   Schematic illustrations of fracture and instability limits.

References p. 233.

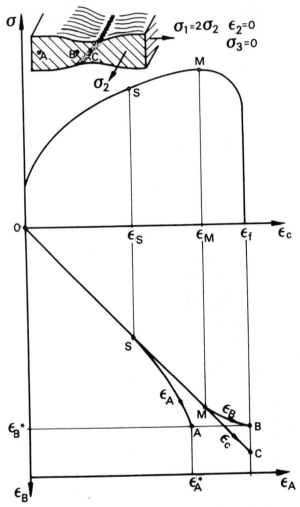

Fig. 18. Strain localization in plane strain.

curve $\sigma(\epsilon)$ correspond to the points of losses of stability of the sheet metal and of the material. Let us assume that for given conditions of forming i.e., for a given set of stabilizing factors, the process of strain concentration in the groove is illustrated by the line OSA, which expresses the relation between the strain outside the neck versus strain in the neck. Let us assume further that the process of strain localization in the shear bands takes place gradually as a result of the action of additional stabilizing factors such as strain-rate sensitivity. This process is illustrated by the curvilinear segment MB expressing the strain at point B versus the strain at point C. Points A, B and C in the graph correspond to the strains in the various areas of the cross-section at the instant of fracture at point C. It can be seen that the position of point A, i.e. the value of limit strain $\epsilon_A{}^*$

outside the groove depends not only on the process of the formation of the groove but also on the position of points B and C connected with the phenomenon of the material instability.

In other words the phenomenon of material instability in the form of shear bands can have in some cases an influence on the limit strain outside the neck.

REFERENCES

[1] M. Azrin and W. A. Backofen, Metall. Trans., *1* (1970), 2857.
[2] J. R. Rice and D. M. Tracey, J. Mech. Phys. Solids, *17* (1969), 201.
[3] H. W. Swift, J. Mech. Phys. Solids, *1* (1952), 1.
[4] P. B. Mellor, J. Mech. Phys. Solids, *5* (1956), 41.
[5] A. Parmer, P. B. Mellor and J. Chakrabarty, Int. J. mech. Sci., *19* (1977), 389.
[6] R. Hill, Trans. ASME, *67* (1967),
[7] Z. Marciniak and K. Kuczyński, Int. J. mech. Sci., *9* (1967), 609.
[8] A. K. Ghosh and W. A. Backofen, Metall. Trans., *4* (1973), 1113.

DISCUSSION

B. Kaftanoglu *(Middle East Technical University, Turkey)*

Thank you for a very interesting paper Professor Marciniak. I would like to report on some of our research carried out on the behavior of tubes under hydraulic pressure. We found—this was on tubes just exactly as you have pointed out—that we wanted to approximate an infinite tube so we used a long tube, and we did obtain, of course, plane-strain condition. We found that the deformation was rather homogeneous and instability occurred almost instantaneously without a secondary bulge forming. So that, in general, the overall homogeneous straining was most important and the local straining was very inhibited. Application of the Hill criterion showed a gross overestimate of the instability strain. Instability was generally found for copper and brass tubes even before maximum pressure was reached exactly as was found by Professor Mellor in his early work. I agree with your argument about the role of frictional stresses in punch stretching. It is certainly an important factor if one is to look at the process because it really is the friction between the tool and the piece which affects the local straining. This raises important questions. Thank you.

Marciniak

The only answer is that this was only an idea presented here that what represents the strength of the localization process depends on so many factors and also upon material properties. So that for each particular case we can get quite different part involvement. Speaking of plane-strain in tubes, wherein the length of the tube doesn't change, we have a sudden change from stable to unstable stage of the process. That is what I have wanted to show.

J. J. Jonas *(McGill University, Canada)*

I think the question of whether you have stable flow or unstable flow is always an interesting one because there are so many ways of trying to decide, or rather define, unstable flow. Now in the talk you gave today you defined unstable flow as the flow that occurs after the forming limit is reached.

Marciniak

Yes. The unstable state is here defined when no strain outside the neck occurs.

Jonas

So, I was going to argue that you could set up an alternative set of definitions based on the strain gradients and on the strain rate gradients as follows. You could say that in the beginning of the process there is really a stable flow where the strain-rate gradients, for example, or the strain gradients are decreasing—this is fully stable flow. Then you come into a range of quasi-stable flow where the strain-rate and strain gradients are increasing, but slowly. Now that happens, I think, before the region that you call quasi-stable flow. At a later stage then the strain-rate gradients and the strain gradients increase quickly so then you have what you could call an unstable flow, but you still have not reached the forming limit. Then at a later point you have the forming limit where outside the groove there is no further deformation and then you have what you could call fully localized flow. The point I'm trying to make is that we can actually divide the experiment into four steps; there is a stable step, there is a quasi-stable step, there is an unstable step, and there is a fully localized step. This is a proposal which I am making.

Marciniak

I agree that this problem of stability can be divided in many ways depending on the stability criterion employed.

J. W. Hutchinson *(Harvard University)*

Professor Marciniak, you made one reference to the effect of sheet thickness on the limit strains. Can you share with us any other insights into the thickness effect? Why it is that the thicker sheets show higher limit strains? None of the analyses seem to get to the bottom of that.

Marciniak

I think that this part of the analysis job is only one among many others. The others are, for example, greater triaxiality, stress, strain, the middle of the neck which is harder when the thickness of the sheet is higher. Friction forces are also higher even though the ratio of the thickness to radius is the same. Finally, bending is higher, and the homogeneity of the material may also be different from thin sheets.

P. B. Mellor *(University of Bradford, U.K.)*

One of your figures showed a sheet stretching over a sort of hemispherical shape on a constant radius with friction—you had a neck at the pole. I would have expected two strain peaks on either side of the pole.

Marciniak

Yes, but in this figure the width of the shape was not shown.

LIMITS TO DUCTILITY SET BY
PLASTIC FLOW LOCALIZATION

A. NEEDLEMAN and J. R. RICE

Brown University, Providence, R.I.

ABSTRACT

The theory of strain localization is reviewed with reference both to local necking in sheet metal forming processes and to more general three dimensional shear band localizations that sometimes mark the onset of ductile rupture. Both bifurcation behavior and the growth of initial imperfections are considered. In addition to analyses based on classical Mises-like constitutive laws, we discuss approaches to localization based on constitutive models that may more accurately model processes of slip and progressive rupturing on the microscale in structural alloys. Among these non-classical constitutive features are the destabilizing roles of yield surface vertices and of non-normality effects, arising, for example, from slight pressure sensitivity of yield. We also discuss analyses based on a constitutive model of a progressively cavitating dilational plastic material which is intended to model the process of ductile void growth in metals. A variety of numerical results are presented. In the context of the three dimensional theory of localization, we show that a simple vertex model predicts ratios of ductility in plane strain tension to ductility in axisymmetric tension qualitatively consistent with experiment. We also illustrate the destabilizing influence of a hydrostatic stress dependent void nucleation criterion. In the sheet necking context, and focussing on positive biaxial stretching, it is shown that forming limit curves based on a simple vertex model and those based on a simple void growth model are qualitatively in accord, although attributing instability to very different physical mechanisms. These forming limit curves are compared with those obtained from the Mises material model and employing various material and geometric imperfections.

INTRODUCTION

A remarkably common observation in highly deformed ductile solids is that a smooth deformation pattern rather abruptly gives way to one involving a band

References pp. 264–265.

(or bands) of localized deformation. Here, we review a theoretical framework which associates this localization of deformation with a material instability. More specifically, we consider an element of a solid, constrained so as to rule out geometric instabilities, subject to loading that *could* give rise to a homogeneous deformation and determine the conditions under which the constitutive relation of the solid permits a highly localized deformation pattern to emerge. Within this framework, the onset of localization can occur as a bifurcation from a homogeneous deformation state or it can be triggered by some initial inhomogeneity, possibly well before a bifurcation would occur if the (small) inhomogeneity were absent. Although not all observed localization phenomena can be expected to fit within this theoretical framework, it is broad enough to encompass descriptions of the localized necking mode characteristic of ductile metal sheets (when these are viewed as two dimensional continua) as well as of the "shear band" localizations that seem to be important precursors to ductile rupture.

As is not surprising, given the point of view adopted here, the onset of localization will depend critically on the assumed constitutive law. We base our analyses on the classical rate independent Prandtl-Reuss equations (suitably modified to account for finite deformations) and departures from this idealized model that include the effects of (i) yield surface vertices, (ii) deviations from plastic "normality" and (iii) the dilational plastic flow (and possible strain softening) due to the nucleation and growth of voids.

The effects of these deviations from the idealized Prandtl-Reuss material model on the predicted onset of flow localization are illustrated and compared. Particular emphasis is given to the example of local necking in metal sheets subject to positive biaxial stretching, where both vertex effects and the weakening induced by ductile void growth appear to provide plausible explanations of observed behavior.

CONSTITUTIVE RELATIONS

The constitutive relations on which we base our discussion are the classical Prandtl-Reuss relations and generalizations of these to allow investigation of (i) yield surface vertex effects as predicted, e.g., from models of the crystalline slip process, (ii) slight pressure-sensitivity of yield and deviations from an associated flow rule, and (iii) dilational plastic flow, and, possibly, strain softening due to micro-rupture by the nucleation and growth of cavities. Although not considered explicitly here, significant departures from isotropy are also important in applications to sheet materials.

We write the rate of deformation $\mathbf{D}$ (symmetric part of $\partial \mathbf{v}/\partial \mathbf{x}$ where $\mathbf{v}$ is the particle velocity) as

$$\mathbf{D} = \mathbf{D}^e + \mathbf{D}^p \tag{1}$$

where, assuming elastic isotropy and small elastic dimension changes, the elastic part, $\mathbf{D}^e$, is

$$\mathbf{D}^e = \frac{1}{2G}\left[\overset{\triangledown}{\boldsymbol{\sigma}} - \frac{\nu}{1+\nu}\mathbf{I}\,\mathrm{tr}(\overset{\triangledown}{\boldsymbol{\sigma}})\right]. \tag{2}$$

Here tr( ) denotes the trace, $\overset{\triangledown}{\boldsymbol{\sigma}}$ is the Jaumann (or corotational) derivative of Cauchy (true) stress and is related to the ordinary material time rate $\dot{\boldsymbol{\sigma}}$ by

$$\overset{\triangledown}{\boldsymbol{\sigma}} = \dot{\boldsymbol{\sigma}} + \boldsymbol{\sigma}\cdot\boldsymbol{\Omega} - \boldsymbol{\Omega}\cdot\boldsymbol{\sigma} \qquad (3)*$$

where the spin $\boldsymbol{\Omega}$ is the anti-symmetric part of $\partial v/\partial x$. The plastic part, $\mathbf{D}^p$, according to the Prandtl-Reuss model, is (e.g., Hill [1])

$$D_{ij}^p = \frac{1}{2h}\frac{\sigma_{ij}'}{\tau_e}\dot{\tau}_e = \frac{1}{4h}\frac{\sigma_{ij}'\sigma_{kl}'}{\tau_e^2}\overset{\triangledown}{\sigma}_{kl} \qquad (4)$$

where $\boldsymbol{\sigma}'$ is the deviatoric part of $\boldsymbol{\sigma}$, h is the plastic hardening modulus in shear and τ_e is the "equivalent" shear stress, related to the equivalent tensile stress σ_e and to $\boldsymbol{\sigma}$ by

$$\tau_e^2 = \frac{1}{3}\sigma_e^2 = \frac{1}{2}\sigma_{ij}'\sigma_{ij}'. \qquad (5)$$

Of course, the above form for $\mathbf{D}^p$ is understood to apply only when the stress state is at yield and the imposed deformations are such as to enforce continued plastic flow; otherwise $\mathbf{D}^p = \mathbf{0}$. The value of the equivalent shear stress τ_e required for continued yielding is considered to be a function of the equivalent "engineering" plastic shear strain γ_e^p, or, alternatively, σ_e is taken to be a function of the tensile equivalent plastic strain ϵ_e^p, where

$$\dot{\gamma}_e^p = \sqrt{3}\dot{\epsilon}_e^p = (2D_{ij}^{p'}D_{ij}^{p'})^{1/2}. \qquad (6)$$

Here, $\mathbf{D}^{p'}$ is the deviatoric part of $\mathbf{D}^p$ (plastic dilatancy is considered subsequently) and the plastic hardening modulus is given by

$$h = d\tau_e/d\gamma_e^p = \frac{1}{3}d\sigma_e/d\epsilon_e^p \qquad (7)$$

as depicted in Fig. 1.

We remark that these constitutive relations involve a $\mathbf{D}^p$ that has a direction "normal" to the yield surface in stress space. Furthermore, this surface is smooth, with a unique normal at the current state. Therefore, the direction of $\mathbf{D}^p$ is independent of that of the stress rate $\overset{\triangledown}{\boldsymbol{\sigma}}$ and, in fact, the yield surface is imagined to expand isotropically, having the mathematical expression $\tau_e = f(\gamma_e^p)$. The criterion for yield is pressure insensitive and the plastic flow is volume preserving (the two go together when normality is assumed).

Now, investigations of the stability of plastic flow against localization (Rudnicki and Rice [2]; Rice [3]) suggest that results can be very sensitive to deviations from the various attributes just discussed. Thus we consider several generalizations of the Prandtl-Reuss relations and, following a formulation of conditions for localization in the next section, we examine their consequences.

First consider the smooth yield surface assumption, with its consequent requirement that the direction of $\mathbf{D}^p$ is not influenced by that of $\overset{\triangledown}{\boldsymbol{\sigma}}$. This assumption is at variance with models for polycrystalline aggregates based on single crystal

---

* $[\boldsymbol{\sigma}\cdot\boldsymbol{\Omega}]_{ij} = \sigma_{ik}\Omega_{kj}$.

References pp. 264–265.

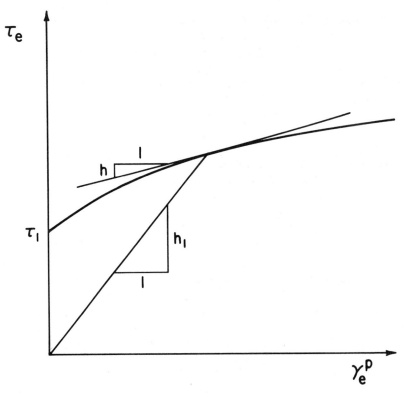

Fig. 1.   Stress-strain curve showing the tangent modulus h and the secant modulus h_1.

plasticity. Hill [4] remarks that the associated discreteness of slip systems in each grain leads to the prediction of a vertex on the yield surface, at least when the yield surface is defined for very small offset plastic strains. Specific calculations for various polycrystal models carried out by Lin [5] and Hutchinson [6] do, in fact, exhibit such yield surface vertices. Unfortunately, no simple mathematical formulation of response at a vertex has yet been developed, but an approximate model, useful when the prior deformation has been carried out at conditions near to proportional stressing, has been developed by Rudnicki and Rice [2] and Stören and Rice [7]. These authors introduce two hardening moduli; h, as introduced before, characterizes the hardening when $\overset{\triangledown}{\boldsymbol{\sigma}}{}'$ merely continues the direction of $\boldsymbol{\sigma}'$. But owing to vertex effects, a new modulus $h_1(>h)$ comes into play to characterize the response to that portion of $\overset{\triangledown}{\boldsymbol{\sigma}}{}'$ directed tangentially to what is taken as the yield surface in the isotropic hardening model. This is illustrated schematically in Fig. 2. The adopted form is

$$D^p_{ij} = \frac{1}{2h}\,\sigma'_{ij}\,\frac{\dot{\tau}_e}{\tau_e} + \frac{1}{2h_1}\left(\overset{\triangledown}{\sigma}{}'_{ij} - \sigma'_{ij}\,\frac{\dot{\tau}_e}{\tau_e}\right) \qquad (8)$$

where it is assumed that $\overset{\triangledown}{\boldsymbol{\sigma}}{}'$ is not strongly different in direction from $\boldsymbol{\sigma}'$. Indeed, on the basis of a simplified plasticity theory based on crystalline slip, Sanders [8] constructed a flow theory, that for stressing paths that were "fully active," in the sense that every slip system, once stressed to yield, is subject to a monotonically increasing shear stress, coincided with the simplest "deformation" theory. Sander's considerations were based on infinitesimal strain and several generalizations of deformation theory to large strains were examined by Stören and Rice. In particular, they showed that if a deformation measure of the type

$$\int \mathbf{D}^p \, dt,$$

but suitably modified to eliminate effects of the spin history Ω on re-orientation of material directions, is taken to be coaxial with $\boldsymbol{\sigma}'$ (again, corrected for rotations), with the ratio between the two depending on τ_e, then the expression (8) for $\mathbf{D}^p$ results with

$$h_1 = \tau_e / \gamma_e{}^p = \frac{1}{3} \sigma_e / \epsilon_e{}^p. \tag{9}$$

That is, the vertex modulus h_1 is the *plastic* "secant modulus" as illustrated in Fig. 1. Stören and Rice mention other possible generalizations of deformation theory and one of these, based on the logarithmic strain tensor, is adopted by Hutchinson and Neale [9] in a companion paper in this volume. As discussed in [23] and [9] the value assumed by the vertex modulus h_1 can, in certain circum-

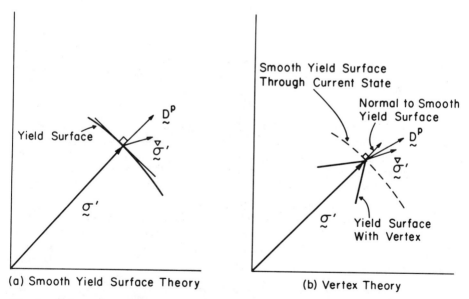

(a) Smooth Yield Surface Theory   (b) Vertex Theory

Fig. 2.   Schematic representations of yield surfaces in deviatoric stress space, showing the directions of the deviatoric stress rate $\overset{\triangledown}{\boldsymbol{\sigma}}'$ and the corresponding plastic strain rate for (a) Flow theory with a smooth yield surface and (b) Flow theory with a vertex at the current state; a hypothetical smooth yield surface passing through the current state is also shown.

stances, depend significantly on the choice of deformation measure on which the deformation theory is based.

Although a deformation theory type formulation of yield vertex effects may be appropriate for the analysis of bifurcations from previously homogeneous, and nearly proportional, deformation paths such a model is undoubtedly not versatile enough to treat finite stress alterations in a direction different from that before bifurcation.

As a means of investigating the effects of dilatant plasticity and pressure sensitive yielding, in a manner that includes the possibility that these do not satisfy an associated flow rule (i.e., "non-normality"), we adopt the simple generalization of the Prandtl-Reuss equations studied by Rudnicki and Rice [2]. In this case we write

$$D_{ij}^p = \frac{1}{h} P_{ij} Q_{kl} \overset{\triangledown}{\sigma}_{kl} \tag{10}$$

where

$$P_{ij} = \frac{\sigma'_{ij}}{2\tau_e} + \frac{\beta}{3} \delta_{ij}, \qquad Q_{kl} = \frac{\sigma'_{kl}}{2\tau_e} + \frac{\mu}{3} \delta_{kl} . \tag{11}$$

Here, **P** gives the "direction" of plastic flow (and in this sense, vertex effects are not included since **P** is assumed independent of $\overset{\triangledown}{\sigma}$) while **Q** gives the outer "normal" to the yield surface in stress space. When β and μ vanish this reduces to the Prandtl-Reuss form (4). More generally, β gives the ratio of plastic dilatancy to equivalent plastic shear,

$$D_{kk}^p = \beta \dot{\gamma}_e{}^p \tag{12}$$

whereas μ measures the pressure sensitivity of yield (as sketched in Fig. 3) in that the amount of a plastic deformation increment depends on

$$d\tau_e + \mu d(\sigma_{kk}/3). \tag{13}$$

Note that plastic "normality" applies only when $\mu = \beta$ (i.e., when **P** = **Q**).

Rudnicki and Rice motivate such a constitutive model with reference to the dilatant, frictional response of rocks. However, there are specific cases in metal plasticity where the same representation seems applicable. First, Spitzig et al. [10] report strength differential (SD) effects in martensitic high strength steels, whereby the compressive yield strength is finitely different than that for tension. Spitzig et al. present evidence that the SD arises from the effect of the mean stress (rather than, say, the third stress invariant) on the value of τ_e required for yield, and hence the effect would seem to fit into the framework of the present formulation. Since the SD is defined as

$$SD = \frac{\sigma_c - \sigma_t}{(\sigma_c + \sigma_t)/2} \tag{14}$$

where σ_c and σ_t are the yield strengths in compression and tension, respectively,

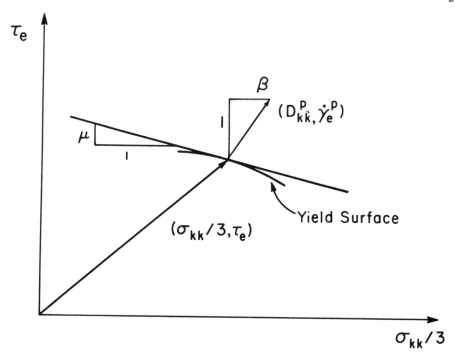

Fig. 3.   Sketch of a yield surface illustrating the geometric interpretations of the pressure sensitivity factor μ and the dilatancy factor β.

the parameter μ is

$$\mu = \frac{\sqrt{3}}{2}(SD), \tag{15}$$

at least if μ is assumed to be essentially independent of the level of σ_{kk} that acts. It might, instead, increase rapidly with large values of σ_{kk} and in such a case (15) will underestimate μ. Spitzig et al. find SD values in the range 0.05 to 0.07 for AISI 4310 and 4330 steels (although the SD value is approximately half for the unaged material and of the order 0.01 to 0.02 for HY-80 [11]; the value would seem to be much lower, near zero, for mild steels). Spitzig et al. also find that the plastic dilatancy is much smaller than required for normality, and according to their experiments, the dilatancy factor β introduced above would have a value of the order of $\mu/15$ and can be taken as zero.

The dilatant and pressure-sensitive plastic flow model described above arises also as a representation of the flow rule for void-containing ductile solids, at least in the approximate development of constitutive relations for this case by Gurson [12, 13]. Based on the approximation of a solid with volume fraction f of voids by a homogeneous spherical body with a concentric spherical cavity, and employing certain other approximations in the rigid-plastic limit analysis of this

References pp. 264–265.

body, Gurson suggests the combined stress yield condition

$$\Phi(\boldsymbol{\sigma}, \bar{\sigma}, f) \equiv \frac{3}{2} \frac{\sigma'_{ij}\sigma'_{ij}}{\bar{\sigma}^2} + 2f \cosh\left(\frac{\sigma_{kk}}{2\bar{\sigma}}\right) - 1 - f^2 = 0 \qquad (16)$$

where σ_{ij} is the macroscopic stress state acting on the voided material and $\bar{\sigma}$ is the tensile flow strength of the matrix material, taken as spatially uniform in the analysis. Note that the yield condition resembles the Mises form (which is assumed in Gurson's model to hold for the matrix material) but that whenever the void volume fraction is non-zero, there is an effect of the mean normal stress on plastic flow as illustrated in Fig. 4. Indeed, McClintock [14] has noted the exponential amplification of hole growth rates over remotely applied plastic strain rates at high values of $\sigma_{kk}/\bar{\sigma}$. Now, as noted by Berg [16], an argument of Bishop and Hill [15] is relevant in this context and, since each element of matrix material is assumed to satisfy the normality rule based on the Mises yield condition, it requires the macroscopic plastic deformation rates of the voided aggregate to satisfy

$$D^p_{ij} = \Lambda \frac{\partial \Phi}{\partial \sigma_{ij}} \qquad (17)$$

where Λ is to be determined in a manner consistent with the strain hardening

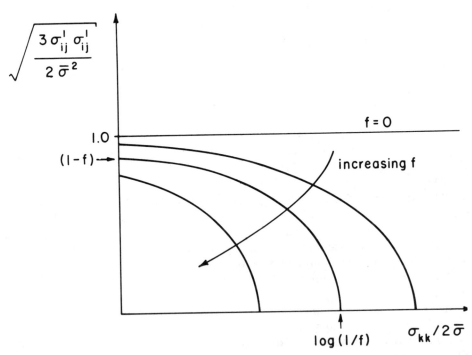

Fig. 4.   Sketch illustrating the dependence of the yield function (16) on the mean normal stress for increasing volume fractions of voids, f.

behavior of the material. Following Gurson, we assume that the yield function remains effectively isotropic (e.g., deviations of the void shape from spherical are neglected), in the sense that (16) continues to describe the yield function and that only the matrix flow strength $\bar{\sigma}$ and void volume fraction f vary during the deformation. The matrix equivalent plastic tensile strain $\bar{\epsilon}^p$ is assumed to vary according to the equivalent plastic work expression

$$(1 - f)\bar{\sigma}\dot{\bar{\epsilon}}^p = \sigma_{ij}D_{ij}^p,$$

so that

$$\dot{\bar{\sigma}} = \bar{h}\,\frac{\sigma_{ij}}{(1 - f)\bar{\sigma}}\,D_{ij}^p \tag{18}$$

where $\bar{h} = d\bar{\sigma}/d\bar{\epsilon}^p$ is the equivalent *tensile* hardening rate of the matrix material (note that h as introduced earlier represented an equivalent *shear* hardening rate). The void volume fraction increases because of the growth of existing voids, which causes f to increase at the rate $(1 - f)D_{kk}^p$, and because of the nucleation of new voids, say by the cracking or decohesion of inclusions or second-phase particles. Accordingly we write

$$\dot{f} = (\dot{f})_{\text{growth}} + (\dot{f})_{\text{nucleation}}$$
$$= (1 - f)D_{kk}^p + A\dot{\bar{\sigma}} + B\dot{\sigma}_{kk}/3. \tag{19}$$

Here the nucleation rate has been represented in terms of the two parameters A and B. Note that if the nucleation of cavities can be correlated exclusively in terms of the equivalent plastic strain (or, since they are related, in terms of $\bar{\sigma}$), as suggested by Gurson's [12, 17] analysis of data by Gurland [18] on a spheroidized carbon steel, then $B = 0$. On the other hand, if the nucleation criterion depends only on the maximum stress transmitted across the particle-matrix interface, as suggested in studies by Argon et al. [19–21], then $B \approx A$ (assuming that $\bar{\sigma} + \sigma_{kk}/3$ is an approximation to this maximum stress). More elaborate nucleation models, based on an extension of Argon et al.'s [21] considerations of the statistics of particle spacing, are discussed by Gurson [12, 17].

Now, since it is assumed that $\Phi(\boldsymbol{\sigma}, \bar{\sigma}, f) = 0$ throughout a program of sustained plastic deformation, the "consistency" condition $\dot{\Phi} = 0$ follows:

$$0 = \frac{\partial \Phi}{\partial \sigma_{kl}}\dot{\sigma}_{kl} + \frac{\partial \Phi}{\partial f}\dot{f} + \frac{\partial \Phi}{\partial \bar{\sigma}}\dot{\bar{\sigma}}$$

$$= \frac{\partial \Phi}{\partial \sigma_{kl}}\overset{\triangledown}{\sigma}_{kl} + \frac{\partial \Phi}{\partial f}\frac{B}{3}\overset{\triangledown}{\sigma}_{kk}$$

$$+ \left[\frac{\partial \Phi}{\partial f}(1 - f)\delta_{ij} + \left(\frac{\partial \Phi}{\partial f}A + \frac{\partial \Phi}{\partial \bar{\sigma}}\right)\bar{h}\frac{\sigma_{ij}}{(1 - f)\bar{\sigma}}\right]\Lambda\frac{\partial \Phi}{\partial \sigma_{ij}}$$

where, in the last version, we have used (18, 19) for $\dot{\bar{\sigma}}$, $\dot{f}$ and also used (17) for D_{ij}^p. From this result we can solve for Λ and the resulting plastic flow rule

References pp. 264–265.

has the form

$$D_{ij}^p = \frac{1}{H} \frac{\partial \Phi}{\partial \sigma_{ij}} \left(\frac{\partial \Phi}{\partial \sigma_{kl}} + \frac{\partial \Phi}{\partial f} \frac{B}{3} \delta_{kl} \right) \overset{\triangledown}{\sigma}_{kl} \tag{20}$$

where

$$H = -\left[\frac{\partial \Phi}{\partial f} (1 - f)\delta_{ij} + \left(\frac{\partial \Phi}{\partial f} A + \frac{\partial \Phi}{\partial \bar{\sigma}} \right) \bar{h} \frac{\sigma_{ij}}{(1 - f)\bar{\sigma}} \right] \frac{\partial \Phi}{\partial \sigma_{ij}}. \tag{21}$$

In fact, given the form of Φ in (16), this plastic flow rule is identical to that introduced in eqs. (10) and (11) provided that the parameters h, β and μ are identified as

$$h = \frac{\bar{h}}{3} \frac{(1 + f^2 - 2fc + f\Sigma s)^2}{(1 - f)(1 + f^2 - 2fc)}$$
$$- \frac{(c - f)\bar{\sigma}}{6(1 + f^2 - 2fc)} \left[3f(1 - f)s + 2A\bar{h} \frac{1 + f^2 - 2fc + f\Sigma s}{1 - f} \right], \tag{22}$$

$$\beta = \frac{\sqrt{3} fs}{2\sqrt{1 + f^2 - 2fc}}, \tag{23}$$

$$\mu = \beta + \frac{(c - f)B\bar{\sigma}}{\sqrt{3}\sqrt{1 + f^2 - 2fc}}, \tag{24}$$

and where we have used the shortened notations

$$\Sigma \equiv \sigma_{kk}/2\bar{\sigma}, \qquad s \equiv \sinh \Sigma, \qquad c \equiv \cosh \Sigma.$$

We note that μ differs from β, so that normality does not apply, whenever $B \neq 0$. That is, for a maximum-stress-dependent criterion for cavity nucleation (so that $B \neq 0$) the plastic flow direction **P** and effective normal **Q** to the yield surface are not coincident. The point is discussed at some length by Gurson [12, 17], who observes that there is no unique direction **Q** when $B \neq 0$. The direction given by (11) and the above μ applies if the stress increments acting serve to continue the void nucleation process. If the stress increments do not cause further nucleation, B can be replaced by zero and then μ and **Q** agree with β and **P** so that "normality" applies.

It may be noted further that the macroscopic hardening rate h in (22) can be negative. This is promoted, particularly, by large triaxial tension so that c and s are large. For example, assume for simplicity that the void nucleation parameter A is zero, that f and even $f \cosh\Sigma$ are small compared to unity, but that the mean normal stress is high so that $\cosh\Sigma \approx \sinh\Sigma \approx e^\Sigma/2$. Then

$$h \approx \frac{\bar{h}}{3} - f \frac{\bar{\sigma}}{8} \exp\left(\frac{\sigma_{kk}}{\bar{\sigma}} \right),$$

and if $\bar{h} = \bar{\sigma}$ (i.e., the matrix is deformed to a strain equal to that for tensile necking) and $\sigma_{kk} = \sqrt{3}(1 + \pi)\bar{\sigma}$, as appropriate to the Prandtl stress triaxiality ahead of a sharp crack in plane strain [22], then h will be negative whenever $f > 0.002$. Of course, when $A > 0$ the softening is yet more pronounced.

ANALYSIS OF LOCALIZATION

The point of view taken here regards the localization of deformation as an instability that can be predicted in terms of the pre-localization constitutive relation of the material. The basic principles were given by Hadamard [25] for elastic solids and extended to elastic-plastic solids by Hill [26], Thomas [44], and Mandel [27]. Not all localization phenomena can be expected to fit within this framework and limitations inherent to this approach have been discussed by Rice [3].

We model the material as rate independent and seek conditions under which the deformations concentrate into a narrow band. Within this framework localization can take place when the constitutive relation allows a bifurcation from a homogeneous deformation field into a highly concentrated band mode. Alternatively, it may be triggered by some initially small nonuniformity of material properties, possibly well before a bifurcation would occur if the initial (small) nonuniformity were absent. Here, particular emphasis is given to Rice's [3] analysis of flow localization in the presence of initial inhomogeneities, which proceeds in the spirit of the Marciniak-Kuczynski [24] approach to localized necking in thin sheets.

We denote the position of a material point in some reference configuration by $\mathbf{x}$ and its position in the current configuration by $\bar{\mathbf{x}}$, where both $\mathbf{x}$ and $\bar{\mathbf{x}}$ are referred to a fixed Cartesian frame. The deformation gradient tensor, $\mathbf{F}$, is given by

$$\mathbf{F} = \frac{\partial \bar{\mathbf{x}}}{\partial \mathbf{x}}, \qquad F_{ij} = \frac{\partial \bar{x}_i}{\partial x_j} \equiv \bar{x}_{i,j} \tag{25}$$

where F_{ij} are the components of the deformation gradient tensor on the fixed Cartesian axes and $(\ )_{,i}$ here, and subsequently, denotes $\partial(\ )/\partial x_i$.

Stress is measured by the nonsymmetric nominal stress tensor $\mathbf{s}$, defined so that $\mathbf{n} \cdot \mathbf{s}$ is the force acting, per unit area in the reference configuration, on an element of surface having (unit) normal $\mathbf{n}$. It satisfies

$$s_{ij,i} = 0 \tag{26}$$

at equilibrium, with body forces neglected. The nominal stress, $\mathbf{s}$, and the true (or Cauchy) stress are related through

$$\boldsymbol{\sigma} = J^{-1}\mathbf{F} \cdot \mathbf{s} \tag{27}$$

where J, the Jacobian, is given by

$$J = \det(\mathbf{F}) \tag{28}$$

and represents the ratio of the volume of a material element in the current configuration to its volume in the reference configuration.

References pp. 264–265.

Now consider the solid subjected quasi-statically to an increment of deformation which in a homogenous (and homogeneously deformed) solid would give rise
to the uniform field $\dot{F}_0$. Suppose, however, that within a thin planar band of
orientation $\mathbf{n}$ in the reference configuration (see Fig. 5) incremental field quantities
are permitted to take on values differing from their uniform values outside this
band. The band is presumed sufficiently narrow to be regarded as homogeneously
deformed.

At the considered instant, the current values of field quantities and material
properties inside the band may be the same as those of corresponding quantities
outside the band, as in a bifurcation analysis. On the other hand, if an initial
imperfection was present, the current values of field quantities and material
properties inside the band will, in general, differ from those outside the band. In
either case two conditions must be satisfied.

First, compatibility requires the displacement increments to be continuous
across the band. Thus (see, for example, [3, 25, 26]),

$$\dot{F}_b = \dot{F}_0 + \dot{q}\mathbf{n} \qquad \text{or} \qquad \Delta\dot{F} = \dot{q}\mathbf{n} \tag{29}$$

where $(\ )_b$ denotes field quantities within the band, $(\ )_0$ denotes corresponding
quantities outside the band, $\Delta(\ ) \equiv (\ )_b - (\ )_0$ and $(\ \dot{}\ )$ denotes the material
derivative.

The second requirement that must be met is that of incremental equilibrium.
This takes the form

$$\mathbf{n}\cdot\dot{s}_b = \mathbf{n}\cdot\dot{s}_0 \qquad \text{or} \qquad \mathbf{n}\cdot\Delta\dot{s} = 0 \tag{30}$$

To proceed further, we assume that the relation between $\dot{s}$ and $\dot{F}$ is piecewise
linear so that

$$\dot{s} = \mathbf{K}:\dot{F} \tag{31}$$

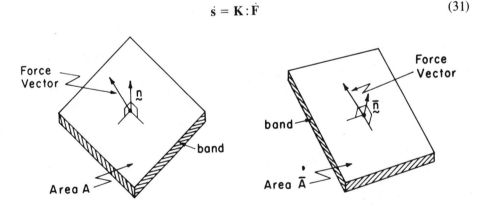

Force Vector $= (\underset{\sim}{\mathbf{n}}\cdot\underset{\sim}{\mathbf{s}})\,A = (\underset{\sim}{\bar{\mathbf{n}}}\cdot\underset{\sim}{\sigma})\,\bar{A}$

a) Reference Configuration   b) Current Configuration

Fig. 5.   Illustration of an element of a band in (a) The reference configuration where
the area is A and the normal is $\mathbf{n}$ and (b) The current configuration where the area is $\bar{A}$
and the normal is $\bar{\mathbf{n}}$.

where the tensor of moduli $\mathbf{K}$ is homogeneous of degree zero in $\dot{\mathbf{F}}$ (for rate independence) but may have a number of branches, the active branch depending on $\dot{\mathbf{F}}$. Combining (29), (30) and (31) gives

$$(\mathbf{n}\cdot\mathbf{K}_b\cdot\mathbf{n})\cdot\dot{\mathbf{q}} = \mathbf{n}\cdot(\mathbf{K}_0 - \mathbf{K}_b)\cdot\dot{\mathbf{F}}_0 \tag{32}$$

where $\mathbf{K}_b$ and $\mathbf{K}_0$ represent the active branches of the tensor of moduli inside and outside the band, respectively. Of course, both $\dot{\mathbf{F}}_b$, which depends on $\dot{\mathbf{q}}$, and $\dot{\mathbf{F}}_0$ must be consistent with the corresponding branches of $\mathbf{K}$.

Localization takes place when

$$\det(\mathbf{n}\cdot\mathbf{K}_b\cdot\mathbf{n}) = 0 \tag{33}$$

Subject to the initial conditions, which specify the imperfection (if any), (32) constitutes a system of three equations for the three components of $\dot{\mathbf{q}}$ which can be solved in an incremental fashion up to the point at which the localization condition (33) is satisfied. At any stage of the loading history the accumulated deformation in the band is simply

$$\mathbf{F}_b = \mathbf{F}_0 + \mathbf{q}\mathbf{n} \tag{34}$$

An alternative formulation, phrased in terms of the Eulerian velocity gradient, $\Gamma = \dot{\mathbf{F}}\cdot\mathbf{F}^{-1}$, and the Cauchy (or true) stress tensor, σ, is sometimes convenient. From (29) and (34), it can be shown that

$$\Gamma_b = \Gamma_0 + \mathbf{Q}\bar{\mathbf{n}} \quad\text{or}\quad \Delta\Gamma = \mathbf{Q}\bar{\mathbf{n}} \tag{35}$$

where $\bar{\mathbf{n}}$ is the unit normal defining the orientation of the band in the current configuration as sketched in Fig. 5 and is given by

$$\bar{\mathbf{n}} = \frac{\mathbf{n}\cdot\mathbf{F}_b^{-1}}{|\mathbf{n}\cdot\mathbf{F}_b^{-1}|} = \frac{\mathbf{n}\cdot\mathbf{F}_0^{-1}}{|\mathbf{n}\cdot\mathbf{F}_0^{-1}|} \tag{36}$$

Whenever (29) holds incremental equilibrium is alternatively expressed by

$$\bar{\mathbf{n}}\cdot\Delta\dot{\sigma} + \dot{\bar{\mathbf{n}}}\cdot\Delta\sigma = 0 \tag{37}$$

The constitutive law, as in the examples discussed in the previous Section, is written in the form

$$\overset{\triangledown}{\sigma} = \mathbf{L}:\mathbf{D} \tag{38}$$

where $\mathbf{D}$ is the symmetric part of Γ, and $\mathbf{L}$ is homogeneous of degree zero in $\mathbf{D}$. Employing (3) and (35) in conjunction with (37) and (38) yields

$$[\bar{\mathbf{n}}\cdot\mathbf{L}_b\cdot\bar{\mathbf{n}} + \mathbf{A}_b]\cdot\mathbf{Q} = \bar{\mathbf{n}}\cdot(\mathbf{L}_0 - \mathbf{L}_b)\cdot\mathbf{D}_0 - (\bar{\mathbf{n}}\cdot\Omega_0 + \dot{\bar{\mathbf{n}}})\cdot\Delta\sigma \tag{39}$$

where

$$\mathbf{A}_b = \tfrac{1}{2}[-\bar{\mathbf{n}}(\bar{\mathbf{n}}\cdot\sigma_b) + (\bar{\mathbf{n}}\cdot\sigma_b\cdot\bar{\mathbf{n}})\mathbf{I} + (\bar{\mathbf{n}}\cdot\sigma_b)\bar{\mathbf{n}} - \sigma_b] \tag{40}$$

and $\mathbf{I}$ is the unit tensor, $(\mathbf{I})_{ij} = \delta_{ij}$. The localization condition becomes

$$\det\{\bar{\mathbf{n}}\cdot\mathbf{L}_b\cdot\bar{\mathbf{n}} + \mathbf{A}_b\} = 0 \tag{41}$$

The alternative formulations (32) and (39) are fully equivalent and the choice between them is solely a matter of convenience (or taste). It does appear to us, however, that (32) is the more convenient formulation to employ when tracing a deformation history corresponding to a given initial imperfection whereas (39) is more suited to a bifurcation analysis. Specifically, if the material is initially homogeneous, the condition governing the onset of bifurcation into the localized band mode is that

$$[\bar{\mathbf{n}} \cdot \mathbf{L}_0 \cdot \bar{\mathbf{n}} + \mathbf{A}_0] \cdot \dot{\mathbf{Q}} = 0 \qquad (42)$$

have a nontrivial solution, where $\mathbf{A}_0$ is given by (40) with σ_b replaced by σ_0, since prior to localization $(\ )_b = (\ )_0$. This presumes that at localization the bifurcating field is associated with the same branch of the tensor of instantaneous moduli as the homogeneous field. For vertex-like yielding it seems possible that in some cases the bifurcating field could take advantage of the moduli associated with a different branch, as suggested by Rice [3].

We emphasize that neither the bifurcation analysis nor the analysis tracing the growth of an initial inhomogeneity require that the tensor of instantaneous moduli, $\mathbf{K}$, be symmetric [3, 27]. However, if the moduli $\mathbf{K}$ are symmetric (the symmetry of $\mathbf{L}$ implying that of $\mathbf{K}$ and vice versa only for an incompressible solid) a direct connection can be made between bifurcation into the band mode and loss of uniqueness in a specific boundary value problem, namely the problem of a solid subject to all around displacement boundary conditions on its surface [26]. Presuming symmetry of the moduli and homogeneity of the deformation field, the band mode is the only bifurcation mode available under these restrictive boundary conditions and is often termed a "material" instability as opposed to "geometric" necking instabilities which these boundary conditions rule out.

Subject to the above conditions and, additionally, assuming incrementally linear material behavior (i.e., $\mathbf{K}$ independent of $\dot{\mathbf{F}}$), a linear stability analysis [3] shows that the onset of bifurcation into the band mode corresponds to the boundary between stability and instability for a displacement mode with a Fourier component having a polarization vector parallel to $\mathbf{n}$. Under less restrictive boundary conditions, "geometric" instabilities (which may include arbitrarily short wavelength modes [23, 28]) invariably precede the band type localization considered here, but the material instability is still available, and sometimes, depending on specific circumstances, at overall deformations not greatly in excess of the geometric modes.

Although here a unified framework for analyzing localization instabilities in a wide class of materials has been presented, the actual onset of localization depends critically on the details of the constitutive description of the solid, as will be illustrated subsequently.

RESULTS FOR PLASTIC FLOW LOCALIZATION

The above formalism for localization has been applied to the various constitutive models discussed in the Section on Constitutive Relations. Consider first the classical Prandtl-Reuss model, eqs. (2)–(7). This emerges as a special case of a calculation by Rudnicki and Rice [2] and for localization the hardening rate must fall to the value

$$h_{\text{crit}} = -2(1 + \nu)GP_{\text{II}}^2 + 0(\tau_e^2/G) \tag{43}$$

where P_{II} is the intermediate principal value of the plastic flow direction tensor **P**, namely

$$P_{\text{II}} = \sigma_{\text{II}}'/2\tau_e . \tag{44}$$

Since τ_e is typically a small fraction of G, the terms $0(\tau_e^2/G)$ are usually negligible and thus this model requires strain softening for localization in all deformation states except for plane strain ($P_{\text{II}} = 0$), in which case $h_{\text{crit}} \approx 0$. The states most resistant to localization are those of axisymmetric extension or compression (or, equivalent, balanced biaxial tension) for which $P_{\text{II}} = \pm 1/2\sqrt{3}$ and

$$h_{\text{crit}} \approx -(1 + \nu)G/6 \approx -0.2G.$$

The prediction of such strongly negative hardening rates for localization may be unrealistic for reasons that can be explored with the help of other constitutive models. Nevertheless, the results are suggestive of significant differences in ductility when plane strain and axisymmetric deformations are compared, e.g., Clausing [29]. In all cases the normal to the plane of localization is in the plane of the greatest and least principal stresses, σ_{I} and σ_{III} respectively. For $\nu = 0.3$ the angle θ between the normal and the σ_{III} axis is 48.7°, 45°, and 41.2° for axisymmetric extension, plane strain (or pure shear), and axisymmetric compression respectively (Rudnicki and Rice [2]).

Now consider the effect of a yield vertex in the form modelled by eqs. (8) and (9). The calculation is rather complicated in general but simplifies when the material is taken to be elastically as well as plastically incompressible, $\nu = 1/2$. The result is most conveniently stated in terms of the moduli

$$h' = \frac{h}{1 + h/G}, \qquad h_1' = \frac{h_1}{1 + h_1/G} \tag{45}$$

which reduce to h and h_1, respectively, for highly ductile materials where we expect $h < h_1 \ll G$. In fact, h' is just the tangent modulus based on the *total* (rather than plastic) strain and, if the vertex modulus h_1 is identified as the plastic secant modulus, then h_1' is the secant modulus based on the total strain.

The critical conditions are established by Rice [3] for cases in which the vertex modulus is large enough to satisfy

$$h_1' > 2(1 - 3P_{\text{II}}^2)h'/(1 - 6P_{\text{II}}^2), \tag{46}$$

with P_{II} given by (44). The inequality reads $h_1' > 2h'$ for plane strain states, and is satisfied for all deformation states if $h_1' > 3h'$. We expect the inequality to be satisfied for materials with light strain hardening. For example, if a pure power law stress-strain relation is satisfied, $\sigma_e \propto (\epsilon_e)^N$, and if h_1' is interpreted as the secant modulus, then the inequality is satisfied for plane strain states whenever $N < 1/2$ and for all states whenever $N < 1/3$.

References pp. 264–265.

In all cases for which the inequality (46) is met h'_{crit} is the value of h' satisfying [3]

$$\tau_e{}^2 = 12P_{\text{II}}^2(h_1{}' - h')^2 + 4h'(h_1{}' - h'), \qquad (47)$$

which includes a plane-strain result of Hill and Hutchinson [28] when we set $P_{\text{II}} = 0$. If we write M for the ratio $h'/h_1{}'$ (noting that $M = N$ for pure power-law materials when the secant modulus definition is adopted for $h_1{}'$), this can be rewritten as the critical condition

$$h'_{\text{crit}} = \frac{M\tau_e}{\sqrt{12P_{\text{II}}^2(1 - M)^2 + 4M(1 - M)}}, \qquad (48)$$

which reduces to

$$h'_{\text{crit}} = \frac{\tau_e}{2}\sqrt{\frac{M}{1 - M}} \quad \text{for plane strain}$$

$$h'_{\text{crit}} = \tau_e \frac{M}{\sqrt{(1 + 3M)(1 - M)}} \quad \text{for axisym. strain}$$

We expect M to have a range similar to that for the hardening exponent, say 0.05 to 0.5, and thus we see from these expressions that localization occurs when h'_{crit} is some *positive*, but typically small, fraction of the equivalent shear stress τ_e. Table 1 gives numerical values and we see that the critical hardening rate under plane strain conditions is always larger than for axi-symmetric conditions, indicating the greater stability of the latter. For pure power-law materials, $\sigma_e \sim \epsilon_e{}^N$, and $h_1{}'$ as the secant modulus (so $M = N$) the above results give the critical strains to localization (expressed in terms of both ϵ_e and the greatest principal strain ϵ_{I}) of

$$(\epsilon_{\text{I}})_{\text{crit}} = \frac{\sqrt{3}}{2}(\epsilon_e)_{\text{crit}} = \sqrt{N(1 - N)}, \qquad \text{plane strain}$$

$$(\epsilon_{\text{I}})_{\text{crit}} = (\epsilon_e)_{\text{crit}} = \sqrt{(1 + 3N)(1 - N)/3}, \quad \text{axisym. strain.} \qquad (49)$$

TABLE 1

Theoretical Predictions of the Critical Hardening Modulus (h'_{crit}) for Localization as a Function of Ratio M of Tangent to Vertex Modulus. The critical strains for localization and the ductility ratio are calculated on the assumption that there is power-law hardening, $\sigma_e \sim (\epsilon_e)^N$, and that the vertex modulus equals the secant modulus (i.e., for "deformation" theory) so that $M = N$.

			For Power-Law Hardening and $N = M$		
	h'_{crit}/τ_e		$(\epsilon_{\text{I}})_{\text{crit}}$		Ductility Ratio
$M \equiv \dfrac{h'}{h_1{}'}$	Plane Strain	Axisym. Str.	Plane Strain	Axisym. Str.	Plane/Axisym.
0.05	0.115	0.048	0.218	0.603	0.361
0.1	0.167	0.092	0.300	0.624	0.480
0.2	0.250	0.177	0.400	0.653	0.612
0.3	0.327	0.260	0.458	0.666	0.688
0.4	0.408	—	0.490	—	—

These values are also shown in the table, as is the ductility ratio

$$\frac{(\epsilon_I)_{\text{crit, pl. str.}}}{(\epsilon_I)_{\text{crit, axisym. str.}}} = \sqrt{\frac{3N}{1 + 3N}}. \tag{50}$$

The last expression is also plotted in Fig. 6.

Clausing [29] reports a decreasing ductility ratio with increasing strength level for a series of seven structural steels; the ratio (based on ϵ_I) varies from 0.72 for a mild steel to 0.17 for a high strength (1770 MPa, or 248 ksi) steel. It is true as a general rule that the hardening exponent decreases with increasing strength level and thus the trend of the results in Table 1 and Fig. 6 seem consistent with Clausing's data, which is summarized in Table 2.

Indeed, comparing Tables 1 and 2 we see that while there is no close numerical

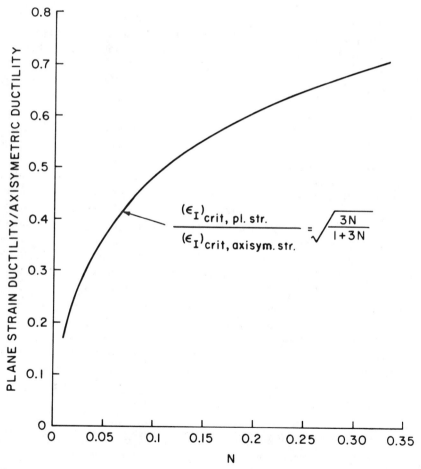

Fig. 6. Ratio of plane strain ductility to axisymmetric ductility as a function of the strain hardening exponent N, as predicted by the vertex model with the assumption that the vertex modulus equals the secant modulus.

TABLE 2

Data on Strain to Fracture by Clausing [29], for Seven Structural Steels Arranged in
Order of Decreasing Strength Level.

Material	σ_{yield} (MPa)	$(\epsilon_1)_{\text{fract}}$ Plane Strain	$(\epsilon_1)_{\text{fract}}$ Axisym. Str.	Ductility Ratio Plane/Axisym.
18 Ni(250)	1770 (248ksi)	0.15	0.89	0.17
10 Ni-Cr-Mo-Co	1300	0.36	1.16	0.31
18 Ni(180)	1270	0.42	1.00	0.42
Hy-130(T)	970	0.54	1.06	0.51
HY-80	610	0.86	1.22	0.70
A302-B	380	0.72	0.98	0.73
ABS-C	280 (39ksi)	0.75	1.04	0.72

agreement, there are some important qualitative similarities. For example, the
range of ductility ratios are somewhat comparable and it is the plane strain
ductility, rather than the axisymmetric ductility, which is most affected by variations in the plastic flow properties. The comparison of this simple vertex model
with Clausing's results is, in fact, encouraging enough that perhaps a great deal
more effort should be directed to the further experimental and theoretical delin-
eation of vertex effects and their role in localization.

Finally, we remark that all of Clausing's steels were materials of high ductility
(at least in axisymmetric tension) and it is plausible that their stability against
flow localization might be limited by some inherent feature of the plastic flow
process, e.g., the vertex yield surface structure that arises from the crystalline
basis for slip. In other alloys, or perhaps in the same alloys under more triaxially
elevated stress states, it is possible that substantial ductility reductions could
occur due to the nucleation and growth of microscopic cavities. It remains an
open question in many systems as to whether ductile fracture arises from some
instability of the plastic flow process, which then concentrates local strains and
leads to cavity nucleation and growth to coalescence, or whether fracture is
traceable instead to some instability brought about by the progressive micro-
rupture process itself. To examine the matter further, we now consider the
plastically dilatant material with pressure sensitive yielding, eqs. (10) and (11).

For this constitutive relation the critical hardening rate is given by Rudnicki
and Rice [2] (we have done some algebraic rearrangement of their eq. (24) in light
of their eq. (19)):

$$h_{\text{crit}} = \left[\frac{1 + \nu}{9(1 - \nu)} (\beta - \mu)^2 - \frac{1 + \nu}{2} \left(\Lambda + \frac{\beta + \mu}{3} \right)^2 \right] G$$
$$+ \frac{(1 + \nu)(\mu - \beta)}{12(1 - \nu)} \sqrt{4 - 3\Lambda^2} \left[\frac{1 - 2\nu}{2} \Lambda \right.$$
$$\left. + \frac{1}{2} \sqrt{4 - 3\Lambda^2} - \frac{1 + \nu}{3} (\mu + \beta) \right] \tau_e + O(\tau_e^2/G). \quad (51)$$

Here Λ (used in place of the symbol N of [2] to avoid confusion with the hardening exponent) is

$$\Lambda = \sigma_{II}' / \tau_e \tag{52}$$

and $\Lambda = -1/\sqrt{3}$ for axisymmetric extension, 0 for a pure shear state of deviatoric *stress*, and $1/\sqrt{3}$ for axisymmetric compression. The solution of (51) applies subject to certain restrictions on the size of μ and β noted in [2]; when these are met the normal to the plane of localization lies in the I–III plane.

We observe first that when normality applies (i.e., when $\beta = \mu$) h_{crit} is non-positive for all stress states (neglecting the $O(\tau_e^2/G)$ terms); it vanishes when $\Lambda = -(\mu + \beta)/3$. But when $\mu \neq \beta$, h_{crit} can be positive for some deformation states.

An important special case is that of a plane strain state. From (11) for the plastic flow direction tensor $\mathbf{P}$,

$$P_{II} = \sigma_{II}'/2\tau_e + \beta/3 = \Lambda/2 + \beta/3, \tag{53}$$

so that $\Lambda = -2\beta/3$ for the case of plane strain. In this instance (51) reduces to

$$h_{crit} = \frac{(1 + \nu)^2}{18(1 - \nu)} \left[(\mu - \beta)^2 G + (\mu - \beta)\sqrt{1 - \beta^2/3} \left(\frac{3}{1 + \nu} \sqrt{1 - \beta^2/3} \right. \right. \tag{54}$$

$$\left. \left. - \mu + \frac{3\nu}{1 + \nu} \beta \right) \tau_e \right] + O(\tau_e^2/G).$$

This last expression may be applied to an analysis of the strength differential effect as discussed in the Section on Constitutive Relations. Using (15) for μ, assuming SD $\ll 1$, and setting $\beta = 0$, one finds

$$h_{crit} = \frac{(1 + \nu)^2}{24(1 - \nu)} \left[(SD)^2 G + \frac{\sqrt{3}}{1 + \nu} (SD) \tau_e \right]. \tag{55}$$

As an example, setting SD = 0.06, $\nu = 0.3$, using $G = 82\ 400$ MPa as appropriate to steel, and assuming that $\sigma_e (= \sqrt{3}\tau_e) = 1400$ MPa (≈ 200 ksi), one obtains

$$h_{crit} \equiv \tfrac{1}{3}(d\sigma_e/d\epsilon_e^p)_{crit} = 42.6 \text{ MPa}(= 6 \text{ ksi}) + O(\tau_e^2/G)$$

or, for comparison with entries in Table 1 based on the vertex model, $h_{crit}/\tau_e \approx 0.05$. But for axisymmetric extension or compression values of h_{crit} computed from (51), for values of the parameters as above, differ little from that of eq. (43) for the classical Prandtl-Reuss model. It is not yet clear as to what the effects of small non-normality, as in the SD effect, and vertex yield would be if present concurrently.

The positive critical hardening rate that may result for states like plane strain, eq. (52), when $\mu \neq \beta$ seems to be typical of other cases with non-normality. For example, Asaro and Rice [30] discuss the localization of plastic flow in ductile single crystals and observe that when there is non-normality, in the sense that stresses other than the Schmid resolved stress enter the criterion for continuing slip, localization can occur with a positive slip system hardening rate. The implications of such ideas for polycrystals have not yet been fully explored.

References pp. 264–265.

As final examples, we consider the constitutive relation based on the Gurson model for porous solids. This has the same form as in (10) and (11) and the localization condition is given as in (51), provided that the parameters h, β, and μ are defined as in eqs. (22)–(24). In these last equations it is to be recalled that $\bar{h}$ and $\bar{\sigma}$ are the equivalent tensile rate of hardening and flow strength of the *matrix* material, f is the void volume fraction, and A and B are parameters entering the expression (19) for the part of the increase of volume fraction f arising from the nucleation of new voids. As remarked earlier, the macroscopic hardening rate h can be negative and, when there is stress-dependent void nucleation ($B \neq 0$), μ can differ from β so that deviations from normality occur.

We limit the analysis to plane strain conditions, so that h_{crit} is given by (54). Further, in practical cases we expect β and μ to be small compared to unity (exceptions may occur at high mean normal stress levels, e.g., in front of a sharp notch or crack) and also we assume that f is small. In these cases we can simplify (22)–(24) to read

$$h \approx \tfrac{1}{3}\bar{h} - \tfrac{1}{2}fcs\bar{\sigma} - \tfrac{1}{3}c\bar{\sigma}A\bar{h}, \quad \text{and} \quad \mu - \beta \approx \frac{1}{\sqrt{3}}cB\bar{\sigma}. \tag{56}$$

Further, we can write $\tau_e = \sigma_e/\sqrt{3} \approx \bar{\sigma}/\sqrt{3}$ in the same circumstances, and thus (54) becomes

$$\tfrac{1}{3}\bar{h} - \tfrac{1}{2}fcs\bar{\sigma} - \tfrac{1}{3}c\bar{\sigma}A\bar{h} = \frac{1+\nu}{36(1-\nu)}\left[\tfrac{1}{3}(cB\bar{\sigma})^2 E + 2cB\bar{\sigma}^2\right] \tag{57}$$

at critical conditions, where E is the elastic tensile modulus.

First we consider the case in which void nucleation is strain controlled,

$$(\dot{f})_{\text{nucleation}} = F\dot{\bar{\epsilon}}^P = (F/\bar{h})\dot{\bar{\sigma}}. \tag{58}$$

Evidently, $A = F/\bar{h}$ and $B = 0$ for this case, and F can be interpreted as the volume fraction of particles converted (effectively) to voids per unit of plastic strain. On the basis of Gurland's [18] data on spheroidized carbides in steel, Gurson [12, 17] suggests that $F \approx 0.3 f_{\text{up}}$, where f_{up} is the volume fraction of unbroken carbide particles. In terms of F, the critical condition becomes

$$(\bar{h}/\bar{\sigma})_{\text{crit}} \approx \tfrac{3}{2}f\cosh\Sigma \sinh\Sigma + F\cosh\Sigma \tag{59}$$

(recall that Σ denotes $\sigma_{kk}/2\bar{\sigma}$). According to this model, and in view of the preceding remarks concerning F, the critical $\bar{h}/\bar{\sigma}$ value is predicted to be of the order of the void volume fraction f or of the order of the volume fraction of uncracked particles, at least for states of only modest stress triaxiality.

In the case of stress dependent nucleation we may write

$$(\dot{f})_{\text{nucleation}} = K(\dot{\bar{\sigma}} + \dot{\sigma}_{kk}/3)/\bar{\sigma} \tag{60}$$

so that K is the volume fraction of particles converted to voids per unit fractional increase in stress. Thus $A = B = K/\bar{\sigma}$ and the critical condition is

$$(\bar{h}/\bar{\sigma})_{\text{crit}} = \frac{1}{1 - K\cosh\Sigma} \left\{ \tfrac{3}{2} f \cosh\Sigma \sinh\Sigma \right.$$
$$\left. + \frac{1 + \nu}{6(1 - \nu)} K\cosh\Sigma \left[1 + \frac{EK\cosh\Sigma}{6\bar{\sigma}} \right] \right\}. \tag{61}$$

We note that in cases for which void nucleation takes place over a comparatively narrow range of stress, K can be several times the volume fraction of void nucleating particles and then the above expression can predict that rather large values of $\bar{h}/\bar{\sigma}$ are required for stability against localization. For example, choosing $\Sigma = \sqrt{3}/2$ as appropriate to the plane strain tension test (without necking effects) and $\nu = 0.3$,

$$(\bar{h}/\bar{\sigma})_{\text{crit}} = \frac{1}{1 - 1.40K} \{2.06f + 0.43K + 0.10K^2 E/\bar{\sigma}\}$$

and when we consider the current void volume fraction f to be negligible compared to K and set $E/\bar{\sigma} = 300$ (e.g., $\bar{\sigma} = 100$ ksi ≈ 700 MPa in steel), we get

$$(\bar{h}/\bar{\sigma})_{\text{crit}} = 0.007, 0.10, \text{ and } 0.40$$

for

$$K = 0.01, 0.05, \text{ and } 0.10$$

respectively.

All the results cited thus far pertain to the onset of bifurcation. However, there may be significant sensitivity to small initial non-uniformities of material properties and this respect of the problem is not yet well explored in general. However, Yamamoto has adopted an approach similar to that outlined in the Section on Analysis of Localization for a material described by the Gurson constitutive model ((10)–(11) and (22)–(24)) but in which the initial void volume fraction, f_0, is slightly larger in a planar slice of material than elsewhere. Yamamoto neglects the nucleation terms ($A = B = 0$) so that instability results only from the porosity increase due to the progressive growth of existing voids.

Some of Yamamoto's results are shown in Fig. 7, for which an initial porosity of $f_0 = 0.01$ is assumed outside the imperfect zone, and the strain ϵ_1 to failure is shown as a function of the initial porosity, $f_{\text{imp.,0}}$, inside the imperfection. The calculations were done for axisymmetric and plane strain tension cases with $\sigma_{\text{III}} = 0$ in all cases. The latter means that the induction of triaxial tension stresses, which occurs as part of the necking process in experiments, was neglected in the calculation. This is a serious defect since, as has been seen, the Gurson dilational plasticity equations are strongly sensitive to the mean normal stress.

Yamamoto's calculations employed the power hardening relation

$$\frac{\bar{\sigma}}{\sigma_y} = \left[\frac{\bar{\sigma}}{\sigma_y} + \frac{3E\bar{\epsilon}^P}{2(1 + \nu)\sigma_y} \right]^N$$

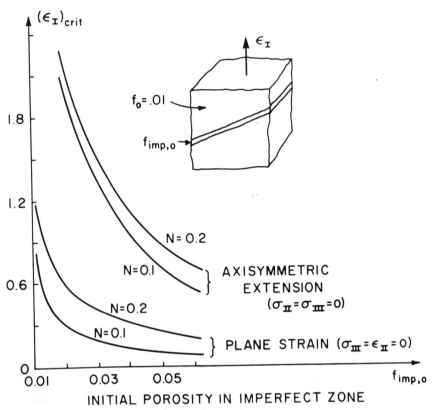

Fig. 7. Curves of the critical strain for localization as a function of the initial void concentration in the band for plane strain tension and axisymmetric tension, from Yamamoto [31].

for the matrix material, with $\nu = 0.3$ and $\sigma_y/E = 0.003$. Results are shown in Fig. 7 for two hardening exponents, $N = 0.1$ and 0.2, and in each case these correspond to a numerical search over orientation angles of the slice to give the least ductility. Although not shown on the figure, Yamamoto also calculates the results for $f_0 = 0$ outside the imperfection and we find that these, when plotted against $f_{imp,0}$, give results which are very similar to those shown. The results indicate that the localization condition for a void-containing material is strongly sensitive to non-uniformities in the porosity distribution. We will see similar effects for sheet metals in the following section. Yamamoto also observes that for a given size imperfection the ductility ratio (plane strain to axisymmetric) increases with decreasing hardening exponent. For example, in calculations for which the size of $f_{imp,0} - f_0$ varies from 0.01 to 0.05, and for $f_0 = 0$ to 0.01, Yamamoto finds a ductility ratio of from 0.13 to 0.20 for $N = 0.1$ and from 0.26 to 0.30 for $N = 0.2$. These seem somewhat low compared to experimental results, but the inclusion of necking effects in the analysis would, presumably, lower the $(\epsilon_I)_{crit}$ for axisymmetric extension more than for plane strain (since far more of the strain to failure shown in Fig. 7 occurs *after* necking for axisymmetric conditions), and this could have the effect of raising the predicted ductility ratios somewhat. This remains a topic in need of further examination.

LOCALIZED NECKING IN BIAXIALLY STRETCHED SHEETS

The mathematical theory of localized necking in thin sheets is a precise two dimensional analog of the three dimensional theory of flow localization outlined in the Section on Analysis of Localization. In that development one needs only replace s and σ by the nominal and true membrane forces, respectively. Indeed, many of the basic principles of the theory were first elucidated within this two dimensional plane stress context [24, 32].

Hill [32] analyzed bifurcations corresponding to localized necking using classical (smooth yield surface, normality) rigid plastic theory. As is well known, this analysis predicts that local necking will not occur in a uniform sheet subject to positive biaxial stretching. Since both practical experience [33–35] and experimental tests [36–39] demonstrate that thin sheets subject to positive biaxial tension can fail by a process of localized necking, there has been considerable interest in attempting to resolve this "paradox". One approach, initiated by Marciniak and Kuczynski [24] (hereafter referred to as M-K), postulates an initial inhomogeneity in the sheet in the form of a localized thickness reduction. This instigates necking by precipitating a drift of the strain state in the neck toward plane strain, while the remainder of the sheet undergoes proportional loading.

Azrin and Backofen [36] carried out experiments aimed at testing the M-K model of localized necking and found that the magnitude of the assumed thickness reductions required to give theoretical predictions of the limit strain, that is the imposed strain at the onset of localized necking, in line with those observed experimentally were much larger than actually measured in the test specimens. Furthermore, although the experiments did show a drift of the strain state in the neck toward plane strain, the dependence of the limit strain on the imposed strain ratio was qualitatively different from that predicted by the M-K model for most of the materials tested. Specifically, for an isotropic material (and most of the materials tested in [36] appear not to have been significantly anisotropic), the M-K analysis predicts that the limit strain should increase with increasing biaxiality, whereas experimentally, for a number of materials, the limit strain was found to be nearly independent of the imposed strain ratio or actually to decrease somewhat with increasing biaxiality.

A different line of attack was initiated by Stören and Rice [7] who showed that a simple model of a material with a vertex on its yield surface, namely a finite strain version of the simplest deformation theory of plasticity, does predict a bifurcation corresponding to localized necking in biaxial tension. The limit strains obtained by means of this analysis are qualitatively more in accord with experiment than those given by the M-K analysis. In particular, for a material exhibiting a pure power law uniaxial hardening curve of the form

$$\sigma = K\epsilon^N \tag{62}$$

where σ is the true stress and ϵ the logarithmic strain in uniaxial tension, Stören and Rice's [7] analysis gives limit strains that monotonically decrease with increasing biaxiality for high hardening materials, $N > \frac{1}{3}$.

Although, in [24] as well as in a number of subsequent investigations, an initial thickness inhomogeneity was employed, this imperfection was assumed equivalent to a local variation in material properties and was not necessarily intended to be regarded as a literal thickness reduction. In some recent work [40, 41] an actual difference in material properties in the incipient neck has been accounted for within the M-K framework. In particular, in [40] Needleman and Triantafyllidis have employed the M-K model in conjunction with Gurson's [12, 13] constitutive relation for porous plastic materials. In [40], as in [31], void nucleation is neglected ($A = B = 0$ in (19)).

Fig. 8 displays limit strains as a function of imposed strain ratio, ρ, where

$$\rho = \epsilon_2 / \epsilon_1 \tag{63}$$

and ϵ_1 and ϵ_2 are, respectively, the major and minor principal in-plane logarithmic strains. The solid curves are the deformation theory results of [7], while the

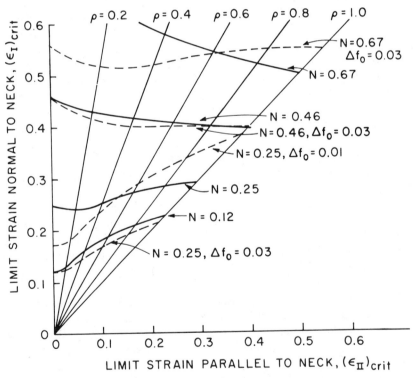

Fig. 8. A comparison of the forming limit curves predicted by the vertex model (solid curves) and the void growth model (dashed curves). Here, Δf_0 is the increased initial porosity in the neck. Based on [7] and [40].

dashed curves are for the void growth model of [40], in which an increased initial void concentration, Δf_0, plays the role of the initial imperfection. Curves are displayed for various values of the strain hardening exponent in (62). It is important to note that in the porous plastic material model this strain hardening exponent pertains to the matrix material and *not* to the void matrix aggregate.

Both theories agree in predicting that for high hardening materials the limit strain will decrease or remain nearly constant with increasing biaxiality whereas for lightly hardening materials the limit strain will increase with increasing biaxiality. By suitably adjusting parameter values, the difference between the predictions of the two theories can be made negligible. This is, of course, a somewhat arbitrary exercise.

The significant point is that the analyses in [7] and [40] suggest that both vertex effects and the weakening induced by incipient ductile rupture provide plausible mechanisms to account for the observed shapes of forming limit diagrams. In essence, the deformation theory model attributes the onset of localized necking to the discrete crystallographic nature of slip whereas the void growth model ties localized necking to the (not well understood) factors responsible for the initiation of ductile rupture.

Of course, vertex effects and incipient ductile fracture are not the only possible physical mechanisms that could be responsible for triggering localizing necking. In Fig. 9 (adopted from [40]) the forming limit curves resulting from various types of inhomogeneities are compared. The curves marked $(\sigma_y{}^B/\sigma_y{}^A) = 0.99$ correspond to a one percent yield stress reduction in the neck with all other initial values being identical inside (region B in Fig. 9) and outside (region A in Fig. 9) the neck. Similarly, the curves marked $(t^B/t^A)_0 = 0.99$ correspond to a one percent initial thickness reduction in the neck with all other initial values being identical inside and outside, and so on. With the exception of the curve marked $\Delta f_0 = 0.01$, the initial void concentration both inside and outside the neck was taken to be identically zero. With an initial void concentration of zero the Gurson constitutive equation reduces to the Prandtl-Reuss equations and, therefore, the void concentration remains zero. The inhomogeneity corresponding to $(N^B/N^A) = 0.99$ gives qualitatively the same behavior as the model incorporating void growth. This is not entirely unexpected since the principal effect of void growth is to increasingly decrease the stiffness of the void matrix aggregate. In contrast, the forming limit curves for $(\sigma_y{}^B/\sigma_y{}^A) = 0.99$ and $(t^B/t^A)_0 = 0.99$ monotonically increase with ρ. On the other hand with $N^A = 0.25$, all the forming limit curves have the same general shape [40].

These results have some bearing on the idea of an "equivalent thickness imperfection." In previous calculations of forming limit diagrams a thickness imperfection was employed as the inhomogeneity and it was hypothesized that this was representative of microstructural inhomogeneities. The results in [40] show that this hypothesis is not necessarily appropriate for high hardening materials. These results also indicate that any microstructural inhomogeneity that has the effect of continually decreasing the effective strain hardening exponent would be expected to give qualitatively similar forming limit diagrams to the void

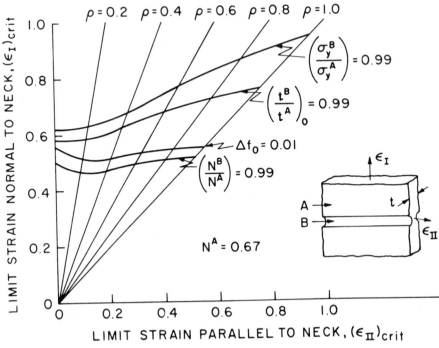

Fig. 9. A comparison of the predicted forming limit curves resulting from various types of initial inhomogeneities with strain hardening exponent $N^A = 0.67$. Here, $(\ )^A$ refers to quantities outside the neck and $(\ )^B$ refers to quantities inside the neck. Adopted from [40].

growth model. For example, in precipatation hardened solids there could be some local region of the sheet which is hardened less than the surrounding material.

This (somewhat obliquely) introduces the important question of the length scale over which it is reasonable to assume inhomogeneities act. Of course, the deformation theory results of [7], being based on a bifurcation analysis, do not require the assumption of an initial inhomogeneity. However, for an M-K type model, this question is a crucial one. Recent work [42] has indicated that the assumption of plane stress is an appropriate one for deformation wavelengths long with respect to the sheet thickness. Thus in the M-K model, the inhomogeneity must be assumed to be one acting over a length scale of at least several sheet thicknesses. If the inhomogeneity is truly local, of the order of the sheet thickness or less, then a plane stress analysis, such as that in [40], may significantly overestimate the rate of growth of the inhomogeneity [42, 43].

Experimental studies directed toward characterizing the type, magnitude and spatial distribution of the inhomogeneities to be expected in sheet materials would be most valuable. As illustrated in Fig. 8, by arbitrarily choosing values of the characterizing parameters for sheet materials, theories based on very different physical mechanisms can yield virtually identical forming limit diagrams. Experimental guidance is required to determine what values of parameters are reasonable. Hopefully, coupled with theoretical analyses of relevant constitutive models, such studies would enable the dominant physical mechanisms responsible for triggering localized necking (which may very well be material dependent) to be identified.

Additionally, direct examination of the material in the neck, at various stages of necking, for evidence of incipient ductile rupture would indicate whether or not this mechanism plays a significant role. Here, it would be particularly important to include specimens with well developed necks, since in [40] it was found that most void growth takes place in the latter stages of necking.

The discussion so far has presumed that the plane stress local necking mode is the one that limits ductility. Although consideration here is focussed on thin sheets, it is by no means certain that a three dimensional band mode, of the type discussed in the previous Section, is ruled out. Somewhat curiously, within the context of the three dimensional theory, the localized necking mode corresponds to a "geometric" instability and not a "material" instability. Whatever the underlying physical mechanism, there is the possibility of the deformation in the necked down region becoming significantly large to precipitate three dimensional flow localization (or to activate a "geometric" instability mode not accounted for in the plane stress theory [23, 28]). The three dimensional localized necking mode, representing an actual material instability, provides an inherent limit, regardless of sheet thickness or other geometric effects, to ductility. Indeed, in the plane strain tension test, e.g. [29], ductility is often found to be limited by highly intense band type localizations, sometimes preceeded by only a small amount of growth of the diffuse "geometric" necking mode.

CONCLUDING REMARKS

The description of localization as a material instability appears to provide a useful framework for analyzing both three dimensional "shear band" localizations and localized necking in ductile metal sheets (when these can be appropriately treated as two dimensional continua). In both contexts, the predicted onset of localization depends critically on the assumed constitutive description. The effects on localization of several deviations from the classical Prandtl-Reuss model have been illustrated. Significant constitutive features not explicitly dealt with here that can be accommodated within the theory include anisotropy, which is particularly important in the sheet metal context, and material rate dependence (which necessitates an analysis with an initial imperfection).

Encouraging results have been obtained from a simple model of a solid with a vertex on its yield surface, both regarding the ratio of plane strain and axisymmetric ductilities and forming limit diagrams. More or less similar results can also be obtained from a simple model of a solid weakened by micro-cavities. Both these constitutive features merit improved modelling in relation to the detailed mechanisms of deformation. A more thorough theoretical and experimental illumination of these aspects of material behavior could shed considerable light on the circumstances under which an inherent instability of the plastic flow process or progressive micro-rupturing is responsible for limiting ductility.

References pp. 264–265.

ACKNOWLEDGEMENT

Support from the Energy Research and Development Administration through contract E(11-1)3084 and the National Science Foundation through grant ENG76-16421 is gratefully acknowledged.

REFERENCES

[1] R. Hill, The Mathematical Theory of Plasticity, Oxford Press (1950).
[2] J. W. Rudnicki and J. R. Rice, J. Mech. Phys. Solids, 23 (1975), 371.
[3] J. R. Rice, Theoretical and Applied Mechanics, Proc. 14th Int. U. Theor. Appl. Mech. Cong. (ed., W. T. Koiter), North-Holland, 1 (1976), 207.
[4] R. Hill, J. Mech. Phys. Solids, 15 (1967), 79.
[5] T. H. Lin, Adv. Appl. Mech., 11 (1971), 255.
[6] J. W. Hutchinson, Proc. Roy. Soc. Lond. A, 139 (1970), 247.
[7] S. Stören and J. R. Rice, J. Mech. Phys. Solids, 23 (1975), 421.
[8] J. L. Sanders, Proc. 2nd U.S. Nat. Congress Appl. Mech. (1954), 455.
[9] J. W. Hutchinson and K. W. Neale, "Sheet necking-II. Time-independent behavior," this volume.
[10] W. A. Spitzig, R. J. Sober and O. Richmond, Acta Met., 23 (1975), 885.
[11] W. A. Spitzig, R. J. Sober and O. Richmond, Metall. Trans. 7A (1976), 1703.
[12] A. L. Gurson, Ph.D. Dissertation, Brown University (1975).
[13] A. L. Gurson, J. Engr. Mater. Technol. (Trans. ASME, H), 99 (1977), 2.
[14] F. A. McClintock, J. Appl. Mech. (Trans. ASME, E), 35 (1968), 363.
[15] J. F. W. Bishop and R. Hill, Philos. Mag., 42 (1951), 414.
[16] C. A. Berg, Inelastic Behavior of Solids (eds., M. F. Kanninen et al.), McGraw-Hill (1970), 171.
[17] A. L. Gurson, Fracture 1977, (ed. D. M. R. Taplin), Univ. Waterloo Press, 2 (1977), 357.
[18] J. Gurland, Acta. Met., 20 (1972), 735.
[19] A. S. Argon, J. Im and A. Needleman, Metall. Trans., 6A (1975), 815.
[20] A. S. Argon and J. Im, Metall. Trans., 6A (1975), 839.
[21] A. S. Argon, J. Im and R. Safoglu, Metall. Trans., 6A (1975), 825.
[22] J. R. Rice and M. A. Johnson, Inelastic Behavior of Solids (eds., M. F. Kanninen et al.), McGraw-Hill (1970), 641.
[23] A. Needleman and V. Tvergaard, J. Mech. Phys. Solids, 25 (1977), 159.
[24] A. Marciniak and K. Kuczynski, Int. J. Mech. Sci., 9 (1967), 609.
[25] J. Hadamard, Lecons sur la Propagation des Ondes et les Equations de L'Hydrodynamique, Paris, Chp. 6 (1903).
[26] R. Hill, J. Mech. Phys. Solids, 10 (1962), 1.
[27] J. Mandel, Rheology and Soil Mechanics (eds., J. Kravtchenko and P. M. Sirieys), Springer-Verlag (1966), 58.
[28] R. Hill and J. W. Hutchinson, J. Mech. Phys. Solids, 23 (1975), 239.
[29] D. P. Clausing, Int. J. Fract. Mech., 6 (1970), 71.
[30] R. J. Asaro and J. R. Rice, J. Mech. Phys. Solids, 25 (1977), 309.
[31] H. Yamamoto, Int. J. Fract. (in press).
[32] R. Hill, J. Mech. Phys. Solids, 1 (1952), 19.
[33] S. P. Keeler, Machinery, 74, Nos. 6-11, February-July (1968).
[34] S. S. Hecker, Sheet Metal Ind. (1975), 671.
[35] S. S. Hecker, J. Eng. Mater. Technol. (Trans. ASME, H), 97 (1975), 66.
[36] M. Arzin and W. A. Backofen, Metall. Trans., 1 (1970), 2857.
[37] A. K. Ghosh and S. S. Hecker, Metall. Trans., 5 (1974), 2161.
[38] A. K. Ghosh and S. S. Hecker, Metall. Trans., 6A (1975), 1065.
[39] M. J. Painter and R. Pearce, J. Phys. D: Appl. Phys., 7 (1974), 992.
[40] A. Needleman and N. Triantafyllidis, J. Eng. Mater. Technol. (Trans. ASME, H) (in press).
[41] A. Needleman, J. Mech. Phys. Solids, 24 (1976), 339.

[42] J. W. Hutchinson, K. W. Neale and A. Needleman, "Sheet necking-I. Validity of plane stress assumptions of the long-wavelength approximation," this volume.
[43] E. J. Appleby and O. Richmond, to be published.
[44] T. Y. Thomas, Plastic flow and fracture in solids, Academic Press (1961).

DISCUSSION

S. Nemat-Nasser (*Northwestern University*)

There are two comments I would like to make. The voids develop at much later stages of deformation, and it seems to me that their effect should be combined with the derivation of the diffuse type of necking from an experimental point of view. In that sense, if you do the calculations using the usual plasticity laws, but including the effects of the voids, you will find that (we have done this calculation) you will develop internal shear bands or large bands of high strain between the voids, and if there are precipitates (as there usually are in these types of materials), then secondary voids (smaller voids) will develop and you will immediately observe that without having to introduce unrealistic numbers of voids that you do get localization effects. Now, I make this comment together with the comment that the conditions of localization that are being used are necessary rather then sufficient. That does not mean that satisfying these conditions necessarily will lead to localization.

Rice

It does once an imperfection is there; that's what comes out in the Marciniak-type analysis.

Nemat-Nasser

You do that for the imperfection, if you do the calculation as we have done with the imperfection, then a diffuse neck forms. Then if you consider the void growth inside, then the shear bands would automatically develop at much lower strain states. Secondly, I would question how you apply this to a round bar because Hadamard's conditions are really applicable to the displacement boundary value problem. Whereas I cannot imagine how it would be applied to a round bar in an axisymmetric displacement state.

Rice

When we look at these shear band conditions, they are for limiting instabilities. The relevance of the displacement boundary conditions is that they are such limiting instabilities that they still take place even if you have complete displacement boundary conditions all around the outer surface of the material.

But I wanted to get to the earlier comment. Your calculation, I would like to remind you again, was a plane-strain calculation. Your void was, in fact, an infinitely long tunnel and in a plane-strain calculation it seems to me hardly surprising that deformation should concentrate along 45° planes. So, I'm not sure that I'd be ready to accept what you say about strain concentrations near existing holes just on the basis of that work.

I think your model probably overestimates the effect. On the other hand, I agree it is important, in fact, to the local strain concentrations near holes and could have a lot to do with the final localizations. You still, however, do not explain why these final localizations occur and that also is an important problem. The cases we're discussing, incidentally, are like those which have been observed by Low and co-workers, especially Low and Cox in 4300 steels I think it was, where you have a hole starting at the large inclusions and then the joining process occurred not as in some materials by these large voids growing together and coalescing, but rather by some shear band forming between these large holes (Nemat-Nasser—Void sheets). That's right, the shear band containing void sheets which probably nucleated from very small carbide particles.

J. W. Justusson (General Motors Research Laboratories)

You referred to the work of Gurson concerning constitutive equations which incorporate the effect of voids. I have the impression that the void shape that he considers is more or less spherical.

Rice

It is more, rather than less. It is spherical. In fact what we've done, in the spirit of isotropic hardening, is to neglect the fact that voids don't remain spherical. The calculation is for a spherical void.

Justusson

The calculations then would not pertain to penny shaped voids or other extreme geometries?

Rice

Of course what happens in low triaxiality stress states is that these things tend to get very much strung out and become much more cylindrical.

R. N. Dubey (University of Waterloo, Canada)

At the beginning of your talk you talked about a vertex modulus. We have two moduli: one you called the plasticity modulus which you can calculate from the tension curve. . . .

Rice

The tangent modulus, I think I called that.

Dubey

Then you had the vertex modulus which you could get with the secant modulus.

Rice

The results I gave in that unreadable table were in terms of the ratio of the vertex modulus to tangent modulus on the left side. On the right side of the table some results were given when you identified the vertex modulus as the secant modulus.

Dubey

My question is: How can you calculate it supposing you don't want to predetermine that?

Rice

The way to calculate it is to do the kind of thing that Sanders and co-workers started years ago and that John Hutchinson has continued; actually trying to calculate the properties of plastic polycrystals. What emerges from those calculations, though not in any simple way, are descriptions of all moduli for all possible directions of deformation from the current state. Some of this sort of thing is in Hutchinson's 1970 Royal Society paper where you can see calculations of the effect of shear moduli for different directions of shearing on the basis of the so-called Self-Consistent Polycrystal Model.

Dubey

I have seen that paper. Can you determine it by a simple test like a tensile test?

Rice

The kind of tests that I guess ought to be done—there's a whole literature on looking for vertices on yield surfaces and it's all terribly inconclusive. Probably the more telling thing would be to deform the material in tension. Then, while you continue to increase the tension, superpose a very small amount of shear. It's got to be small because you want to make sure that the loading is still what's called "fully active", that it continues to activate all currently active slip systems. Then, if you measure the shear strain response to that small superimposed shear stress that would tell you what the vertex modulus is, if indeed it exists at all.

SHEET NECKING-III. STRAIN-RATE EFFECTS*

J. W. HUTCHINSON

Harvard University, Cambridge, Massachusetts

K. W. NEALE

Université de Sherbrooke, Sherbrooke, Quebec, Canada

ABSTRACT

The effect of material strain-rate dependence on necking retardation is examined for biaxially-stretched sheets. Rate-dependent versions of both flow theory and deformation theory are employed in an analysis of the growth of long-wavelength nonuniformities. Material strain-rate sensitivity is seen to substantially increase the predicted limit strains beyond their corresponding values for time-independent material response. We also discuss the influence of strain-rate dependence on imperfection-sensitivity and forming limit curves.

INTRODUCTION

In this third and final Part of our investigation of sheet necking, we concentrate on the influence of material strain-rate dependence on necking retardation. A relatively small amount of strain-rate dependence is known to lead to substantially increased straining prior to necking [1–3]. This phenomenon was discussed in [1] for an axisymmetric bar under uniaxial tension. Ghosh [2, 3] has also studied the effects of strain-rate sensitivity on necking. In particular, he has collected experimental data for flat strips under uniaxial tension which shows that the maximum amount of overall axial strain attained is strongly dependent on the material strain-rate sensitivity characteristics. In addition, he has developed an approximate analysis for this strip problem which agrees very well with the trend of experiments.

As emphasized in [1], necking in strain-rate sensitive materials is inherently a nonlinear process. While classical linearized analyses, such as those discussed in Part I, can provide useful information regarding the early development of

---

* The material in this three-part paper was presented orally in Session II under the title "Constitutive Relations for Sheet Metal" and Session IV under the title "Sheet Necking: Influence of Constitutive Theory and Strain-Rate Dependence".

References p. 283.

nonuniformities, they do not give meaningful estimates for limit strains. Alternatively, an approximate nonlinear analysis based on the long-wavelength simplification discussed in [1] for axisymmetric bars under uniaxial tension does reproduce the essential details of the phenomenon, and will be applied herein.

As in Part II, we shall examine in detail here the basic differences between the results obtained using J_2 flow theory of plasticity and those determined with a rate-dependent J_2 deformation-type theory. Marciniak et al [4] have considered the effects of strain-rate sensitivity on localized necking in sheets; their analysis employs J_2 flow theory only and is restricted to the biaxial tension range. The present analysis is valid for a wider range of strain states. We shall examine imperfection-sensitivity and, in particular, how material strain-rate dependence alters forming limit curves.

LONG-WAVELENGTH(M-K) ANALYSIS

As mentioned previously, it is essential that nonlinearities be properly accounted for in an analysis of the necking process when the material response is time-dependent. Here, our analysis is based on long-wavelength (M-K) simplification. This approximation was applied in Part II to examine the influence of nonhomogeneities on sheet necking. An examination of some of its limitations was given in Part I.

Throughout this study we shall employ the same notation as that used in Part II. The derivation of the basic relations here closely parallels the detailed development given in the Section on Long-wavelength Analysis of II. To avoid unnecessary repetition, we shall simply discuss those steps of that analysis which must be modified to account for strain-rate dependent material behavior. The assumptions of material incompressibility and initial isotropy also apply here.

As in Part II, we consider a thin sheet which is initially uniform except for the presence of a nonhomogeneity concentrated in a narrow band across the sheet (Fig. 1), and examine the growth of this initial nonuniformity as the uniform section of the sheet is loaded.

To describe time-dependent material response, the following true stress-natural strain relation replaces (II: 11)*

$$\sigma_e = K\epsilon_e{}^N\dot{\epsilon}_e{}^m \qquad \text{(deformation theory)} \qquad (1)$$

$$\sigma_e = K\bar{\epsilon}^N\dot{\bar{\epsilon}}^m \qquad \text{(flow theory)} \qquad (2)$$

Here, σ_e and ϵ_e (or $\bar{\epsilon}$) are the effective stress and effective strain, equal to the true stress and true (logarithmic) strain, respectively, in uniaxial tension, and $\dot{\epsilon}_e$ (or $\dot{\bar{\epsilon}}$) is the effective strain-rate. N and m denote the strain hardening and strain-rate hardening exponents. As in II, we shall employ two distinct symbols and definitions for effective strain. In the deformation theory analysis ϵ_e is used and is defined in total form as $\epsilon_e = (2\epsilon_i\epsilon_i/3)^{1/2}$, where ϵ_i are the principal values of logarithmic strain. The effective strain-rate $\dot{\epsilon}_e$ is the time rate of change of ϵ_e.

This denotes equation (11) of Part II.

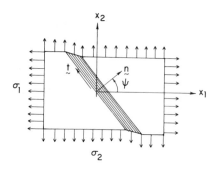

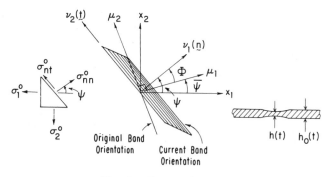

Fig. 1. Conventions.

Conversely, in the flow theory analysis we define the effective strain-rate in differential form as $\dot{\bar{\epsilon}} = (2\dot{\epsilon}_{ij}\dot{\epsilon}_{ij}/3)^{1/2}$, and the effective strain $\bar{\epsilon}$ is then the integral of $\dot{\bar{\epsilon}}$ with respect to time. In any monotonic proportional loading history $\epsilon_e = \bar{\epsilon}$.

Equation (1) or (2) brings in the effect of the strain-rate on the effective stress-strain relation in the simplest possible way. The constants K, N and m are those commonly determined from uniaxial test data. For multiaxial states for the rate-dependent "deformation theory", (1) is still supplemented by (II: 8), i.e.,

$$\epsilon_i = \mu s_i \tag{3}$$

where now μ depends on $\dot{\epsilon}_e$ through (1). The flow theory relation (II: 43), i.e.,

$$d\epsilon_{ij} = \frac{3}{2}\frac{d\bar{\epsilon}}{\sigma_e}s_{ij} \tag{4}$$

still goes with (2). For $m = 0$ both versions reduce to their counterpart relations used in Part II. Neither relation can be expected to adequately represent rate-dependent material behavior for arbitrary histories of stress or strain. But they can be expected to be reasonable generalizations of their respective time-independent limits for histories likely to be experienced in sheet-metal forming when stresses and strains increase monotonically with relatively slowly varying strain-rates. The rate-dependent flow theory, (2) and (4), is the same as that used by Marciniak et al [4]. We will refer to (1) and (3) as a rate-dependent "deformation theory'' in order to keep in mind its connection with the true deformation theory of Part II. In general, the rate-dependent version does not yield path-independent stress-strain behavior unless the effective strain-rate $\dot{\epsilon}_e$ is held constant.

The analysis producing equations (II: 33) to (II: 39) involves only equilibrium and the definition of strains so that these equations remain valid for the present investigation. However, in view of (1), (II: 40) is modified as follows, for deformation theory,

$$\frac{\sigma_{nn}^0/\sigma_e^0}{\sigma_{nn}/\sigma_e}\left(\frac{\dot{\epsilon}_e^0}{\dot{\epsilon}_e}\right)^m = (1 - \xi)\left(\frac{\epsilon_e}{\epsilon_e^0}\right)^N \exp(\epsilon_3 - \epsilon_3^0) \tag{5}$$

with ϵ_e, ϵ_e^0 and their rates replaced by $\bar{\epsilon}$, $\bar{\epsilon}^0$ and their rates for flow theory. Equation (II: 41) for the ratio σ_{nn}^0/σ_e^0 and (II: 42) for ϵ_3^0 are unchanged.

J_2 **Flow Theory Analysis**—The flow theory relations (II: 43) to (II: 48) and (II: 50) to (II: 52) apply for the present analysis. However, since (5) replaces (II: 40), the expression analogous to (II: 49) becomes

$$(1 - B - G)^{1/2}H\left[1 - B\left(\frac{d\bar{\epsilon}^0}{d\bar{\epsilon}}\right)^2\right]^{-1/2}\left(\frac{d\bar{\epsilon}^0}{d\bar{\epsilon}}\right)^m$$

$$= (1 - \xi)\left(\frac{\bar{\epsilon}}{\bar{\epsilon}^0}\right)^N \exp(C\bar{\epsilon}^0 + \epsilon_3) \tag{6}$$

where (II: 51)

$$\frac{d\epsilon_3}{d\bar{\epsilon}^0} = -\frac{A}{H}\left[1 - B\left(\frac{d\bar{\epsilon}^0}{d\bar{\epsilon}}\right)^2\right]^{1/2} - D\frac{d\bar{\epsilon}^0}{d\bar{\epsilon}} \tag{7}$$

The parameters A, B, C, D, G and H here depend only on the imposed strain ratio ρ and current band orientation ψ. For $\psi = 0$, the above equations reduce to those given by Marciniak et al [4].

From a straightforward incremental solution of (6) and (7), we can numerically determine the groove deformation $\bar{\epsilon}$ as a function of the prescribed uniform deformation $\bar{\epsilon}^0$. This numerical procedure is described more fully in II. Of particular interest here is the limit strain $\bar{\epsilon}^{0*}$, i.e., the maximum attainable strain in the uniform region of the sheet. As will be seen in the results which follow, the corresponding strain in the groove becomes unbounded, i.e., $\bar{\epsilon} \to \infty$, as $\bar{\epsilon}^0$ reaches its maximum value when the material response is strain-rate dependent.

J_2 **Deformation Theory Analysis**—The deformation-theory expressions (II: 53) to (II: 67) and (II: 69) are also valid for the present analysis. From (5), the modified form of (II: 68) becomes

$$\left[\frac{(\rho + 2)\cos^2\psi + (2\rho + 1)\sin^2\psi}{\sqrt{3}(1 + \rho + \rho^2)^{1/2}}\right]\left(\frac{d\epsilon_e^0}{d\epsilon_e}\right)^m = \frac{(1 - \xi)}{\lambda_1\lambda_2}\left(\frac{\epsilon_e}{\epsilon_e^0}\right)^N \exp(C\epsilon_e^0)\frac{Y}{\epsilon_e} \tag{8}$$

A step-wise numerical solution of this equation immediately gives the increments of ϵ_e in terms of the prescribed increments of ϵ_e^0. At each step the current values of principal stretches λ_i, groove angle ψ and principal axes orientation θ must be updated. The quantities λ_i and θ, however, depend on the parameters δ, γ, η and ϕ in (II: 53). From the matching conditions (II: 56), the current values of ϕ and γ are readily calculated. We compute δ and η in an incremental fashion by expressing (II: 65) and (II: 69) in rate form.

For the case $\bar{\psi} = \psi = 0$, the present analysis reduces to

$$(1 - B)^{1/2} \left[1 - B \left(\frac{\epsilon_e^{\,0}}{\epsilon_e} \right)^2 \right]^{-1/2} \left(\frac{d\epsilon_e^{\,0}}{d\epsilon_e} \right)^m = (1 - \xi) \left(\frac{\epsilon_e}{\epsilon_e^{\,0}} \right)^N \exp(C\epsilon_e^{\,0} + \epsilon_3) \qquad (9)$$

where ϵ_3 is given by (II: 72). The above relation (9) is analogous to (II: 71).

An interesting feature of the present analysis, both for flow theory and deformation theory, is that the relationship between $\epsilon_e^{\,0}$ and ϵ_e (or between $\bar{\epsilon}^0$ and $\bar{\epsilon}$) is independent of the load history experienced by the sheet and, in particular, is independent of the rate $\dot{\epsilon}_e^{\,0} (\equiv \dot{\bar{\epsilon}}^0)$ at which the sheet is being deformed. Thus the limit strain for prescribed ρ and $\bar{\psi}$ is determined solely by the parameters m, N and ξ.

PLANE STRAIN CASE ($\rho = 0$)

For the plane strain case ($\rho = 0$), $\bar{\psi} = \psi = 0$ and, as in II, flow theory and deformation theory give identical predictions. In the long-wavelength analysis of the previous Section, $B = D = G = 0$, $H = 1$ and $A = C = \sqrt{3}/2$. Equations (7) or (II: 72) give $\epsilon_3 = -\sqrt{3}\epsilon_e/2 = -\epsilon_1$ and the relations (6) and (9) can be written as

$$(\epsilon_1^{\,0})^{N/m} \exp(-\epsilon_1^{\,0}/m)d\epsilon_1^{\,0} = (1 - \xi)^{1/m} \epsilon_1^{\,N/m} \exp(-\epsilon_1/m)d\epsilon_1 \qquad (10)$$

This relation is identical to the expression obtained in [1] for the axisymmetric bar under uniaxial tension, with ϵ_1 identified as the axial strain. The results and conclusions of that study are therefore directly applicable to the present problem. For $m = 0$, (10) reduces to (II: 74).

Typical results using (10) taken from [1] are shown in Fig. 2, where curves of $\epsilon_1/\epsilon_1^{\,0}$ are plotted against $\epsilon_1^{\,0}/N$ for an initial geometric nonuniformity $\xi = .005$. For the case $m = 0$, the curves are identical to those given in Fig. 4 of Part II. Of particular interest here is the limit strain, i.e., the maximum value of $\epsilon_1^{\,0} (= \epsilon_1^{\,0*})$ attained in the uniform region of the sheet. From (10) and this figure it can be seen that, when the material response is strain-rate dependent, the uniform strain $\epsilon_1^{\,0}$ reaches a maximum as the strain in the groove, ϵ_1, becomes unbounded. In contrast, for $m = 0$ the limiting value, which will be denoted by $\bar{\epsilon}_1^{\,0*}$, is attained when $\epsilon_1 = N$ (see Part II).

The curves of Fig. 2 indicate that material strain-rate dependence greatly influences the maximum uniform strain that can be achieved. Fig. 3 illustrates this phenomenon for small values of strain-rate exponent ($m \leq .05$). Here,

$$\delta\epsilon_1^{\,0*} = \epsilon_1^{\,0*} - \bar{\epsilon}_1^{\,0*} \qquad (11)$$

is the increase, due to strain-rate dependence, of the limit strain in uniform region above the corresponding limit strain for a time-independent material ($m = 0$).

References p. 283.

J. W. HUTCHINSON, K. W. NEALE

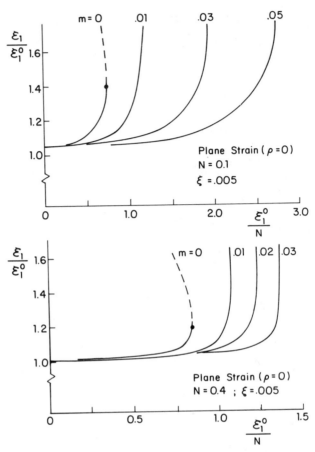

Fig. 2. Effect of strain-rate sensitivity index m on development of the strain in neck ϵ_1 as indicated by growth of ratio $\epsilon_1/\epsilon_1{}^0$, where $\epsilon_1{}^0$ is the strain outside the neck.

As discussed in [1–3], very small values of m can lead to relatively large increases in $\epsilon_1{}^{0*}$. It is also evident that $\delta\epsilon_1{}^{0*}$ decreases with increasing initial nonuniformity ξ, and increases with increasing strain-hardening exponent N. In fact, the numerical results depicted in Fig. 3 indicate that $\delta\epsilon_1{}^{0*}$ is nearly proportional to $\sqrt{N}$ for small m, as suggested by the asymptotic formula (12) given below.

In Fig. 4 the increase in limit strains for $m = .05$ are plotted against the initial imperfection ξ. (Curves of limit strain $\bar{\epsilon}_1{}^{0*}(m = 0)$ vs. ξ for these cases can be found in Fig. 5 of Part II.) As mentioned above, the delay in necking $\delta\epsilon_1{}^{0*}$ decreases as the imperfection ξ increases. A basic difference between necking in time-dependent and time-independent materials can also be seen in Fig. 4. As $\xi \to 0$, the limit strain $\bar{\epsilon}_1{}^{0*}$ approaches a finite value $(=N)$ when $m = 0$. In contrast, when the material response is time-dependent $(m \neq 0)$, $\epsilon_1{}^{0*} \to \infty$ as $\xi \to 0$. It is for this reason that $\delta\epsilon_1{}^{0*}$ becomes infinite at $\xi = 0$.

The numerical results of Figs. 2–4 indicate that very small values of strain-rate exponent m can substantially increase the limit strain $\epsilon_1{}^{0*}$ beyond its time-inde-

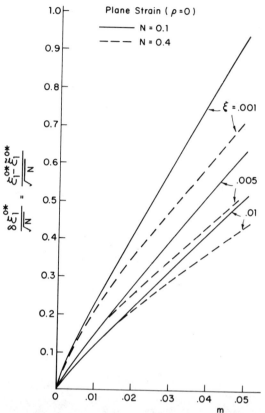

Fig. 3. Necking retardation due to small amounts of strain rate sensitivity as measured by the increase of the limit strain outside the neck over the time-independent value at m = 0.

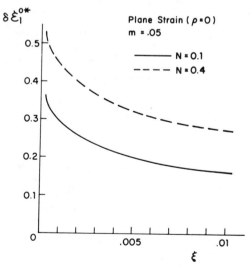

Fig. 4. Imperfection-sensitivity of additional limit strain in plane strain with $m = 0.5$. (Deformation theory and flow theory predictions coincide for $\rho = 0$.)

References p. 283.

pendent value $\tilde{\epsilon}_1{}^{0^*}$. An analysis was carried out in [1] to determine an exact asymptotic relation for the influence of very small m on $\epsilon_1{}^{0^*}$. The result of that analysis, also valid for the present problem, is

$$\frac{\delta\epsilon_1{}^{0^*}}{\sqrt{N}} \simeq \frac{m}{2\sqrt{2\xi}} \ln\left(\frac{4\pi\xi}{m}\right) \tag{12}$$

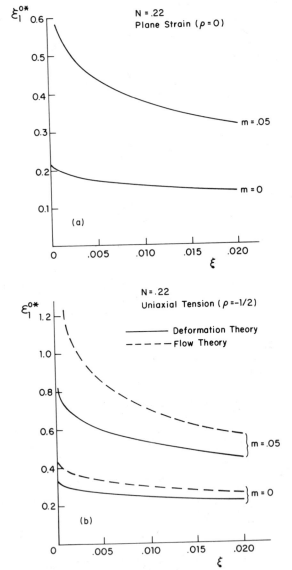

Fig. 5. Imperfection-sensitivity for two plasticity theories for time-independent limit $(m = 0)$ and for $m = .05$: (a) plane strain, (b) uniaxial tension, (c) equibiaxial tension.

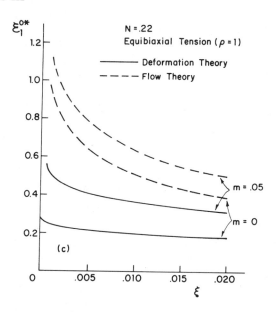

Here it is assumed that $\xi \ll 1$, $m < 2\xi$ and that m/N is small. This asymptotic relation has a rather restricted range of applicability. Nevertheless, it does reveal that the slope of the $\delta\epsilon_1^{o*}$ -- m relation is infinite at $m = 0$, thereby illustrating a very strong dependence on m. It also indicates that $\delta\epsilon_1^{o*}$ is proportional to $\sqrt{N}$ for small m, as do the numerical results in Fig. 3. Furthermore, equation (12) implies that the amount of increase $\delta\epsilon_1^{o*}$ is quite sensitive to small variations in imperfection values ξ. We expect this since, as seen in Fig. 4, $\delta\epsilon_1^{o*}$ becomes infinite as $\xi \to 0$.

RESULTS AND DISCUSSION FOR FULL RANGE OF ρ

The long-wavelength relations (6), (7) for flow theory, and (8) for deformation theory, together with the condition ϵ_e (or $\bar{\epsilon}$) $\to \infty$ at $\epsilon_e^o = \epsilon_e^{o*}$, were used to investigate the influence of strain-rate sensitivity on limit strains. The range of strain ratios $-1/2 \leq \rho \leq 1$ was considered. Computations were carried out with various values of initial band orientation $\bar{\psi}$ to find the critical angle $\bar{\psi}^*$ giving the minimum limit strain. In the biaxial tension range ($\rho \geq 0$) this angle is $\bar{\psi}^* = 0$.

In Figs. 5(a)–(c) the variation of limit strain ϵ_1^{o*} with initial nonuniformity ξ is shown for $\rho = 0$, $-1/2$ and 1, respectively, and $N = .22$. Time-independent results ($m = 0$) taken from Part II are included here for comparison. A strain-rate hardening exponent $m \leq .05$ is typical of most common sheet metals at room temperature.

For the plane strain case ($\rho = 0$, Fig. 5(a)) there is no distinction between the predictions of flow theory and deformation theory, as mentioned in the previous Section. The curve for $m = 0$ exhibits the imperfection-sensitivity characteristics discussed in II. A much stronger imperfection-sensitivity is observed for the strain-rate dependent results ($m = .05$) since, as stated earlier, ϵ_1^{0*} becomes infinite as $\xi \to 0$. We also see that the limit strains for $m = .05$ are considerably higher than their corresponding values for $m = 0$.

The curves of ϵ_1^{0*} vs. ξ for uniaxial tension ($\rho = -1/2$, Fig. 5(b)) also exhibit

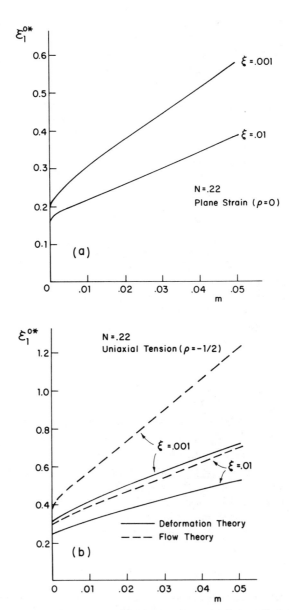

Fig. 6.   Dependence of limit strain outside neck on strain-rate index for two plasticity theories: (a) plane strain, (b) uniaxial tension, (c) equibiaxial tension.

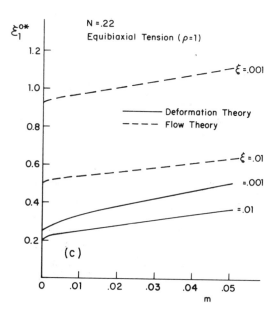

imperfection-sensitivity characteristics. Here, the solid and dashed curves refer to the deformation theory and flow theory predictions, respectively. For $m = 0$, the flow theory limit strains are somewhat higher and slightly more imperfection-sensitive than the corresponding deformation theory results as discussed in II. Both theories predict approximately the same *relative* increase in ϵ_1^{0*} due to strain-rate dependence ($m = .05$).

For the equibiaxial tension case ($\rho = 1$, Fig. 5(c)), a much larger discrepancy between the flow theory and deformation theory predictions is observed. As discussed in II, the flow theory limit strains for $m = 0$ are substantially higher than those based on deformation theory, and there is also considerably more imperfection-sensitivity associated with flow theory. With material strain-rate dependence ($m = .05$) the increase in limit strain for a given imperfection ξ is greater for deformation theory than for flow theory. Nevertheless, strain-rate effects do not raise the deformation theory predictions to the level of the time-independent flow theory results.

In Figs. 6(a)–(c) the dependence of limit strain ϵ_1^{0*} on strain-rate exponent m is illustrated for $\xi = .01$, .001 and $\rho = 0$, $-1/2$, and 1. Fig. 6(a) for plane strain shows the same trend as Fig. 3, namely, that strain-rate effects substantially increase limit strains beyond their time-independent values and that the amount of increase diminishes with increasing ξ. Figures 6(b) and 6(c) bring out the discrepancies between flow theory and deformation theory discussed previously. The uniaxial tension results of Fig. 6(b) indicate that the difference between both theories is greater for the lower values of ξ and higher values of m. The enormous discrepancy for the equibiaxial tension case is illustrated in Fig. 6(c).

Figs. 7(a) and 7(b) depict the manner in which the limit strain ϵ_1^{0*} under uniaxial tension varies with initial band orientation $\bar{\psi}$, as well as current groove orientation ψ^* at necking. Again, solid curves represent deformation theory predictions and dashed lines refer to flow theory. In these figures the values given in parentheses adjacent to each curve refer to m and ξ, respectively. The time-independent curves ($m = 0$) here are taken from II and, as mentioned there, they indicate that the flow theory limit strains are much more sensitive to variations in $\bar{\psi}$ and ψ^* than the predicted deformation theory results. With strain-rate dependence ($m = .05$) this sensitivity is accentuated for flow theory, particularly for smaller ξ, while the limit strains according to deformation theory are still rather unaffected by small changes in $\bar{\psi}$ or ψ^*.

In Fig. 7(a) it can be seen that the critical value of $\bar{\psi}$ which minimizes $\epsilon_1^{o^*}$ with flow theory is greatly decreased by material strain-rate dependence, especially as the imperfection ξ is lowered. This effect, counteracted by the tendency of limit strain to increase as m is varied from $m = 0$ to $.05$, gives current minimizing values of ψ^* that do not vary substantially (see Fig. 7(b)). The corresponding variations in critical ψ^* for different m and ξ according to deformation theory are somewhat larger.

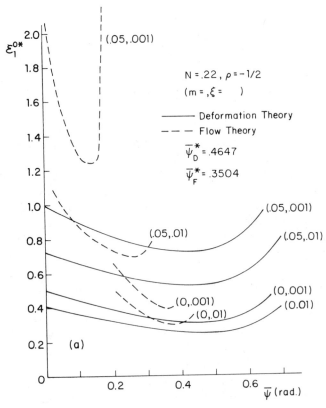

Fig. 7.   Dependence on limit strain outside neck on orientation of necking band: (a) plotted against initial orientation angle $\bar{\psi}$, (b) plotted against final orientation angle ψ^*.

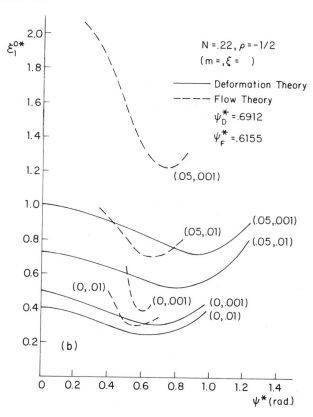

The effect of material strain-rate dependence on forming limit curves is depicted in Figs. 8 and 9 for strain-hardening exponents $N = .22$ and $.50$, respectively, and an imperfection level $\xi = .01$. Part II results for time-independent material response $(m = 0)$ are also included here.

In the range $-1/2 \leq \rho \leq 0$, the shapes of the forming limit curves in Figs. 8 and 9 for $m = .05$ closely resemble the time-independent curves. The effect of increasing m in this range is essentially to just shift the forming limit curve upwards. In the biaxial tension range $(0 \leq \rho \leq 1)$, however, there is a flattening of the curves due to strain-rate dependence. Nevertheless, the flow theory curves still rise rather steeply for $m = .05$ and both N-values, whereas the corresponding forming limit curves with deformation theory fall with increasing ρ.

A detailed discussion of the theoretical forming limit curves and their relationship to published experimental data was presented in Part II of this paper which dealt with time-independent material behavior. The outcome of the discussion on the issue of flow theory vs. deformation theory was the contention that deformation theory seems to give better qualitative agreement with experiments than does flow theory for the overall straining histories considered here. In particular for $\rho \geq 0$, where the discrepancies between theories are greatest, only deformation theory predicts the experimental trends of (i) a slightly rising forming limit

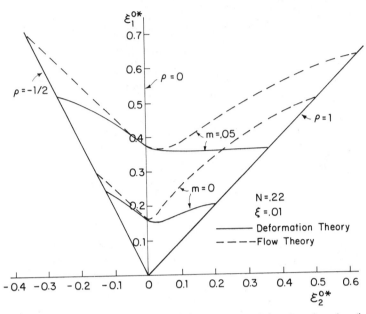

Fig. 8.   Forming limit curves for $N = .22$ for two plasticity theories showing influence of strain-rate dependence.

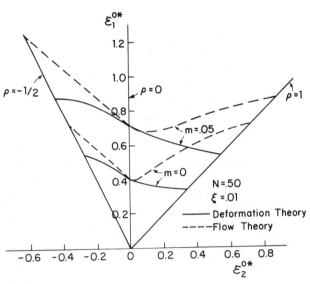

Fig. 9.   Forming limit curves for $N = .50$ for two plasticity theories showing influence of strain-rate dependence.

curve for the lower N-values ($\leq.25$), or (ii) a curve which falls somewhat for the higher N-values ($\geq.50$). The results of the present study indicate that, even when strain-rate effects are incorporated in the analysis, simple flow theory does not predict this tendency.

ACKNOWLEDGMENT

The work of J.W.H. was supported in part by the Air Force Office of Scientific Research under Grant AFOSR 77-3330, the National Science Foundation under Grant ENG76-04019, and by the Division of Applied Sciences, Harvard University. The work was conducted while K.W.N. was on leave at Harvard University. The support of the Faculty of Applied Sciences at the University of Sherbrooke is gratefully acknowledged.

REFERENCES

[1] J. W. Hutchinson and K. W. Neale, Acta Met., 25 (1977), 839.
[2] A. K. Ghosh, J. Eng. Mater. Technol. (Trans. ASME, H), 99 (1977), 264.
[3] A. K. Ghosh, Research Publication GMR-2323, General Motors Research Laboratories, Warren, Mich. (1977).
[4] Z. Marciniak, K. Kuczyński and T. Pokora, Int. J. Mech. Sci., 15 (1973), 789.

DISCUSSION

A. K. Ghosh *(Rockwell International)*

Why did you choose Roger Pearce's data on brass for the negative side—why not our data? The way Roger Pearce conducted these tests, he takes a wide strip, cuts a slot in the strip for the negative side (slots of various width to height ratios) and the grid size is fixed. So as he moves closer to plane strain, because the grid size is fixed, you get strain gradients. And this is why I feel the negative side is not as steep as our results by punch stretching. He comes off to much higher plane-strain values.

Neale

In fact, I tried to qualify the discussion. There are a number of things which you do in an experiment which don't correspond to the very simple analysis that I've given here: we assume, for example, proportional straining in all regions of the sheet.

The other thing is that even though some experimental results, in the negative ρ range, might agree very well with Hill's calculations, I think it's hard sometimes to reconcile these results with the flow theory of plasticity because often when we do a test we are restricting the angle at which the neck eventually forms. Now as I showed on the slide (although I didn't discuss it in relation to experiments) flow theory analysis results are highly sensitive to the angle at which necking does occur. So, if, in a test, we were constraining the angle quite a bit, to be zero for example, and if you work out the flow theory prediction it will give a prediction of limit strain much higher than Hill's results. However, deformation theory, as you saw, is fairly insensitive to this variation in angle.

S. P. Keeler (National Steel Company)

It bothers me that you use an fcc for a high N (in your notation) and a bcc for low N. What would happen if you had identical N values, let's say 0.2, and comparing a ferrous behavior versus a brass behavior? As Ghosh, Hecker, Azrin, and others have shown, you get two different behaviors with the same N value. How do you handle that problem?

Neale

In this analysis, we are using the simple, classical constitutive laws of plasticity. The only variable, at least in this analysis, is the strain hardening parameter N. We can use refined constitutive laws (continuum plasticity laws) as Prof. Hutchinson discussed yesterday, which take into account anisotropy, or we can develop models such as the ones that Prof. Rice discussed earlier, to model microscopic fracture. I think that's probably the only way. This analysis is based on a certain constitutive law, a certain continuum approach and if you want to model other types of material behavior, you will have to use a different constitutive law.

U. S. Lindholm (Southwest Research Institute)

We're new to the forming area but we've done a lot of multiaxial testing on tubular-type specimens and plotted many strain-to-failure diagrams which are essentially equivalent to the forming limit diagram if you assume the instability point and the failure strain are close approximations to one another. For ductile titanium, our data would seem to indicate in both quadrants (less than zero and greater than zero) that forming is more closely approximated by the flow theory. This is also true for beryllium but there the failure is more of a brittle failure than ductile failure, particularly if a more biaxial (1 to 1) strain ratio is imposed.

P. B. Mellor (University of Bradford, U.K.)

I think we are making great progress in learning how ignorant we are. I must, I'm afraid, complicate the situation even more. We've just completed some tests on brass using the Marciniak technique and we get a rising curve from plane strain to balanced biaxial tension which agrees fairly closely with the work which has been done in brass in that quadrant on curved specimens. So, I don't know how we determine that. We have been subjected to considerable discussion of strain rate during recent days, and I have just one query about the use of the expression for stress, strain and strain rate. Really it doesn't affect the result of this theoretical paper but when we're actually getting the N value to feed into there, how can we get the N value? Because the N value you get from your stress strain curve itself for aluminum or steel does depend on the speed of testing, in fact on the strain rate. So I'm not quite sure. . . .

Neale

Well, I think it's the other term in the expression (the m value for strain rate sensitivity) which indicates the way your stress-strain curve does vary with speed of testing.

Mellor

Yes, but on the experimental side, I've not really seen any great data from which the N and m values have been calculated.

John Hutchinson (Harvard University)

I'm sure some of the experimentalists here could answer this much better than I. My understanding is that there is a fairly standard series of tests and it's a series of tests precisely. You take a tensile specimen say, and pull it at some fixed strain rate. Then you repeat the test at slightly higher strain rates and do a sequence of such tests. Then on two log-log plots, you pick the N and the M which best fit the slopes of the corresponding log-log plots. I believe that's what done.

PLASTIC FLOW PROPERTIES IN RELATION TO LOCALIZED NECKING IN SHEETS

A. K. GHOSH

Rockwell International, Thousand Oaks, California

ABSTRACT

Considerable interest exists in understanding the levels of maximum useful strains achievable prior to localized necking (forming limits) and their dependence on the imposed stress-state. Marciniak's model of imperfection growth has successfully explained the rise in forming limit as the imposed strain-ratio (ϵ_2/ϵ_1) is increased from zero (plane strain) toward unity (balanced biaxial tension). Experimental studies on formability of various materials have, however, revealed basic differences in behavior, such as the "brass-type" and the "steel type", exhibiting respectively, zero and positive dependencies of forming limit upon the strain-ratio. Such results cannot be reconciled without proper attention to the details of strain hardening and strain-rate hardening behaviors of these materials, particularly as functions of strain and strain-ratio. A review of these properties for several materials will be presented in an attempt to show their importance on the necking behavior. Furthermore, the dependence of the patterns of behavior upon the mode of stretching (in-plane and punch stretching) is discussed. Some results of a Marciniak-type model of *material imperfection* are also considered.

INTRODUCTION

In this paper, we shall review some important experimental results on limiting strain levels in sheet metal forming. The theories advanced toward understanding and predicting such strains under complex loading conditions will be described. Of particular interest is the analysis of Marciniak and Kuczyński [1]. Calculations based on this analysis will be examined in the light of observations and more accurate description of material's plastic flow properties. The scheme of presentation will be somewhat chronological in nature, bringing in refinements in the characterization of flow behavior for better quantitative prediction. Finally some discussions on imperfection sensitivity will be presented for a number of different imperfection types.

References p. 311.

Forming Limit Diagrams—As early as 1963, a study of failure in biaxially stretched sheets by Keeler and Backofen [2] showed the existence of what is known today as the Forming Limit Diagram (FLD). The main discovery was that the largest principal strain before any localized thinning in a sheet increased as the degree of biaxiality increased. They had tested several materials including steel, copper, brass and aluminum sheets by stretching over solid punches (i.e., the principal strains ϵ_1 and $\epsilon_2 > 0$.) An aggregate of limit strain data, ϵ_1^* for these materials in the annealed condition exhibited a general upward trend with increasing ϵ_2. Although their data did contain suggestions of different failure patterns for different materials, it was not until 1970 that Azrin and Backofen [3] showed the existence of basically two different types of failure limits—one for steel exhibiting a rapid rise in limit strain with increasing biaxiality, and the other for brass with negligible dependence (Fig. 1). In addition, a decreasing forming limit pattern was shown for austenitic stainless steel; however, this behavior was complicated, according to the authors [3], by martensitic transformation occuring

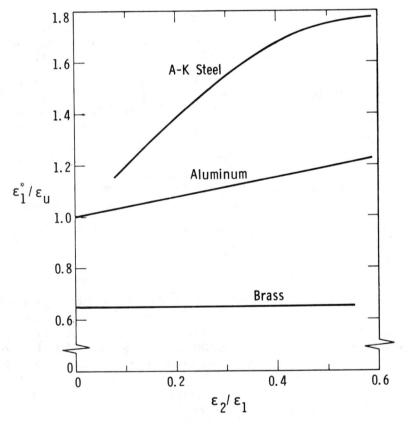

Fig. 1. Limit strain, ϵ_1^*, for in-plane stretching normalized with respect to the instability strain in uniaxial tension, ϵ_u, is plotted as a function of strain ratio, ϵ_2/ϵ_1, for three materials—aluminum-killed steel, 1100 aluminum and 70-30 brass. (Data from [3] and [4].)

during deformation. FLD for aluminum was determined later on and was found to fall in between the behaviors exhibited by steel and brass [4].

Azrin and Backofen's experiments involved stretching of flat sheets with no punch contact and, therefore, were free from the friction and curvature effects present in the previous work. (It will be subsequently shown that these factors do have a pronounced influence on the forming limit diagram). Their main conclusion was that the forming limit diagrams for different materials differed not only in magnitude but also in general character. Subsequently, punch-stretch experiments by Hecker [5] showed that different aluminum alloys have different FLDs, some of which are about half the level of that for low carbon steel (Fig. 2). In these diagrams both positive and negative values of minor strain are included and the forming limit is found to decrease rather sharply from the high values in uniaxial tension toward low values in plane strain ($\epsilon_2 = 0$). This behavior is shown universally by all materials, even though the slope of the curve might change somewhat.

Understanding the Basic Behavior—*Flat Sheet Stretching*—For the negative minor strain regime Hill [6] explained how, during uniform deformation of a sheet, a

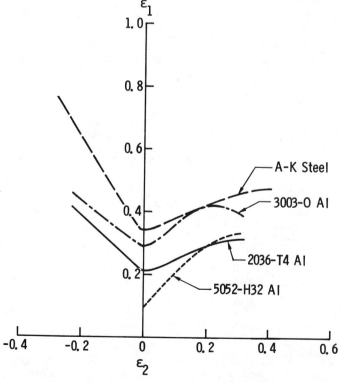

Fig. 2.   Forming limit diagrams for a number of aluminum alloys (Aluminum Association of America designations) compared with that of aluminum-killed steel. (From Hecker [5].)

References p. 311.

localized shear zone can develop along a direction of zero-length-change ($d\epsilon_2 = 0$). This happens when the plastic work increment within the zone becomes less than that for uniform deformation. On this basis the forming limit would change from a value of $2n$ in uniaxial tension to n in plane strain, where n = strain-hardening exponent in the parabolic stress-strain law: $\sigma = K\epsilon^n$. Hill's theory did not, however, explain localized necking for positive minor strains, since a direction with $d\epsilon_2 = 0$ does not exist in the sheet plane under these conditions.

A major development in the understanding and predicting the levels of FLD for $\epsilon_2 > 0$ was made by Marciniak and Kuczyński [1] (the M-K model.) They postulated that strain non-uniformity can develop from local regions of weakness in the material (Fig. 3). For convenience, the weakness was represented by a linear thinned region (trough) perpendicular to the direction of major principal strain. Imposing the same ϵ_2 inside and outside it while proportional straining continued outside, these regions were shown to deform faster than the rest of the sheet (i.e., with increasing accumulation of ϵ_1 within them). Eventually a condition of plane strain ($d\epsilon_2 = 0$) can develop if the homogeneous region of the sheet stops deforming before the weak region reaches fracture strain. However, the attainment of plane strain in a macroscopic sense is not a necessary condition when fracture process terminates useful flow.

In any event, strain accumulated in the homogeneous region is called forming limit. Because the terminal condition is near plane strain, a greater change in strain-state is required when initial strain paths are further from it, thereby accumulating larger strains before failure. This explained why forming limit increased on either side of plane strain state. Additionally, the analysis was capable

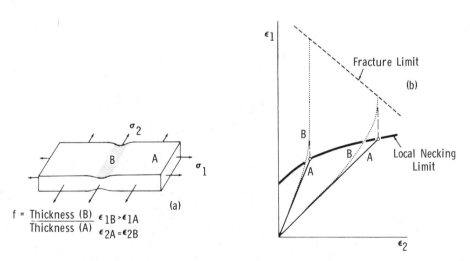

Fig. 3. (a) An imperfection (long thinned region, B) bounded by homogeneous material, A, undergoing biaxial stretching according to Marciniak and Kuczyński's model [1]. (b) Proportional strain paths in region A (solid line) and those in region B (dashed line) are shown in biaxial strain space. When region B reaches fracture limit (or plane strain state), the strain accumulated in region A defines the forming limit.

of predicting forming limit diagrams (for $\epsilon_2 > 0$) that are independent of stress-state (or even decreasing with increasing biaxiality) when large imperfection size and/or low fracture strain were used in the analysis.

In the M-K model the presence of an imperfection is thus necessary to explain failure in biaxial stretching. The types of possible imperfections and the details of their growth can, however, have large flexibility. Van Minh, Sowerby and Duncan [7] have assumed that a material containing voids deforms like a thinned region. Yamaguchi and Mellor [8] have associated thickness variation with grain size effect and used M-K analysis to calculate FLDs as a function of grain size. Several other types of imperfections will be treated later on in this paper.

Recently, Stören and Rice [9] have taken quite a different approach to explain localized necking during biaxial stretching. In contrast to the smooth yield surface assumed in the M-K model, these authors proposed that an instability in the constitutive relationship, such as a vertex on the yield surface, can develop during deformation. The treatment follows from Hill's theoretical development [10], which predicts the development of a yield vertex as a natural consequence of deformation and explains bifurcation from a state of uniform deformation occuring when the vertex becomes sufficiently sharp to satisfy a plane strain condition. This approach thus assumes that fracture is always preceded by localized necking on a microscopic scale. By considering an elastic-plastic, strain hardening material, Stören and Rice have shown that an increase in the strain hardening exponent can change the shape of FLDs from one that increases with biaxiality ratio (ϵ_2/ϵ_1) to that of a decreasing type.

Punch Stretching—Both analytical models discussed above should relate to deformation in a flat sheet and must be applied with caution to stretching of sheets over rigid punches such as in actual production, or even in hydraulic bulging. The reason for this is that these deformation processes give rise to strain gradients in the sheet, either through the combined frictional and curvature effects as in punch stretching, or due to constraints imposed at the boundary of the deforming sheet as in hydraulic bulging. As long as a strain peak develops and progressively sharpens during deformation, it is not necessary to invoke either a constitutive instability or a material weakness to understand localized necking.* In fact, with smooth yield surface and no imperfection, force calculations based on experimentally determined strain history show that at a certain stage during punch stretching all material elements near the pole area of the dome stop deforming, and strain concentrates within a narrow zone [11].

Experimental Variations—In support of the discussion related to punch stretching and in-plane deformation of flat sheets, the experimental FLDs obtained by these methods are presented in Fig. 4 for three test materials [12]. In in-plane stretching, a reduced gage section is stretched in a flat plane without punch contact. The

---

* This does not, however, mean that such phenomena do not occur. During strain localization it is possible that either imperfection growth, or vertex formation, or fracturing process can actually terminate useful flow.

References p. 311.

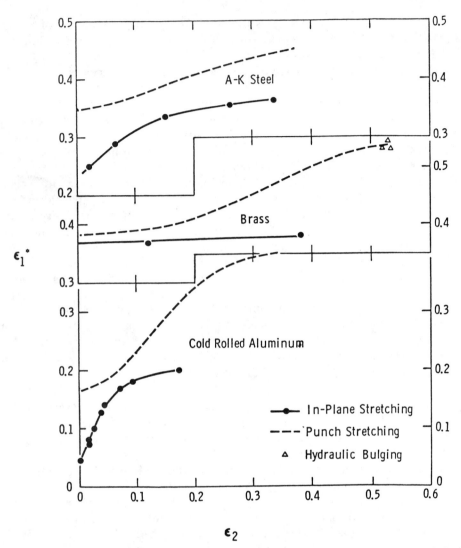

Fig. 4.   FLDs from in-plane and punch stretching for three different test materials: A-K steel, 70-30 brass and cold-rolled aluminum (from [12]).

section geometry is altered to obtain different ϵ_1 and ϵ_2. In punch stretching, clamped sheets are stretched directly over a hemispherical punch. Decreasing friction produces greater ϵ_2—a solid, dry punch is used to obtain the plane strain end while a rubber punch is used near balanced biaxial tension. Bending strain component and normal pressure are rather small in these experiments and cannot explain the large increase in FLD arising out of punch contact. The increment is different for different materials and contains some influences of material flow properties as well.

More detailed strain measurements show that the development of strain peak

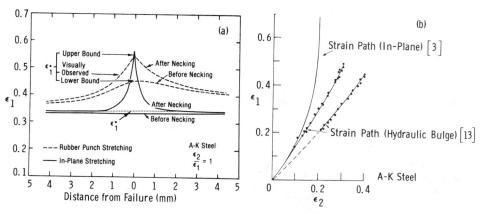

Fig. 5. (a) Distributions of major strain in the vicinity of failure showing a gradual strain peak for rubber bulging and a sharp one for in-plane stretching. The latter one seems to occur abruptly within a narrow region from a state of uniform strain. (b) Strain paths (ϵ_1 vs. ϵ_2) of the failure location in in-plane stretching (from [3]) and hydraulic bulging (from [13]), showing an early departure from proportional straining in the case of in-plane stretching.

is rather abrupt for in-plane stretching [12], which is further associated with a sharp change in strain path [13] (Fig. 5). Punch stretching and hydraulic bulging of sheets, on the other hand, promote a slower change in strain path, and tend to raise the FLD. A reexamination of Azrin and Backofen's in-plane strain measurements reveals that appreciable deviation from proportional loading occurs even outside the failure site (see Fig. 9 in [3] for example). This can be caused by a lack of strong external constraints necessary to enforce proportional loading in a flat plane.* Therefore, calculated FLD based on proportional loading would be expected to be larger than in-plane FLD and might even approach punch FLD. This possibility should exist at least in the neighborhood of balanced biaxial tension where hydraulic bulging or rubber punching produces very small strain gradients.

Furthermore, during punch stretching, friction retards the deformation over the punch and causes a thinning front to progressively pass through different material elements in the sheet [11]. As a consequence, strain cannot concentrate at one imperfection site in a continuous manner as it can in a flat sheet. This is schematically illustrated in Fig. 6 in terms of the M-K concept of imperfections. Assuming periodic distribution of weak and strong regions in the sheet and superimposing their individual strain distributions (greater thinning in the weak regions) on the normal strain distribution due to the punch, it becomes clear that as this latter peak passes through elements 1, 2 and 3, it can enforce excess

---

* In in-plane experiments, there is a strain-state variation from the specimen shoulder (near plane strain) toward the center of the gage section. As more elements near the shoulder unload (or their strain rates approach zero), a state closer to plane strain begins to spread into the gage section. An analogue of this effect for tensile specimens has been shown in [14].

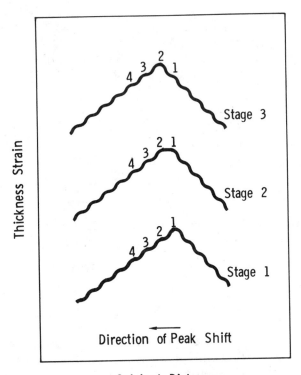

Fig. 6.   A hypothetical illustration of the small strain nonuniformities from material imperfections superimposed on the large wave length peak due to frictional (or constraint) effects. The movement of the latter peak through stages 1, 2 and 3 brings different strong and weak regions to rest at the peak location.

deformation in the stronger areas as well. This can retard the development of a thinning difference between adjacent weak and strong regions and allow a larger limit strain. However, this phenomenon can operate only as long as the strain peak can shift. For the state of plane strain the peak cannot shift after the force maximum, since thinning concentrates essentially in one element (unless the material possesses a positive strain-rate sensitivity). Rate-sensitivity promotes more even strain distribution within a gradient and resists localization [11]. More will be said about this subject later.

COMPARISON BETWEEN THEORY AND EXPERIMENT

Marciniak's original model used an empirical strain hardening law:

$$\bar{\sigma} = K(\epsilon_0 + \bar{\epsilon})^n \tag{1}$$

where $\bar{\sigma}$ = effective stress, $\bar{\epsilon}$ = effective strain, ϵ_0 = a prestrain term, n = strain-hardening exponent and K = strength constant. The imperfection was assumed to be a thickness variation or a variation in K times thickness. Imperfection size f, was thus defined as $\Delta(Kt)/Kt$ where t = thickness. The analysis basically consists of satisfying force equilibrium along the major principal direction (for equal ϵ_2) between the imperfection and outside. Using a straightforward numerical procedure the computation is carried out until the strain outside the imperfection does not appreciably change. (For details on the analysis, see [1]). Fig. 7a shows calculated FLDs for aluminum-killed steel (for two realistic values of f) and compares them with the experimental results. Fig. 7b compares the predictions of Stören and Rice [9] and Hill [6] with the same experimental data.

The interesting fact about the calculated FLDs in Fig. 7a is that the plane strain value is always less than $n - \epsilon_0$ ($\epsilon_0 \simeq 0$ for steel), while the experimental (both in-plane and punch stretching) values are greater than $n - \epsilon_0$. The curve slope can be reduced by using larger values of f, but, since this also reduces the plane strain value substantially, it can be concluded that calculated FLDs are too low near plane strain ($\epsilon_2 \simeq 0$) and too high near balanced biaxial tension ($\epsilon_2 \simeq \epsilon_1$). Also the predicted slopes on both sides of $\epsilon_2 = 0$ are too high.

While the predicted FLDs in Fig. 7a represent the attainment of plane strain within the neck, it appears that, in reality, fracture terminates flow in the vicinity of balanced biaxial tension since the fracture strain for A-K steel [15], [16] is found to be lower in this region. In the absence of a suitable model for fracture, this experimental fracture strain is used as a terminal condition for the neck. The predicted FLD now approaches the punch FLD in this region. Experiments with hydraulic or rubber bulge also produces a fracture-like failure, with little visible necking. Calculations in [3] based on M-K model suggested that a lower value of plastic anisotropy parameter*, r, can raise the FLD near balanced biaxial tension. However, if fracture terminates flow in this region, the influence of r may not be verified unless the dependence of fracture strain on r is also known. Fig. 7b shows that vertex theory predicts a plane strain limit of $n(\epsilon_0=0)$. The predicted FLD is somewhat lower in comparison to the experimental ones, even though it gives the correct trend for $\epsilon_2 > 0$. For $\epsilon_2 < 0$, however, it bends back toward the pure shear state ($\epsilon_2 = -\epsilon_1$) leading to an increasing discrepancy. Hill's prediction, also shown in Fig. 7b is found to be 10–20% lower than the experimental FLD for the range of $\epsilon_2 < 0$ shown here.

Figs. 8a and b show similar results for 70-30 brass. The value of $n = 0.56$ used here is the best linear fit to the log $\bar{\sigma}$ vs. log $\bar{\epsilon}$ data for uniaxial tensile test. While this does not describe both small and large strain ranges adequately ($n = 0.68$ with $\epsilon_0 = 0.06$ as in [3] may be a better description) the conclusions we will draw are not influenced by it. Fig. 8a shows that the predicted plane strain limit is less than n, but in contrast to the case of steel it is considerably greater than the experimental limit. Again, the calculated FLD rises rapidly on either side of plane strain which is also in disagreement with the experimental FLDs. However, when

---

* This parameter enters the analysis through its association with the flow rule, and the effective stress and strain relations in plasticity. Hill-modified effective stress and strain formulations are used (see [1] for details).

References p. 311.

experimental fracture strain limit is incorporated into the analysis, a closer agreement with the punch FLD (near balanced biaxial tension) is obtained.

Fig. 8b shows the predictions of the vertex theory [9] for $n = 0.50$ and Hill's prediction [6] for $n = 0.56$. Hill's prediction is substantially greater than the measured values for $\epsilon_2 \leq 0$. Yield vertex theory also predicted a plane strain limit of n; however, the FLD for $\epsilon_2 > 0$ is predicted well by their analysis, and particularly so for the in-plane FLD. For negative ϵ_2, however, Stören and Rice predictions deviate from both Hill's prediction and experimental data, and approach a lower value in pure shear.

Even though some appropriate trends are predicted by the bifurcation analysis

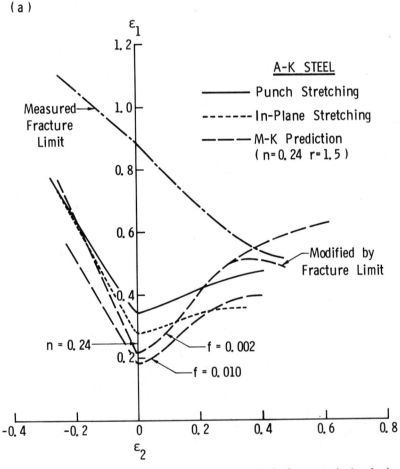

Fig. 7. (a) FLDs calculated according to M-K analysis for a strain hardening model with imperfection sizes of 0.002 and 0.010 and compared with the experimental results on A-K steel. Experimental fracture limit curve (from [16]) assists a better prediction. (b) Predictions from Hill [6] and Stören and Rice's yield vertex model [9] are also compared with experimental FLDs.

(b)

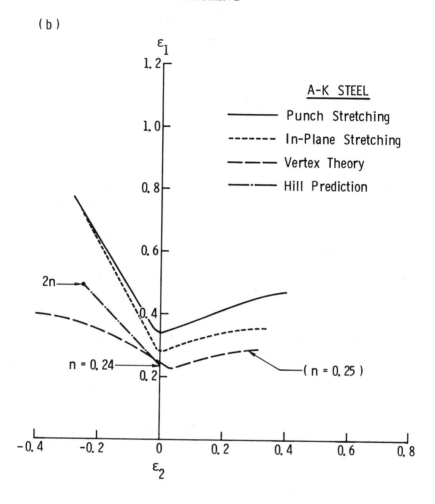

of Stören and Rice it is unclear why the predicted FLD trends for $\epsilon_2 < 0$ are so drastically different from the experimental ones. Furthermore, this analysis predicts that the angle of localized neck in a tension test measured with respect to the tensile axis would decrease with increasing strain-hardening exponent, n. This is not validated by experiments. In fact, brass with a very high n (~0.56) and cold-rolled aluminum with a very low n (~0.04) exhibit local necks at almost 90° to the tensile axis, while A-K steel with an intermediate n (~0.24) forms necks at about 50° to the tensile axis. Diffuse necking cannot be held responsible for this since steel shows a great deal of diffuse necking while brass and aluminum do not. A correlation between the angle of neck and r-value has been observed experimentally—the angle increasing with decreasing values of r($r_{\text{steel}} \sim 1.5$, r_{brass} ~ 0.9, $r_{\text{Al}} \sim 0.6$). Such behavior is, however, expected from normality conditions for Hill's anisotropy-modified smooth yield surface. Finally, the fracture-like failure in balanced biaxial tension cannot be rationalized by the development of

References p. 311.

a yield vertex. Thus, even though vertex formation has been experimentally observed in certain cases, it could not be concluded to be a preferred mechanism for localized necking without further experimental and analytical support.

Refinements in the Constitutive Relations—*Rate Sensitivity*—Considerable help in explaining the FLD levels is derived from a better description of the plastic flow properties. The importance of even a small positive strain-rate sensitivity was explained when Marciniak et al. [15] incorporated strain-rate sensitivity through a simplified law:

$$\bar{\sigma} = K(\epsilon_0 + \bar{\epsilon})^n \dot{\bar{\epsilon}}^m \tag{2}$$

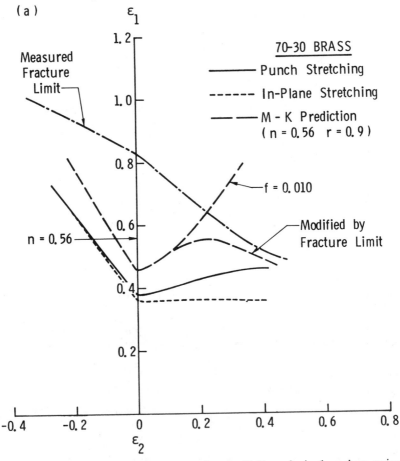

Fig. 8. (a) Partial FLD calculated according to M-K analysis (based on uniaxial $\bar{\sigma}$–$\bar{\epsilon}$ data for brass) for an imperfection size of 0.010 is compared with the experimental FLDs for brass. (b) Predictions from Hill [6] and Stören and Rice's yield vertex model [9] are also compared with experimental FLDs.

(b)

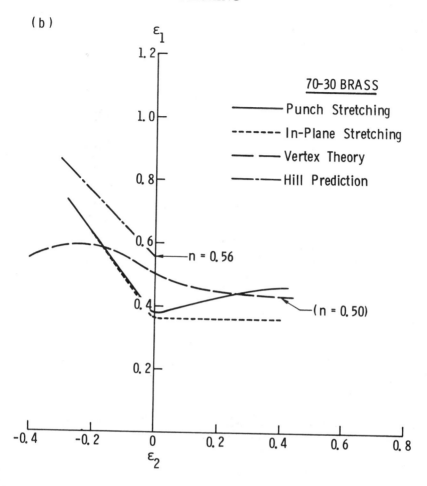

into their analysis, where $\dot{\bar{\epsilon}}$ = effective strain rate and m = strain-rate sensitivity index. While the small $m(\sim0.012)$ for steel was ignored before this time, calculations including this are found (Fig. 9) to bring the predicted FLDs in close agreement with experimental ones. It should be noted that the agreement is better for the gradient-free and proportional straining regions (i.e., plane strain region for the in-plane FLD and the region near balanced biaxial tension for the punch FLD).* The importance of such small m on the capacity for large diffuse necking elongations in a tensile test and cup depths in a punch test has also been established [14], [17]. It is realized now that nearly one half of steel's excellent formability arises due to this effect.

---

* As mentioned in the section on experimental variations, the presence of strain-rate sensitivity allows A-K steel to take advantage of the strain peak shift during punch stretching to elevate the FLD (near plane strain) in comparison to in-plane stretching. Since thickness gradient aids this effect, the punch FLD is also raised by an increase in sheet thickness. Such effects are miniscule for brass because its strain-rate sensitivity is zero.

References p. 311.

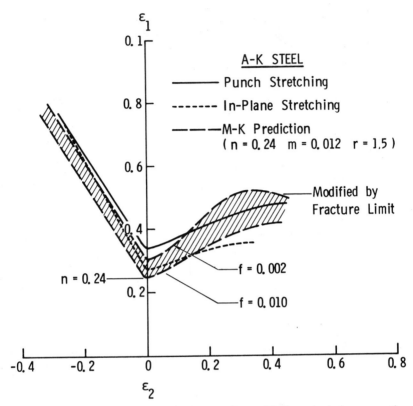

Fig. 9. FLDs calculated for A-K steel according to M-K analysis incorporating a rate-sensitive constitutive law (equation (2)). A small *m* of 0.012 for A-K steel is used in the simulation.

Equation (2) may not be the best constitutive description for the sheet stamping materials, where an overstress model [18] might describe the behavior somewhat better. An equation of this type, used satisfactorily in the modeling of the tensile test is given by:

$$\bar{\sigma} = K(\bar{\epsilon} + \epsilon_0)^n + \Delta\sigma \qquad (3a)$$

where the overstress $\Delta\sigma$ is a function of strain-rate alone

$$\Delta\sigma = Km' \ln(\dot{\bar{\epsilon}}/\dot{\bar{\epsilon}}_0) \qquad (3b)$$

and where m' = strain-rate sensitivity index and $\dot{\bar{\epsilon}}_0$ is a reference strain rate at which rate-sensitive part of the flow stress is negligible. This kind of constitutive equation would be expected on the basis of thermally activated motion of dislocations, for which strain hardening forms the athermal part and overstress the thermally activated part of the total flow stress. For the small strain-rate sensitivity observed in steel, however, the results are not too strongly influenced by either equation (2) or (3), unless temperature, strain and strain-rate regimes are altered such that the behavior changes appreciably.

Biaxial Hardening—In the case of brass it is surprising that the plane-strain limit is considerably less than n. While rate-sensitivity is zero for brass, accurate measurement of uniaxial strain hardening shows that the instantaneous slope of log $\bar{\sigma}$ vs. log $\bar{\epsilon}$ curve (i.e. instantaneous n) falls to as low as 0.47 before maximum load in tensile test. This is shown in Fig. 10. Since strain hardening at large strains is of interest in the prediction of necking, $n = 0.47$ may be a better approximation than the best fit value of 0.56. However, this strain-dependence and the low plane-strain limit led us to question whether the strain hardening behavior for brass was invariant with stress-state. In the following paragraphs we will present several experimental evidences to the contrary.

Fig. 11 shows the effective stress-strain curves for brass measured from three different tests: (i) uniaxial tension ($\epsilon_2/\epsilon_1 \sim -0.5$), (ii) plane strain tension ($\epsilon_2/\epsilon_1 \sim 0$), and (iii) a punch stretch test ($\epsilon_2/\epsilon_1 \sim 0.10$). Sample geometry for plane strain tension is shown in the figure. Incremental deformation of many individual samples produced the aggregate of data points (see [4] for details). The set-up for punch test is also sketched here. Again incremental loading of many samples was used. Major and minor strains in the sheet at the boundary of punch contact were read from a photogrid previously printed on the sheet. From measurements of the load (P) and the diameter (d) and the average sheet thickness (t) at the punch contact boundary the meridional stress, σ_r, is calculated from the equation

$$\sigma_r = P\rho/\pi d^2 t \tag{4}$$

where $\rho =$ punch radius. Strain ratio (ϵ_2/ϵ_1) in this test was maintained near 0.1, while that in the plane strain test was near zero, remaining approximately constant during the test. The plane strain data points are shown by open circles and punch test results by closed circles. Since all these data fall together, a single curve is drawn through them.

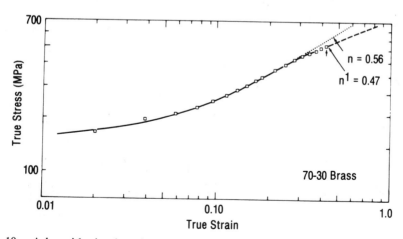

Fig. 10.   A logarithmic plot of true stress vs. true strain from uniaxial tension test of brass exhibits a drop in instantaneous n from 0.56 to 0.47 at large strain.

References p. 311.

A. K. GHOSH

Significant differences in the stress level and curve slope are observed between the uniaxial and biaxial tests in Fig. 11. While stress differences are more when the curves for the major axial direction are considered, the slope differences remain even after plotting the effective stress-strain curves.* Load calculations in deformation processing are not significantly influenced by such finding, however, the slope differences have serious implications on necking instability. Results similar to this were obtained earlier by prestraining sheets in biaxial tension and subsequently performing tensile tests [4].

The absence of an invariant strain hardening curve suggests that isotropic hardening model is not really justified, and perhaps a nonconventional yield function needs to be employed. However, from an engineering standpoint it may not be unreasonable to adjust the empirical hardening law while maintaining conventional yielding theory. Again best fit to log $\bar{\sigma}$ vs. log $\bar{\epsilon}$ plot for these data indicate $n = 0.43$. This value is substantially lower than the best fit uniaxial n of 0.56 and even less than the terminal uniaxial n of 0.47.

A number of authors have found similar departures from the uniaxial stress-strain curves for brass [19]–[21], copper [20], [21] and several aluminum alloys [19], [20], [22] when tested in biaxial tension. They have performed hydraulic bulge tests on circular sheets clamped at the periphery ($\epsilon_2/\epsilon_1 \sim 1.0$). Biaxial extensometers developed by Duncan and Johnson [23] have facilitated the automatic recording of the bulge curvature and strain in the sheet. From these and the records of bulge pressure $\bar{\sigma}$–$\bar{\epsilon}$ curves can be calculated quite routinely [22]. This test is not only capable of producing stress-strain information in biaxial tension but it can do so up to a larger strain, since necking does not occur until an effective strain of about ~0.80. The stress-strain behavior at these large strains is therefore of interest in the necking analysis.

Fig. 12 shows the instantaneous value of n as a function of $\bar{\epsilon}$ for brass from Yoshida et al [20]. Both uniaxial and biaxial n are found to change as a function of $\bar{\epsilon}$. As in Fig. 11, initial hardening is greater in biaxial tension, however, biaxial n soon drops below that of uniaxial and continues to drop to rather low values with continued deformation. It is quite possible that in the neighborhood of the limit strain, an ideally plastic state would be approached, where even the slightest imperfection can trigger a neck. Lack of space does not permit showing more data here; however, the evidence is rather strong (see [4], [20], [22] for example)

---

* The effective stress and strain values are obtained from Hill's equations for anisotropic plasticity, with the use of associated flow rule to assess the stress-ratio from strain-ratio. These equations are:

$$\bar{\sigma} = \sqrt{\frac{3(1 + r)}{2(2 + r)}} \{1 + (\sigma_2/\sigma_1)^2 - 2r(\sigma_2/\sigma_1)/(r + 1)\}^{1/2}\sigma_1 \tag{5}$$

$$\bar{\epsilon} = \sqrt{\frac{2(1 + r)(2 + r)}{3(2r + 1)}} \{1 + (\epsilon_2/\epsilon_1)^2 + 2r(\epsilon_2/\epsilon_1)/(r + 1)\}^{1/2}\epsilon_1 \tag{6}$$

and

$$\sigma_2/\sigma_1 = \frac{(1 + r)(\epsilon_2/\epsilon_1) + r}{1 + r + r(\epsilon_2/\epsilon_1)} \tag{7}$$

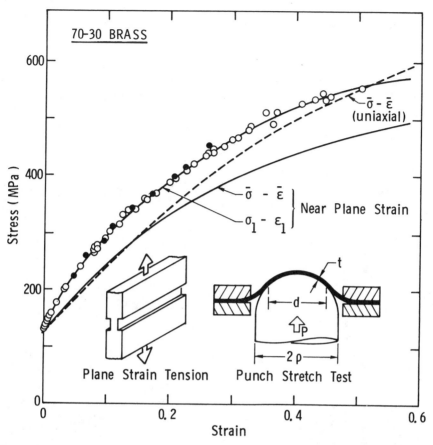

Fig. 11.   Axial stress-strain data points from two biaxial tests on brass: (1) plane strain tension ($\epsilon_2/\epsilon_1 \sim 0$), indicated by open circles, and (ii) punch stretch test ($\epsilon_2/\epsilon_1 \sim 0.1$), indicated by solid circles. The effective stress-strain curves drawn from these data exhibit substantial difference in the rate of hardening from that of uniaxial tension.

that biaxial n continues to drop with strain well below the uniaxial value for brass, copper, Al-Cu alloy, Al-Mg alloy, austenitic stainless steel, etc.

A thorough characterization of the strain- and stress-state dependence of strain hardening behavior is required for accurate predictions of FLD and the strain distributions in forming. Furthermore, in order for a new theory for deformation and yielding to be successful it must incorporate such effects. Realistic models of fracture also need such input. However, before such things could be accomplished it may be worthwhile to repeat the M-K analysis for brass with at least the plane strain n-value of 0.43. The effect of this partial correction is shown in Fig. 13. The plane strain region of the predicted FLD now shows good agreement with the experimental one. As discussed before, punch stretching and in-plane stretching produce about the same plane strain limit for brass due to the absence of strain-rate hardening. The predictive capability of M-K model is thus good

References p. 311.

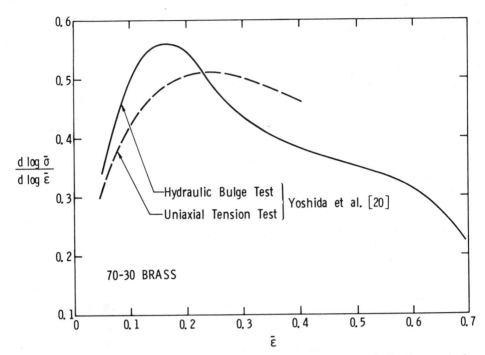

$\dfrac{d \log \bar{\sigma}}{d \log \bar{\varepsilon}}$

Fig. 12.   Instantaneous value of $n(d \log \bar{\sigma}/d \log \bar{\varepsilon})$ as a function of effective strain for uniaxial tension and hydraulic bulge test ($\epsilon_2/\epsilon_1 \sim 1.0$) (from Yoshida et al [20].)

when realistic constitutive relations, such as biaxial strain hardening, rate-sensitivity and measured fracture strains are used. Even though no attempt has been made here to predict fracture, which has been introduced only in an empirical way, the overall trends are certainly encouraging.

IMPERFECTION SENSITIVITY

A major issue that has not yet been addressed is the strong imperfection sensitivity predicted by the M-K model, but not supported by experimental observations. Of particular interest is the work of Azrin and Backofen [3] who studied the effect of small grooves of various depths on the in-plane stretching of steel. They concluded that the groove depth had to be at least 1.3% of the sheet thickness ($f = 0.013$) before the FLD is affected by its size. The M-K model predicted, on the other hand, that the FLD would drop very significantly if f is raised from 0.001 to 0.01. In this section we will show that this can be explained on the basis of a level of imperfections of metallurgical origin pre-existing in the material.

Material Imperfections—A material imperfection is defined as a local variation in flow properties or constituents that can influence the flow and fracture properties of a material. These could originate from small variations in thermomechanical processing history or chemistry of the metal and be microscopically present as local variations in grain size, grain preferred orientation, amount of residual cold work, inclusions and second phase content, as well as variations in their size and spacing. They can manifest as variations in the plastic flow parameters, K, n, m, r, ϵ_0, and yield strength. A variation in inclusion density can cause differential microvoiding and weakening before failure. Needleman and Triantafyllidis [24] have recently made some interesting calculations based on a void growth model that predict "brass" and "steel" type behaviors (for in-plane stretching). One of their results shows that nearly all of the void growth occurs within a few percent of the limit strain, which may be the reason why metallographic examination of necked samples does not reveal extensive voiding. However, neither the punch FLD (near balanced biaxial tension) nor the observed fracture limits can be

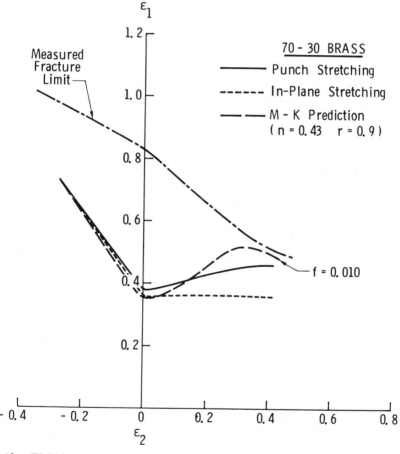

Fig. 13.   FLD for brass recalculated on the basis of the plane strain n-value of 0.43 by the use of M-K analysis and compared with experimental data.

References p. 311.

explained on the basis of their model. It may be necessary, therefore, to refine the model by incorporating the effects of void coalescence.

Fig. 14 schematically defines several types of material imperfections in terms of their effects on the log $\bar{\sigma}$–log $\bar{\epsilon}$ curves. Parabolic hardening is used here for simplicity. The inequalities in plasticity parameters for the homogeneous (A) and imperfect (B) regions are chosen such that the imperfection is the weaker site. K-type imperfections give an increasing true stress differential between regions A and B with increasing strain, and are therefore rather detrimental. Their effect is identical to that of thickness imperfections. n-type imperfections can be viewed in, at least, two different ways. For equal K inside and outside the imperfection, the imperfection should have a higher n (as in (ii)), while for equal yield strength (as in (iv)) the imperfection should have a lower n. In terms of the plasticity parameters we have discussed so far, this last imperfection type is a combination of n-type and K-type acting "additively" to cause the flow stresses to diverge, unlike (ii). This imperfection is therefore more detrimental than even the K-type.

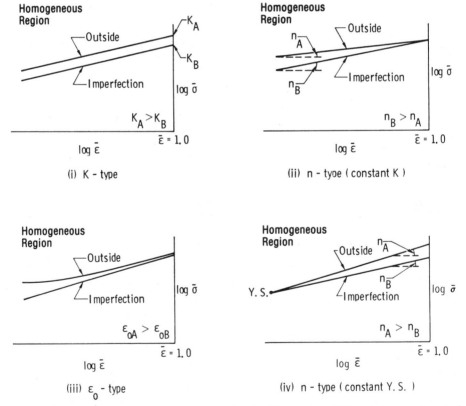

Fig. 14.   Schematic log $\bar{\sigma}$ vs. log $\bar{\epsilon}$ curves for illustrating different types of material imperfections.

A prestrain-type imperfection (type (iii)) demands a larger prestrain outside the imperfection site than the one inside to render the outside stronger. However, the effect of such an imperfection is reduced with increasing strain. An imperfection in m according to equation (2) is also similar to K-type except that m_B needs to be greater than m_A, while according to equation (3) $m_A' > m_B'$. In either case, they are of smaller magnitudes than K-type imperfections. While real material imperfections could be mixtures of all these types, we shall discuss the effects of only single imperfection types for simplicity. Combined types of imperfection may produce more detrimental effects than its single constituents if they act in an "additive" manner, or a beneficial effect if they act in a "subtractive" manner.

Limit Strains for Material Imperfections—Fig. 15 shows the theoretical influence of several single (material) imperfection types on the plane strain limit of A-K Steel in the gradient-free case. Imperfection size, f, is defined as $\Delta x/x$, where x denotes the parameter of interest. K-type or thickness imperfections are found

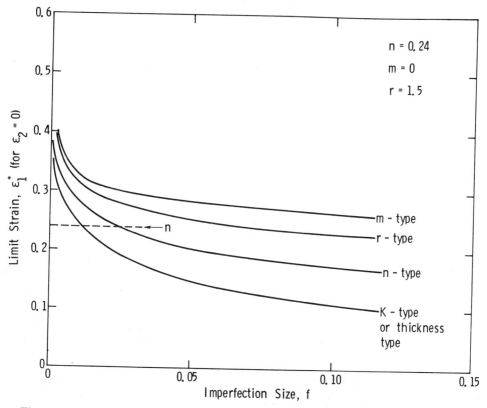

Fig. 15. Calculated influence of imperfection size on the plane strain limit of A-K steel for different types of material imperfections.

References p. 311.

to be the most detrimental; however, even here f has to be at least 0.01 before limit strain can drop below n. The next worse imperfection shown here is the n-type, which requires an $f > 0.025$ to reduce the limit strain below n. Imperfections in r and m are rather mild and permit relatively high FLDs even with $f > 0.05$. Although the influence of prestrain imperfections are not shown here, one could guess that it would be milder than an n-type imperfection. This would also be true of an imperfection which constitutes a constant flow stress differential between the two regions (in their $\bar{\sigma}$–$\bar{\epsilon}$ plot). Furthermore, if an imperfection exists in that parameter which is mostly responsible for the ductility of a material, it constitutes a severe imperfection. For example, an n-type imperfection is more severe than an m-type for steel $(n \gg m)$, while the reverse is true for a material whose $m > n$.

On the basis of this modified M-K analysis, the observed in-plane strain limits can now be rationlized as arising from an $f = 0.006$ if the imperfection is K-type, or an $f = 0.012$ if it is n-type, or $f = 0.03$ if it is r-type, or even $f = 0.07$ if it is m-type. Imperfection sizes of this magnitude are quite likely to be present in the as-received material. As a further exercise of predictability from n-type imperfections, the entire steel FLD is shown in Fig. 16. The in-plane FLD is predicted

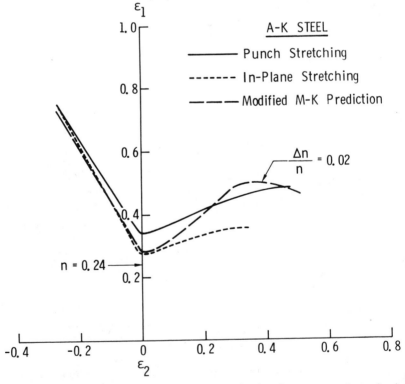

Fig. 16. FLD for A-K steel calculated on the basis of an n-type imperfection and compared with experimental results.

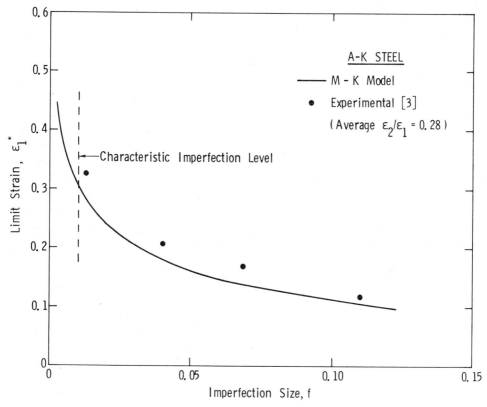

Fig. 17. The calculated dependence of limit strain (for $\epsilon_2/\epsilon_1 \sim 0.28$) on thickness imperfection size, compared with experimental results on A-K steel from [3].

well near plane strain, while the agreement with the punch stretching FLD is good near balanced biaxial tension. For previously explained reasons, this agreement is excellent considering the fact that real material imperfections may be considerably more complex.

The measured FLD level can thus be associated with a characteristic level of material imperfection (see Fig. 17) representing the largest of such imperfections within a specimen. Azrin and Backofen's experiment on the effect of groove depth can now be re-examined in this light. Fig. 17 shows their experimental data and M-K prediction on limit strain (for $\epsilon_2/\epsilon_1 \sim 0.28$) as a function of thickness imperfection size. The data point for the smallest f (~ 0.013) does not represent a drop in the limit strain, while the others do. For a machined groove to act as an imperfection, either it has to be larger than the characteristic level of material imperfections, or it has to be situated exactly at the site of the largest material imperfection. The probability for the latter is rather low.* It is thus reasonable

---

* Even if this were to happen, a small amount of cold work introduced during machining can reduce the severity of such an imperfection.

References p. 311.

that the smallest imperfection did not influence the FLD level. The f-dependence of limit strain for all the other data points appear to be predicted well by M-K model in view of the crudeness of both analysis and experiment. A possible reason for the higher experimental limit strains is a reduction in the effective f, due to (i) larger material-type imperfections outside the groove than inside, and/ or (ii) slight hardening of the groove from the cold work during machining.

The presence of a characteristic level of material imperfection is quite realistic, although to establish the actual imperfection size and type may require a considerable amount of materials testing. For example, Needleman and Triantafyllidis [24] have used $f = 0.01$ for an n-type imperfection (type (iv) in Fig. 14) to predict a "brass" type FLD. However, this f is equivalent to a combined n- and K-imperfection in our terminology and amounts to a rather large imperfection. Nevertheless, the discussion on material imperfections does provide a rational base and does demonstrate the potential for a modified M-K model in explaining the observed FLD levels. Future challenges include an understanding of the fracture process and the development of more accurate constitutive laws, both of which appear to be of extreme importance.

CONCLUSIONS

1. Reasonably good predictions of sheet metal forming limits are obtained by using Marciniak and Kuczyński's model of imperfection growth, provided an accurate constitutive law is used and imperfections of metallurgical origin are accounted for.
2. The presence of a small positive strain-rate sensitivity ($m \sim 0.012$) permits aluminum-killed steel to have a plane strain forming limit which is greater than n, and a uniaxial forming limit which is greater than $(r + 1)n$, where n = strain hardening exponent and r = plastic anisotropy parameter.
3. The relatively low FLD in stretching ($\epsilon_2 \geq 0$) for brass appears to be due to a decrease in its strain hardening capacity with both increasing biaxiality (ϵ_2/ϵ_1) and increasing effective strain ($\bar{\epsilon}$).
4. Relaxation of external constraints during deformation may be responsible for the early loss of proportional straining during in-plane stretching; by contrast, stronger external constraints permit a higher FLD in punch stretching. Strain peak shift further enhances punch FLD (over in-plane FLD) provided rate-sensitivity is positive.
5. Fracture strain influences the punch FLD near balanced biaxial tension ($\epsilon_1 = \epsilon_2$), while instability limits the flow for other strain-states.
6. The model of Marciniak and Kuczyński appears to predict the sensitivity to thickness imperfections reasonably well if the material is assumed to contain a certain level of material imperfections to start with, such as variations in K, n, m, r, yield strength, inclusion density etc.

ACKNOWLEDGMENTS

The author would like to thank Dr. S. S. Hecker and Mr. R. Pearce for their permission to publish some of their results. This work was performed under the Rockwell Independent Research and Development Program.

REFERENCES

[1] Z. Marciniak and K. Kuczyński, Int. J. Mech. Sci., 9 (1967), 609.
[2] S. P. Keeler and W. A. Backofen, Trans. ASM, 56 (1963), 25.
[3] M. Azrin and W. A. Backofen, Metall. Trans., 1 (1970), 2857.
[4] A. K. Ghosh and W. A. Backofen, Metall. Trans., 4 (1973), 1113.
[5] S. S. Hecker, J. Eng. Mater. Technol. (Trans. ASME, H), 97 (1975), 66.
[6] R. Hill, J. Mech. Phys. Solids, 1 (1952), 19.
[7] H. van Minh, R. Sowerby and J. L. Duncan, Int. J. Mech. Sci., 17 (1975), 339.
[8] Y. Yamaguchi and P. B. Mellor, Int. J. Mech. Sci., 18 (1976), 85.
[9] S. Stören and J. R. Rice, J. Mech. Phys. Solids, 23 (1975), 421.
[10] R. Hill, J. Mech. Phys. Solids, 6 (1958), 236.
[11] A. K. Ghosh and S. S. Hecker, Metall. Trans., 6A (1975), 1065.
[12] A. K. Ghosh and S. S. Hecker, Metall. Trans., 5 (1974), 2161.
[13] M. J. Painter and R. Pearce, J. Phys. D: Appl. Phys., 7 (1974), 992.
[14] A. K. Ghosh, Met. Trans., 8A (1977), 1221.
[15] Z. Marciniak, K. Kuczyński and T. Pokora, Int. J. Mech. Sci., 15 (1973), 789.
[16] A. K. Ghosh, Metall. Trans., 7A (1976), 523.
[17] S. S. Hecker, Met. Eng. Q., 14 (1974), 30.
[18] L. E. Malvern, J. Appl. Mech. (Trans. ASME) 18 (1951), 203.
[19] R. Pearce, Int. J. Mech. Sci., 13 (1971), 299.
[20] K. Yoshida, K. Yoshii, H. Komorida, M. Usuda and H. Watanabe, Scientific Papers, Inst. of Physical and Chemical Research, Tokyo, 64 (1970), 24.
[21] H. J. Kleemola and A. J. Ranta-Escola, Metall. Trans., 7A, (1976), 595.
[22] D. J. Lloyd, B. D. McLaughlin and H. Sang, Scr. Metall., 11 (1977), 297.
[23] J. L. Duncan and W. Johnson, Sheet Met. Ind., 42 (1965), 271.
[24] A. Needleman and N. Triantafyllidis, Brown University Report NSF-ENG76-16421/1, June 1977 (to be published in J. Eng. Mater. Technol.).

DISCUSSION

C. L. Magee (Ford Motor Company)

I'd like to thank Dr. Ghosh for bringing the experiments that people have talked about into better focus, except perhaps for the one neglected variable— the sheet thickness. That seems to not even appear on those forming limit diagrams despite all work showing its importance. I'd like to mention some recent results that we've obtained on aluminum which show in fact much smaller sheet gage effects on the FLD than those recorded for steels. I'm curious to know of any other measurements in other materials in which sheet thickness effects in particular have been studied. Also, what kinds of reasons might lead to these variations?

Ghosh

Do you mean, the variations in the effect of thickness on the level of the forming limit curve for different metals?

Magee

First of all, the effect of thickness on the forming limit curves and secondly, studies on different materials.

Ghosh

If you try to take into account sheet thickness, the assumptions of the plane stress analysis that we have done here breakdown, as has been pointed out before. What really needs to be included in any analysis concerned with the thickness effect is thickness strain and σ_3. If you do this, you find that the thicker sheets give a larger gradient in strain over a larger distance—distance over which you measure the strain is larger, and this gradient effect is again helped by the influence of strain rate sensitivity. Whenever there is a gradient in strain or strain rate, the effect of strain rate sensitivity will spread it even further. Therefore, I would expect that for the thicker sheet, because it has a gradient distributed over a larger distance, the measured forming limit will be somewhat higher because of this gradient effect. The effect will also be greater in the case of steel as compared to aluminum, which does not have any strain rate sensitivity. We have made some calculations on punch stretching including this gradient effect and we are able to predict (this a very crude calculation) as a function of increasing thickness, the data that Keeler has shown of the effect of gage on the plane strain FLD for AK steel. We can duplicate that fairly well.

Magee

But you need rate sensitivity to do so?

Ghosh

Yes.

J. L. Duncan (*McMaster University, Canada*)

I wanted to make a comment on the previous paper but the competition was too fierce. Amit Ghosh has introduced some experimental results and I've always found that there's nothing like fact to kill a good argument. My comment applies to both contributions, whether it's worthwhile to direct attention to forming on the far left hand side of the FLD that is to strain, for example, more in the shear region than you get in uniaxial tension. I mention this because practitioners of sheet metal forming do observe failures in this region, particularly in corners of square panels and things like that and there would seem to be quite a disparity between behavior of materials in that region.

SESSION V

ANALYSIS OF DEFORMATION
IN STAMPING OPERATIONS

Session Chairman
B. BUDIANSKY

Harvard University
Cambridge, Massachusetts

DEFORMATION ANALYSIS OF LARGE SIZED PANELS IN THE PRESS SHOP

H. ISHIGAKI

Toyota Motor Company, Japan

ABSTRACT

This paper describes experimental results of forming limits for steel sheets and deformation states in large-sized autobody panels together with examples of applying these results to analysis of actual press forming operations. The forming limit is defined as the amount of strain at which a pronounced roughening in surface finish occurs. This roughening signals the onset of necking as determined by means of stylus instrument measurement and visual observation. Based on this definition, forming limit curves were obtained for both simple and complex deformation paths. An example of using these data in the forming of a new quarter panel is given.

The deformation states in large-sized autobody panels are defined by classifying the measured strains in the formed panels into several deformation ranges depending on the magnitude and ratio of strains. The result is, for each panel, a map-like diagram (strain composition diagram) from which the forming severity of panels may be evaluated. Several examples of using such a diagram in actual forming operations to aid in understanding the relations among deformation defects, deformation ranges and die geometry are given.

INTRODUCTION

Forming analysis of large sized autobody panels has been made on both the most critical areas for avoiding fracture and on the whole formed panel for avoiding surface defects due to press forming (deformation defects) such as minute unevenness, wrinkles and springback. The strain states of the most critical areas are defined by using the scribed circle test developed in recent years [1], [2]. By comparing these strain states with the fracture limit strain obtained experimentally, it has become possible to make rough estimates of forming severity relating to fracture [3]–[5].

References pp. 337–338.

In addition, it has been necessary to investigate the forming limit for cases of complex deformation paths which are often encountered in actual press forming operations [6], [10]. Most laboratory fracture limit studies deal with simple deformation paths (constant strain ratio). Further, the occurrence of necking should be avoided in order to have a truly successful forming operation. It may be reasonable to assert that the amount of deformation at the onset of necking is a measure of the forming limit [11]–[14]. Experimentally, however, it is found difficult to detect the onset of necking accurately. In this paper, the onset of necking is defined as the occurrence of a pronounced roughening in surface finish during straining. With this definition, forming limits were obtained for both simple and complex deformation paths [15], [16].

On the other hand, in order to avoid deformation defects, it may be necessary and useful to analyze the deformation state over the whole formed panel and to investigate the relations among the deformation state, deformation defects and the die geometry. Avoidance of deformation defects has so far depended mostly on skillful techniques [17], [18]. It is also expected that the most suitable steel sheet may reasonably be selected by these investigations. Therefore, an attempt is made to understand to a reasonable degree the changes in the character of the strain distributions and the deformation behavior with the progress of press forming by using the scribed circle technique. As a consequence of this defor-

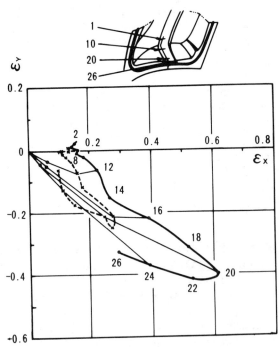

Fig. 1.   Deformation diagram (Front Fender Panel $(F)_1$).

mation analysis over the whole formed panel, some interesting relations among the deformation state, deformation defects and the die geometry are observed.

FORMING LIMIT OF STEEL SHEET AND THE DEFORMATION STATE IN THE MOST CRITICAL AREA

Deformation State in the Most Critical Area—The largest logarithmic strain (ϵ_x) and the logarithmic strain perpendicular to the ϵ_x direction in the most critical area are measured by using scribed circles of 6.4 mm diameter [19]. The measured strains are expressed as the deformation diagram in the $\epsilon_x - \epsilon_y$ rectangular coordinates [1]. As an example, Fig. 1 shows the development strains in the most critical area in a front fender panel $(F)_1$. From the bold-line curve which expresses the strain distribution in the finished stage, we can see that ϵ_x grows the largest at the material position numbered 20 under a nearly uniaxial tensile deformation state. Also, from the thin-line curves which express the deformation paths at each material position, we can see that most positions have monotonic deformation paths. However, a few positions, such as numbers 12 and 16, have remarkably varied deformation paths during the forming process.

Figure 2 shows the deformation diagrams for five critical areas in front fender

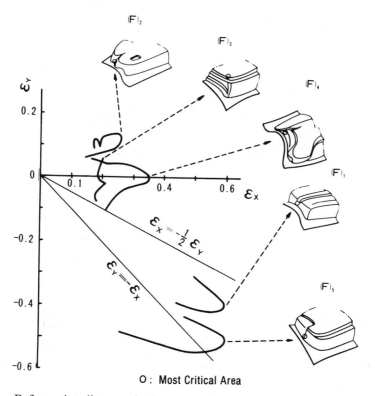

O : Most Critical Area

Fig. 2.   Deformation diagrams in the most critical areas in Front Fender Panels.

References pp. 337–338.

panels, $(F)_1-(F)_5$. The amount and ratio of strains in the most critical areas are seen to vary widely depending on the shapes of punch and die and the press forming techniques.

Forming Limits and the Strain at Critical Surface Roughness—The roughening behavior in surface profile was investigated experimentally. Figure 3 shows the various strain ratios investigated and the general shapes of tools and blanks used in this experiment. The strain ratio β is equal to ϵ_y/ϵ_x. The surface finish is measured along the ϵ_x direction by means of stylus instrument. The material is an aluminum-stabilized steel (K1) as shown in Table 1.

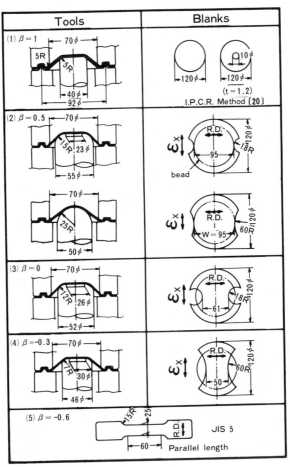

ϕ : Diameter
R : Radius
t : Blank Thickness

Fig. 3.   Geometry of tools and blanks used in experiment. All dimensions in mm.

Step (1)   Before deformation

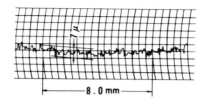

Step (2) $\varepsilon_x = 0.41$ $\varepsilon_y = -0.14$

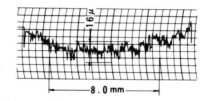

Step (3) $\varepsilon_x = 0.47$ $\varepsilon_y = -0.14$

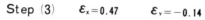
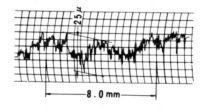

Step (4) $\varepsilon_x = 0.53$ $\varepsilon_y = -0.17$

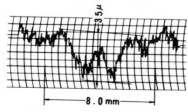

Fig. 4.   Change in surface profile during deformation ($\beta = -0.32$, sheet K1).

TABLE 1

Mechanical Properties of Aluminum-Stabilized Steel Sheets

Materials	Yield point (kg/mm²)	Tensile strength (kg/mm²)	Elongation (%)	n value	r value
K 1	16.3	30.5	47.7	0.242	1.53
K 2	15.3	29.9	45.5	0.236	1.61

(0.8mm thick)

Figure 4 shows an example of the measured profiles of surface finish in sheet K1 stretched under $\beta = -0.32$. Step (1) is the profile before deformation. The strain ϵ_x increases gradually through steps (2) to (4). At step (2), the surface roughness increases uniformly all over the measured range. But, harmonic waves occur at step (3), and then, some of the waves grow into the deep valleys,

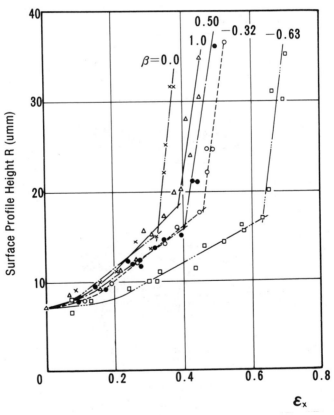

Fig. 5.   Relations between R and ϵ_x (sheet K1).

presumably necking, at step (4). It is obvious that some waves at step (3) grow rapidly into the valleys which immediately develop into fracture with little additional stretching. This figure suggests that the amount of deformation which begins to form the waviness at step (3) should be considered as the practical forming limit, because additional stretch-forming is not stable. As for the other strain ratios, though the profile shape varies with strain ratio, the same kind of waviness as at step (3) in Fig. 4 is observed just before the onset of necking.

In order to express these results quantitatively, the profile height of surface roughness $R(\mu m)$ is used. As shown in Fig. 4, the maximum peak-to-valley height R is determined by the deviation of roughness profile from the basic wave line of the largest period within the length of 8.0 mm. This approach separates the effect of waviness from warp in the sheet or the gradient of thickness distribution.

Figure 5 shows the experimental results in the $\epsilon_x - R$ rectangular coordinates. For every strain ratio, R initially increases in proportion to ϵ_x, but as soon as ϵ_x

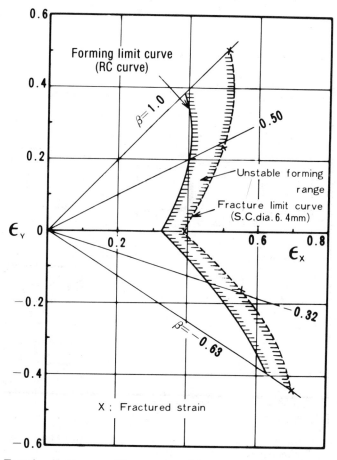

Fig. 6.   Forming limit curve (Rc curve) and fraction limit curve (sheet K1).

References pp. 337–338.

exceeds a critical value, R begins to grow explosively. At peak heights exceeding
this critical value, the local decrease of thickness takes place so rapidly that the
press forming operation is unstable. This critical amount of deformation should
therefore be considered as the forming limit in actual press forming, and is called
here the strain at critical surface roughness (ϵ_{xc}, ϵ_{yc}).

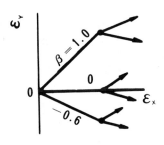

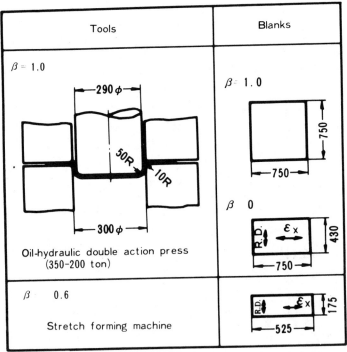

ϕ : Diameter
R : Radius

Fig. 7.   Deformation paths investigated; Geometry of tools and blanks used for initial
deformation. All dimensions in mm.

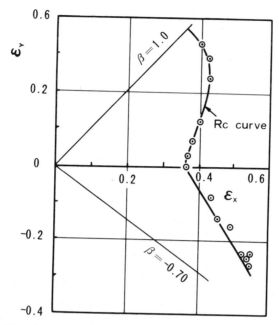

Fig. 8.   Forming limit curve in simple deformation path (sheet K2).

In order to understand the effect of strain path more accurately, it is possible from Fig. 5 to draw the forming limit curve (R_c curve) as a function of the amount and ratio of strains in the $\epsilon_x - \epsilon_y$ rectangular coordinates, Fig. 6. It is seen that the forming limit curve has the smallest ϵ_x under plane-strain stretching. The fracture limit curve measured by means of the scribed circles of 6.4 mm in

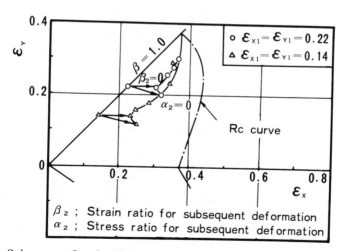

Fig. 9.   Subsequent forming limits after initial equibiaxial stretching (sheet K2).

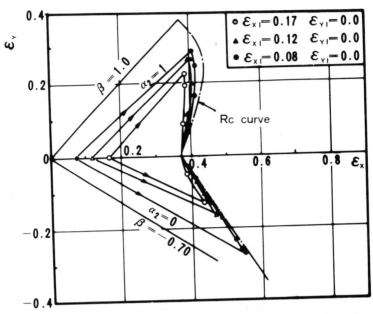

Fig. 10. Subsequent forming limits after initial plane strain stretching (sheet K2).

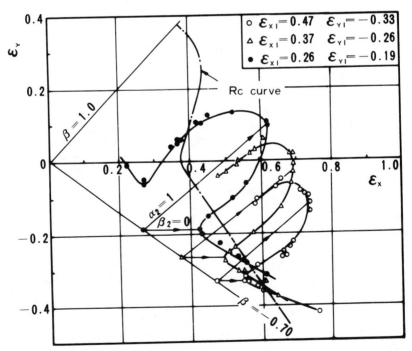

Fig. 11. Subsequent forming limits after initial uniaxial tension (sheet K2).

diameter is also shown in Fig. 6. The forming limit curve is smaller by 0.07 to 0.12 in ϵ_x than the fracture limit curve. It is obvious from Fig. 6 that the range between the forming limit curve and the fracture limit curve corresponds to deformation states in pressed parts which cannot be obtained successfully.

One can conclude from these investigations into the roughening behavior of surface profile that the critical amount of deformation at which the surface profile height begins to grow explosively should be considered as the forming limit in actual press forming operations.

Forming Limits Under Complex Deformation Paths—The forming limit, defined above as the strain at a critical surface roughness, was investigated in cases of not only the constant strain-ratio forming (the simple deformation path), but also involving a varying strain ratio in the forming process (the complex deformation path). In order to obtain a simple forming limit curve, the same size of spherical punch and blanks as shown in Fig. 3 were used, where the width of blank W was regulated to obtain the required wide range of strain ratios. For uniaxial tension, the specimen geometry shown in Fig. 3 was also used. To distinguish the initial and subsequent deformation paths in the complex case, number subscripts are added to the characters respectively, such as ϵ_{x1}, ϵ_{x2}, β_1, β_2.

Figure 7 shows the deformation path investigated and the shapes of tools and blanks used for the initial deformation. The subsequent deformation is added by using the same spherical punch and blanks as shown in Fig. 3 where blanks are

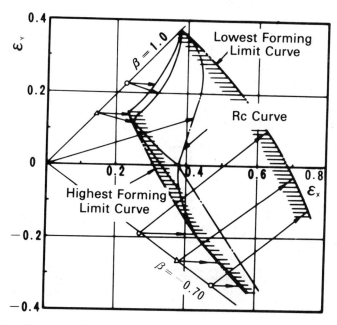

Fig. 12.   Variable range of forming limit depending on deformation path (sheet K2).

References pp. 337–338.

taken from the preformed parts. The material is an Aluminum-stabilized steel K2 as shown in Table 1. In this experiment, the critical condition in surface roughness for the forming limit is assessed by visual observation.

Figure 8 shows the forming limit curve (R_c curve) of sheet K2 obtained for simple deformation paths. Figure 9 shows the forming limits obtained for complex deformation paths in which the initial path is equibiaxial stretching. These forming

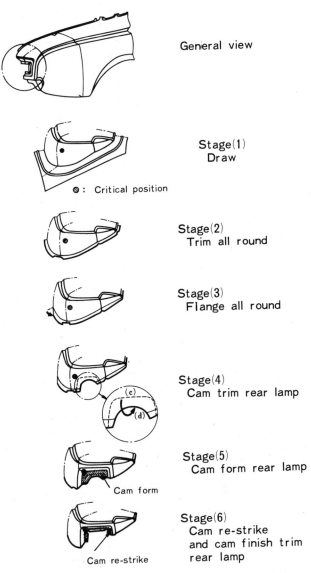

General view

Stage(1)
Draw

⊘ : Critical position

Stage(2)
Trim all round

Stage(3)
Flange all round

Stage(4)
Cam trim rear lamp

Stage(5)
Cam form rear lamp

Cam form

Stage(6)
Cam re-strike
and cam finish trim
rear lamp

Cam re-strike

Fig. 13.   Stamping sequence of Rear Quarter Panel $(Q)_1$.

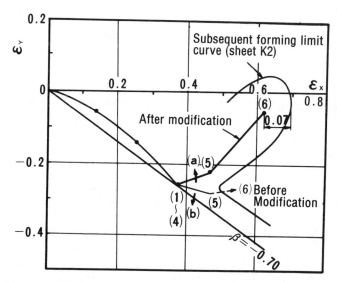

Fig. 14.   Deformation path of a critical position in panel $(Q)_1$.

limits are located at much smaller ϵ_x than the case of the simple deformation path (R_c curve) and have the smallest critical strain ϵ_{xc} for subsequent stretching in plane strain or uniaxial tension. Figure 10 shows the results for initial plane-strain stretching. It is seen that the forming limits are at slightly smaller ϵ_x than the R_c curve. Figure 11 shows the results for initial uniaxial stretching. In this case, the forming limits highly exceed the R_c curve in subsequent equibiaxial stretching, but are located at smaller ϵ_x than the R_c curve for subsequent plane strain stretching. It is obvious from these curves that large amounts of additional deformation are obtainable for cases involving initial uniaxial tension and subsequent equibiaxial stretching.

In these investigations, the variable range of forming limit dependence on the deformation path was obtained. The forming limit was found to vary widely over the range between the lowest forming limit curve and the highest forming limit curve shown in Fig. 12, depending on the deformation path.

Application of Forming Limit Curves to Press Forming Operations—The above experimental results can be applied to actual press forming operations. An example is the tail portion of a rear quarter panel $(Q)_1$ deformed through six stages as shown in Figure 13. Fracture occurred occasionally at stage (5) and always at stage (6) during the initial tryout operation. Figure 14 shows the deformation path of the position which tended to fracture (encircled in stage (4) of Fig. 13). The position was stretched initially under uniaxial tension at stages (1–4) and subsequently subjected to nearly plane strain stretching at stage (5). At stage (6), according to the shape of die and the flow of metal, the deformation path of the fractured position was expected to be along the arrow line indicated by "before modification." The forming limit curve of sheet K2 as illustrated in Fig. 14 suggested that the occurrence of fracture might be avoided by changing the deformation path to the direction of either bold arrow line (a) or (b).

In an attempt to change the direction of deformation path upward, the geometry of cam trim line at stage (4) in Fig. 13 was modified from the wide shape (c) to the narrow shape (d) as shown. This modification successfully changed the deformation path onto the line marked "after modification" (a), and consequently, made a safety margin of 0.07 in ϵ_x at stage (6). As a result, the occurrence of fracture was avoided.

DEFORMATION STATE OF FORMED PANELS AND DEPENDENCE ON DIE GEOMETRY

Strain Composition Diagram—As described in [21], the deformation state of a whole formed panel may be expressed approximately by the profile elongation, the amount of draw-in in the representative vertical sections, the increase of surface area of sheet formed in the die cavity and the amount of draw-in from the die periphery into the cavity. Frequently, however, this approximation has not been effective for avoiding deformation defects. In order to provide a more effective means, it is necessary to know the strain distribution in detail over the whole formed panel and the strain increment behavior with the progress of press forming. Therefore, the largest engineering strain e_x and the engineering strain e_y perpendicular to the e_x direction in the sheet plane of large-size autobody panels were measured by using scribed circles 40 mm and 6.4 mm in diameter.

The histograms which show the appearance frequency of e_x and e_y within certain ranges of strain on a panel were obtained as shown in Fig. 15. The appearance frequency of e_x in roof panel (R) or front door panel (FD) is maximum in the range of e_x smaller than 0.02. The frequency of e_x in front fender panel (F), or rear quarter panel (Q), or trunk panel (T) is maximum in the range of e_x larger than 0.02, as also shown in Fig. 15. The frequency of e_y in each panel is maximum in the range between -0.02 and 0.02.

As a consequence of these investigations, all the measured strains were classified into six types of deformation ranges, $A-F$, as shown in Fig. 16. A map-like expression of strain distributions is thus obtained by using a picture-like pattern based on this classification. This map-like expression is called here the "strain composition diagram."

As an example, Fig. 17 illustrates the strain composition diagram over the whole body surface of a car K. The deformation range F which represents a very small amount of deformation occurs in large areas of panels $(FD)_k$ and $(R)_k$. As a consequence of these investigations, the characteristics of the deformation state of each panel can be expressed clearly by the strain composition diagram.

Figure 18 shows how a modification of the die geometry for a front fender panel $(F)_5$ affects the deformation state in the formed panel. The modification consisted of changing a front portion of the die geometry from flat to curved as indicated by the feature lines in Fig. 18. As a result of this modification, the

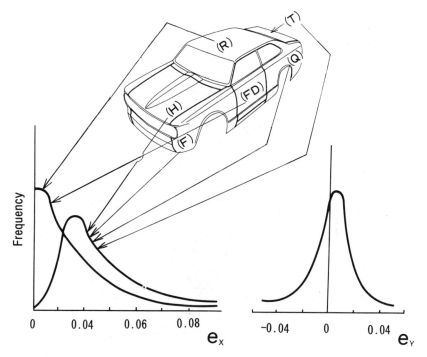

Fig. 15.   Appearance frequency of e_x and e_y for various panels.

deformation ranges C, D and E which appeared previously in the upper portion of the front curves of the fender are decreased. In the same time, the deformation ranges B and F which are usually seen in stretch formed flat panels, are increased. This example clearly shows that the effect of die geometry on the deformation state can be revealed by using the strain composition diagram.

Deformation Behavior During a Press Forming Stroke—We have often observed that surface warp and body wrinkles grow remarkably at the earlier stages of a press forming stroke, but disappear gradually near the end of the stroke. According to these experiences in the press shop, it may be expected that the strain in each portion will not grow monotonically during a press forming stroke. The successive change of deformation state in panels from the beginning to the end of one press stroke was investigated by using the strain composition diagram, in order to understand the deformation behavior over the whole formed surface.

As an example, Fig. 19 shows the change of strain composition diagram during a forming stroke for panel $(F)_6$ formed with the punch having a somewhat conical curved surface and the die similar to the shape of the punch surface. It is easily observed that the deformation range B occurs on the left side portion at step (1), then spreads gradually toward the right side portion through steps (2) and (3),

References pp. 337–338.

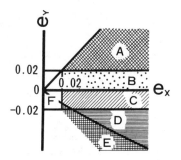

Deformation range		Range of e_x and e_y
A	↟	$e_x \gtrsim 0.02$ $e_x \gtrsim e_y \gtrsim 0.02$
B	↕	$e_x \gtrsim 0.02$ $0.02 > e_y \gtrsim 0$
C	↕	$e_x \gtrsim 0.02 \quad 0 > e_y > -0.02$ $e_y \gtrsim \{(1+e_x)^{-\frac{1}{2}} - 1\}$
D	⟊	$-0.02 \gtrsim e_y \gtrsim \{(1+e_x)^{-\frac{1}{2}} - 1\}$
E	⟊	$e_x \gtrsim 0.02$ $\{(1+e_x)^{-\frac{1}{2}} - 1\} > e_y \gtrsim e_x$
F	•	$e_x < 0.02$ $|e_y| < 0.02$

Fig. 16. Classification of strains into deformation ranges.

and finally covers most of the panel, and that the deformation range A grows rapidly near the corners at steps (3) and (4).

Figure 20 shows the change of strain composition diagram for panel $(Q)_2$ formed with a punch which has a surface relatively flatter than the surface of panel $(F)_6$ and the die also somewhat flatter. It is obvious that unlike the case of panel $(F)_6$, the deformation range B spreads toward the inside portion from the circumferential portion.

These examples show that the growth of strain in each area and the characteristics of deformation behavior, which are governed mostly by the shapes of the punch and the die during a press forming stroke, are observed clearly in the strain composition diagram.

The Relation Between Deformation Defects and Deformation Behavior—In order to reasonably assess the severity of forming and provide some technical means for avoiding defects, it is necessary to investigate the relations between the deformation defects and the deformation behavior.

To do this, a correlation between the commonly observed deformation defects and the deformation ranges in the same portion of the formed panels (as illustrated in Fig. 17) was made, Fig. 21.

But, in order to more accurately assess severity, the effects of the changing behavior in the strain composition diagram on the surface shape of a panel must be studied in more detail. Therefore, the strain composition diagrams of panel $(F)_7$ were determined for five steps of a forming stroke, including the blank holding and the finished stroke. These are compared with the geometrical characteristics of the same panel surfaces obtained by the projection of a special lattice pattern film for distinguishing the surface warp in the panel, as shown in Fig. 22. A noticeable warp occurs on the left side portion at step (2), corresponding to the occurrence of the deformation range B in the same area. But, through steps (3) and (4), the warp becomes smaller as the deformation range B spreads over the warped portion. It may be suggested that the pattern of deformation range B surrounded by range F is related to the occurrence of warp, and its

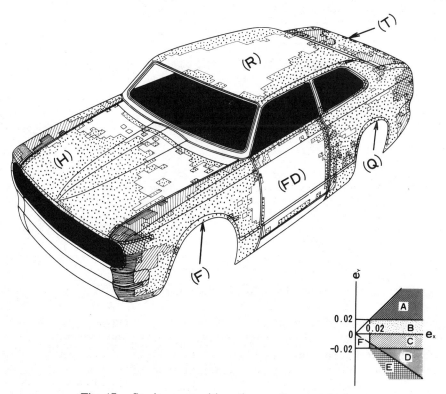

Fig. 17.   Strain composition diagram for a car body.

References pp. 337–338.

Before modification ·
(panel (F)₅)

After modification
(panel (F)₆)

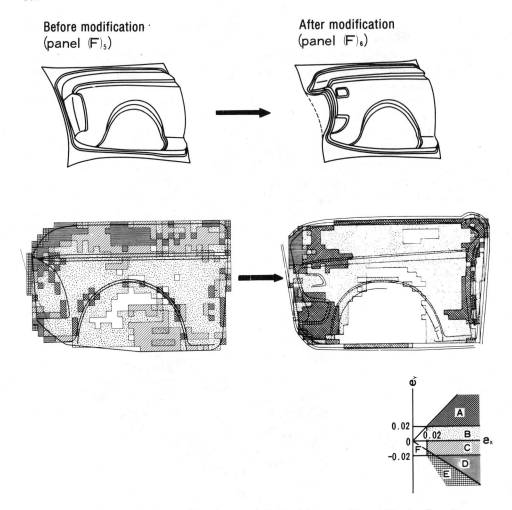

Fig. 18.   Effect of die modification on deformation state (Front Fender Panel).

change to a pattern mostly covered with range B is related with the disappearance of warp.

Next, the punch contacting zone in the panel is observed for understanding the effect of the shapes of the punch and the die on the above relation as shown in Fig. 22. The punch contacting zone at step (2) is nearly consistent with the area where the noticeable warp occurs and the range B is present.

These results obtained from Fig. 22 suggest a correlation among the deformation defects, the strain composition diagram and the punch contacting zone, and raise the possibility of predicting the occurrence of deformation defects and

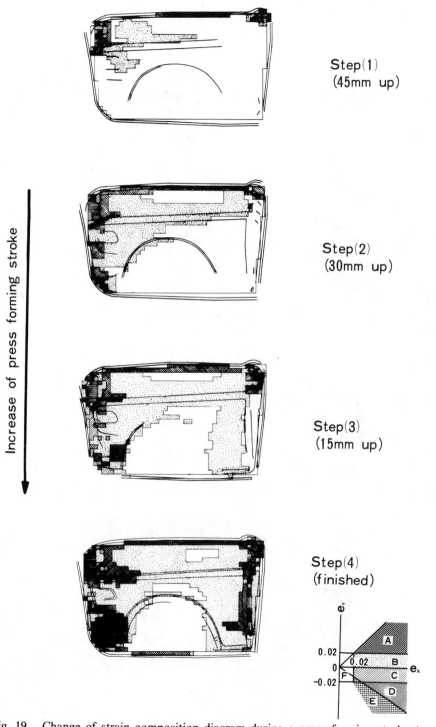

Step(1)
(45mm up)

Step(2)
(30mm up)

Step(3)
(15mm up)

Step(4)
(finished)

Increase of press forming stroke

Fig. 19.   Change of strain composition diagram during a press forming stroke (panel
$(F)_6$).

References pp. 337–338.

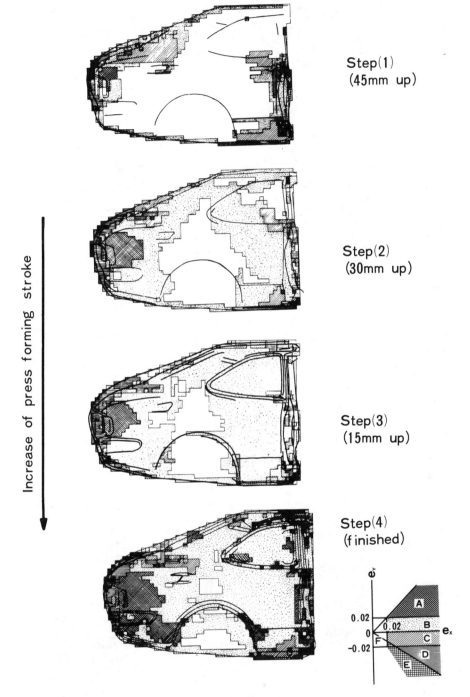

Fig. 20.   Change of strain composition diagram during a press forming stroke (panel $(Q)_2$).

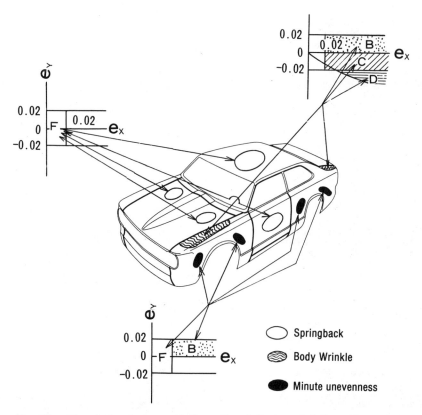

Fig. 21.   Correlation between deformation defects and deformation ranges.

providing some technical means for avoiding them. For predicting the occurrence of deformation defects more accurately, it may be necessary and useful to investigate some basic forming phenomena appearing in Fig. 22, such as the initial shape of the panel resulting from the blank holding stroke, the shape of warp caused by punch contact, the growth of warp and its disappearance during the press forming stroke, [22] and the minute unevenness due to the residual effects of warp.

CONCLUSION

The forming limits of steel sheet and the deformation state of the most critical area are investigated in order to avoid fracture in press forming of large-sized autobody panels. A pronounced roughening in the surface profile is detected immediately before the onset of necking by the stylus instrument measurement or by visual observation, and is reasonably defined as the forming limit. The

References pp. 337–338.

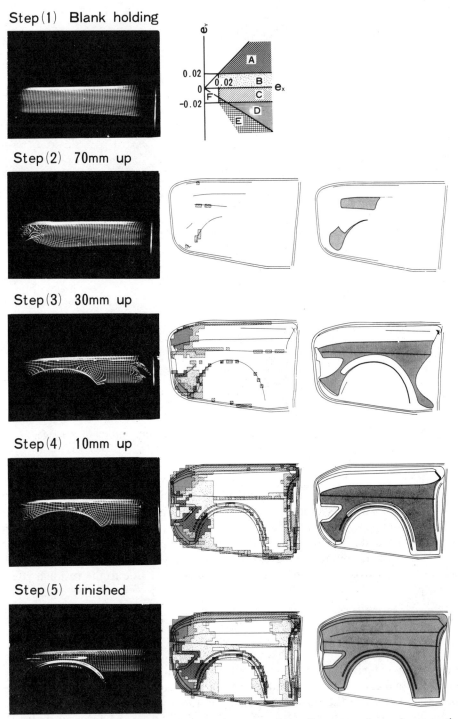

Fig. 22.   Changes of surface shape, strain composition diagram and punch contacting
zone during a press forming stroke (panel $(F)_7$).

forming limit is determined experimentally with respect to the strain ratio and its variation during forming process. By comparing the deformation state with the forming limit, it is possible to provide some technically effective means for avoiding fracture.

On the other hand, the deformation state over the whole formed panel is investigated in order to avoid surface deformation defects.

The character of the deformation state is delineated by classifying the strains into several ranges depending on the amount and ratio of strains. Based on these classifications, one obtains a map-like diagram. By using this type of diagram, the effects of die modification on the deformation state and the change of deformation state during a press forming stroke are understood. As a consequence of these investigations, there appear some possibilities of correlation among the deformation state, the deformation defects and the die geometry.

ACKNOWLEDGMENTS

The author would like to thank Dr. K. Yoshida and Mr. K. Miyauchi, The Institute of Physical and Chemical Research for their useful suggestions in preparing this paper. The author also would like to thank Dr. J. Kokado and Dr. M. Ohyane, The Kyoto University, and Mr. T. Kobayashi, Toyota Motor Co. Ltd., for their valuable guidance and encouragement.

REFERENCES

[1] K. Yoshida, Reports, Inst. of Physical and Chemical Research, 44 (1968), 169.
[2] T. Kobayashi, H. Iida and S. Sato, Proc. Int. Conf. Sci. Tech. Iron and Steel, Tokyo, (1971), 837.
[3] S. P. Keeler, Sheet Met. Ind., 42 (1965), 683.
[4] G. M. Goodwin, SAE Paper No. 680093, (1968).
[5] K. Nakazima, T. Kikuma and K. Renko, Preprint of 19th Joint Meeting for Plastic Working, (1968), 277.
[6] K. Yoshida, K. Miyauchi, A. Tajikawa, T. Iwasaki and S. Mizunuma, Scientific Paper, Inst. of Physical and Chemical Research, 61 (1967), 119.
[7] F. Negroni, S. Kobayashi and E. G. Thomsen, Trans. ASME 67-WA/Prod-12.
[8] H. Kubotera and Y. Ueno, Reports of Tech. Res. Inst., Nippon Kokan, 45 (1969), 45.
[9] Y. Tozawa, M. Nakamura and I. Shinkai, Int. Conf. Sci. Tech. Iron and Steel, Tokyo, (1971), 936.
[10] T. Kobayashi, K. Murata, H. Ishigaki and T. Abe, J. Japan Soc. Tech. Plasticity, 11 (1970), 495.
[11] Z. Marciniak and K. Kuczynski, Int. J. Mech. Sci., 9 (1967), 609.
[12] K. Yoshida, K. Abe, K. Hosono and A. Takezoe, Reports, Inst. of Physical and Chemical Research, 44 (1968), 128.
[13] Y. Hayashi, Preprint of 19th Joint Meeting for Plastic Working, (1968), 165.
[14] M. Azrin and W. A. Backofen, Met. Trans., 1 (1970), 2857.
[15] O. Kienzle, Werkstattstechnik, 56 (1966), 542.
[16] M. Ohyane and K. Kosakada, Trans. JSME, 36 (1970), 1017.
[17] T. Kobayashi, K. Murata and H. Ishigaki, J. Japan Soc. Tech. Plasticity, 9 (1968), 451.
[18] A. S. Kasper and P. J. VanderVeen, SAE Paper No. 720019, (1972).
[19] T. Kobayashi, H. Ishigaki and T. Abe, Proc. 7th Biennial Congress 7th IDDRG, (1972).
[20] K. Miyauchi, Preprint of 20th Joint Meeting for Plastic Working, (1969), 249.

[21] T. Kobayashi, K. Murata, H. Ishigaki and T. Abe, Proc. Int. Conf. Sci. Tech. Iron and Steel, Tokyo, (1971), 841.

[22] K. Yoshida, K. Abe and M. Usuda, Reports, Inst. of Physical and Chemical Research, Tokyo, 46 (1970), 153.

DISCUSSION

J. V. Laukonis *(General Motors Research Laboratories)*

In one of your slides you show the effect of prestrain on forming limit curves. I gathered that the highest forming limit curve was for prestrain along the uniaxial path.

Ishigaki

Yes, the highest forming limit curve was obtained by uniaxial tension followed by balanced biaxial stretching.

Laukonis

Is this for steel material such as aluminum-killed steel?

Ishigaki

Yes.

Laukonis

Would you expect the same kind of behavior for the same prestrain for 2036-T4 aluminum which is also a material for exterior panels? We studied balanced biaxial prestrain for both materials, and found that subsequent formabilities are quite different. We have not finished the study of the uniaxial prestrain path. I was wondering if you have any information on that?

Ishigaki

I have investigated only aluminum-killed steel.

R. A. Ayres *(General Motors Research Laboratories)*

I wondered, when you were able to change the strain distribution on panels, did you change the strain state at failure or simply the location where failure occurs? At what state of strain do most failures occur? Speaking in terms of histogram, did most of the failures occur at plane strain?

Ishigaki

Do you mean the strain distribution in the whole panel surface or in the most rupturable portion?

Ayres

When you change the strain distribution, does the strain state at failure always shift to plane strain?

B. Budiansky *(Harvard University)*

I guess maybe the question is: what would a histogram of the most-critical distributions look like?. . . .You talk about distributions of e_x and e_y, but what Dr. Ayres really wants, I guess, is: what is the maximum frequency of the most critical combinations of strains?

Ishigaki

These hisotgrams as shown are completely general. We obtain the histogram on many types of panels.

D. P. Koistinen *(General Motors Research Laboratories)*

In one of the slides you show the surface roughness as measured by stylus instrument and there is a 35 μm dip on the roughness curve at a considerable strain. Would that be a general measurement of the surface or would it be at the particular area over the neck?

Ishigaki

That is around the neck section.

Koistinen

So you knew at earlier steps where the neck would be.

Ishigaki

Yes. When we are measuring the surface, we can find the initiation of an explosive increase of roughness so we measure around these portions.

DEFORMATION ANALYSIS OF AXISYMMETRIC SHEET METAL FORMING PROCESSES BY THE RIGID-PLASTIC FINITE ELEMENT METHOD

S. KOBAYASHI and J. H. KIM

University of California, Berkeley, California

ABSTRACT

The paper describes the development of a finite-element model for analyzing the sheet-metal forming processes. Materials are assumed to be rigid-plastic with the view that the usefulness of an analysis method depends largely on solution accuracy and computation efficiency.

First, the variational formulation applicable to sheet metal forming is described by considering solution uniqueness and the effect of geometry change involved in the forming processes. From this variational formulation, a finite-element process model based on the membrane theory is developed. Then, three basic sheet-metal forming processes, namely, the bulging of a sheet subject to the hydrostatic pressure, the stretching of a sheet with a hemispherical head punch, and deep drawing of a sheet with a hemispherical head punch, are solved. The solutions by the rigid-plastic finite-element method are compared with existing numerical solutions and the experimental data. The agreement is generally excellent and it is concluded that the rigid-plastic finite-element method is efficient for analyzing the sheet-metal forming problems with reasonable accuracy.

INTRODUCTION

The metal-forming processes basically involve large amounts of plastic deformation, and, due to the complexities of plasticity, the exact analysis of a process is infeasible in most of the cases. Thus, a number of approximate methods have been suggested, with varying degrees of approximation and idealization. Among these, techniques using the finite-element method take precedence because of their flexibility, ability to obtain a detailed solution, and the inherent proximity of their solutions to the exact one.

A prime objective of mathematical analysis of metalworking processes is to

References pp. 362–363.

provide necessary information for proper design and control of these processes. Therefore, the method of analysis must be capable of determining the effects of various parameters on metal-flow characteristics. Furthermore, the computation efficiency, as well as solution accuracy, is an important consideration for the method to be useful in analyzing metalworking problems.

With this viewpoint, successful efforts have been carried out in analyzing various deformation processes, such as compression, heading, piercing, extrusion and drawing by the rigid-plastic finite element method [1]–[7].

This paper describes the development of a rigid-plastic finite-element model for analyzing the sheet-metal forming processes. First, the variational formulation applicable to sheet metal forming is described by considering solution uniqueness and the effect of geometry change involved in the forming processes. From this variational formulation, a finite-element process model based on the membrane theory is developed. Then, three basic sheet-metal forming processes, namely, the bulging of a sheet subject to the hydrostatic pressure, the stretching of a sheet with a hemispherical head punch, and deep drawing of a sheet with a hemispherical head punch, are solved. The solutions by the rigid-plastic finite-element method are compared with existing numerical solutions and the experimental data.

FORMULATION

The variational formulation used in the previous rigid-plastic analyses [1]–[7], is inadequate for analyzing sheet metal forming processes. This is due to nonuniqueness of deformation mode for the quasistatic deformation of a rigid-plastic solid under certain types of boundary conditions [8]–[11]. Furthermore, out-of-plane sheet metal forming processes involve large geometrical change during deformation [12]. These two points must be taken into consideration in deriving proper formulation of process modeling of sheet metal forming.

Variational Principle—At the generic stage of deformation two configurations are of interest: the current configuration B_1 of the body and the future configuration B_2 which it will assume in the nearest future. In this presentation B_1 is sometimes called the undeformed configuration and B_2 the deformed configuration. Associated with B_1 is a system of orthogonal curvilinear coordinates with coordinates X^I and covariant base vectors G_I. Curvilinear coordinates x^i and covariant base vectors g_i are associated with B_2. Incremental displacement vector field u describes the motion from B_1 to B_2.

Now, consider the virtual work done by the external forces in B_2 under a virtual displacement δu, which is given by

$$\delta W = \int_{S_2} T^i \delta u_i \, da \qquad (1)$$

where T^i is the true traction in B_2; da is area element; S_2 the surface of B_2 where traction is prescribed. Body forces are assumed to be negligible. The integral is with respect to B_2, but since we do not know configuration B_2 in advance, we need to transform Eq. (1) into an integral with respect to B_1. Using Nanson's formula and the relationship between Cauchy stress and Kirchhoff stress [13], we have

$$\delta W = \int_{S_1} x^i|_I \tau^{IJ} N_J \delta(G_i{}^N u_N)\, dA, \tag{2}$$

where the vertical bar "$|$" denotes the covariant differentiation, dA the area element of B_1, N_J the unit normal vector to the surface S_1 of B_1, u_N covariant component of incremental displacement with respect to configuration B_1, τ^{IJ} the Kirchhoff stress, and the Shifter $G_i{}^N = \mathbf{G}^N \cdot \mathbf{g}_i$.

Transforming the surface integral to a volume integral by the Gauss-Green theorem, eq. (2) becomes

$$
\begin{aligned}
\delta W &= \int_{B_1} [x^i|_I \tau^{IJ} \delta(G_i{}^N u_N)]|_J\, dV \\
&= \int_{B_1} [x^i|_I \tau^{IJ}]|_J \delta(G_i{}^N u_N)\, dV + \int_{B_1} x^i|_I \tau^{IJ} [\delta(G_i{}^N u_N)]|_J\, dV.
\end{aligned} \tag{3}
$$

Since the first integral vanishes because of the equilibrium condition and the Kirchhoff stress is symmetric, eq. (3) is expressed as

$$\delta W = \int_{B_1} \tau^{IJ} \delta(E_{IJ})\, dV, \tag{4}$$

where $E_{IJ} = \frac{1}{2}(u_I|_J + u_J|_I + u^K|_I u_K|_J)$ and E_{IJ} is called Lagrangian strain. Since the Kirchhoff stress tensor is referred to the fixed base vectors in B_1, the following additive decomposition can be used:

$$\tau^{IJ} = \sigma^{IJ} + d\tau^{IJ}, \tag{5}$$

where σ^{IJ} is the Kirchhoff stress tensor at configuration B_1 and is numerically equal to the Cauchy stress since at B_1 these two stress measures are indistinguishable. $d\tau^{IJ}$ is the increment of the Kirchhoff stress tensor. Substitute eq. (5) into eq. (4) to obtain

$$\delta W = \int_{B_1} (\sigma^{IJ} + d\tau^{IJ}) \delta(E_{IJ})\, dV. \tag{6}$$

Expressing eq. (1) in terms of nominal traction $\tilde{\mathbf{T}} = \mathbf{T}\dfrac{da}{dA}$, we have

$$\delta W = \int_{S_1} \tilde{T}^I \delta(u_I)\, dA. \tag{7}$$

Equating eqs. (6) and (7) and rewriting the equation in terms of physical components,

$$\int_{B_1} (\sigma_{IJ} + d\tau_{IJ}) \delta(E_{IJ})\, dv = \int_{S_1} \tilde{T}_I \delta(u_I)\, dA. \tag{8}$$

To denote that E_{IJ} is small, we replace E_{IJ} by dE_{IJ}. Now the constitutive

References pp. 362–363.

equation for a rigid plastic solid can be derived as [14]

$$h \, dE_{IJ} = \frac{\tau'_{IJ} \, d\bar{\sigma}}{\bar{\sigma}} \tag{9}$$

where $\bar{\sigma} = (\frac{3}{2}\tau'_{IJ}\tau'_{IJ})^{1/2}$, if τ'_{IJ} is the deviatoric component of τ_{IJ}. The effective stress $\bar{\sigma}$ is the function of plastic work, given by $\bar{\sigma} = H(\int dE)$ where $d\bar{E}$ is defined by $d\bar{E} = (\frac{2}{3}dE_{IJ} \, dE_{IJ})^{1/2}$ and $h = \frac{2}{3}\frac{dH}{dE} = \frac{2}{3}H'$. Substituting eq. (9) into eq. (8) and following the usual argument for proof, it follows the variational principle; among all the kinematically possible modes (starred (*) terms) the actual one minimizes the functional π:

$$\pi = \int \bar{\sigma}(d\bar{E}^*) \, dV + \frac{1}{2} \int H'(d\bar{E}^*)^2 \, dV - \int \hat{T}_I(u_I^*) \, dA. \tag{10}$$

It should be mentioned that Wang [15] derived a similar expression in his work on a variational method for large plastic deformation of metal sheets.

Finite Element Formulation—Finite element modeling is achieved by approximating the functional π by a function ϕ as

$$\pi \simeq \phi = \Sigma \phi^{(m)}(\mathbf{u}^{(m)}) \tag{11}$$

where $\mathbf{u}^{(m)}$ is the incremental displacement vector at nodes associated with the m-th element. The formulation is given for axisymmetric plane stress problems based on the membrane theory. The function $\phi^{(m)}$ for the m-th element in eq. (11) is obtained as follows:

The stress-strain increment relation for a sheet with normal anisotropy is given, using the Hill's criterion [16], by

$$\frac{dE_r}{(1 + R)\sigma_r - R\sigma_\theta} = \frac{dE_\theta}{(1 + R)\sigma_\theta - R\sigma_r} = \frac{d\bar{E}}{(1 + R)\bar{\sigma}}, \tag{12}$$

where R is the normal anisotropy parameter which is the ratio of the width strain to the thickness strain in uniaxial tension. The effective stress and the effective strain are defined as

$$\bar{\sigma} = \sqrt{\sigma_\theta{}^2 - \frac{2R}{1 + R}\sigma_r\sigma_\theta + \sigma_r{}^2}, \tag{13}$$

$$d\bar{E} = \frac{1 + R}{\sqrt{1 + 2R}}\sqrt{dE_r{}^2 + \frac{2R}{1 + R}dE_\theta \, dE_r + d\bar{E}_\theta{}^2}. \tag{14}$$

Subscripts r and θ are the meridian and the circumferential directions, respectively. $d\bar{E}$ may be written in matrix form as

$$d\bar{E} = \sqrt{\tfrac{2}{3}}[d\,\mathbf{E}^T D \, d\mathbf{E}]^{1/2}, \tag{15}$$

where $d\mathbf{E} = (dE_r, dE_\theta)^T$ and

$$D = \frac{3(1 + R)}{2(1 + 2R)} \begin{vmatrix} 1 + R & R \\ R & 1 + R \end{vmatrix}. \tag{16}$$

Approximating the sheet geometry by a series of conical frustums, as shown in Fig. 1, the incremental displacement vector at nodes is given by $u^{(m)} = \langle v_1, w_1, v_2, w_2 \rangle^T$ for a representative element m, where v_i, w_i are the radial and the axial components of incremental displacement of the i-th node. Then the incremental displacement field inside the element may be written as

$$\mathbf{u} = \begin{Bmatrix} v \\ w \end{Bmatrix} = \begin{bmatrix} \dfrac{1 + t'}{2} & 0 & \dfrac{1 - t'}{2} & 0 \\ 0 & \dfrac{1 + t'}{2} & 0 & \dfrac{1 - t'}{2} \end{bmatrix} \begin{Bmatrix} v_1 \\ w_1 \\ v_2 \\ w_2 \end{Bmatrix} \tag{17}$$

$$= N\mathbf{u}^{(m)}$$

where t' is the local coordinate varying from the value of -1 at node 2 to $+1$ at node 1. Due to this incremental displacement field an element of length s_0,

$$s_0 = \sqrt{\{(r_0)_1 - (r_0)_2\}^2 + \{(z_0)_2 - (z_0)_1\}^2}, \tag{18}$$

is stretched to a new length s,

$$s = \sqrt{\{(r_0)_1 - (r_0)_2 + v_1 - v_2\}^2 + \{(z_0)_2 - (z_0)_1 + w_2 - w_1\}^2}$$
$$= \sqrt{(r_1 - r_2)^2 + (z_2 - z_1)^2} \tag{19}$$

where $(r_0)_i$, $(z_0)_i$ are the radial and the vertical positions of the i-th node at the undeformed configuration and $(r)_i$, $(z)_i$ at the deformed configuration. Since the element is straight, any point of t' in the local coordinate is shown to have a

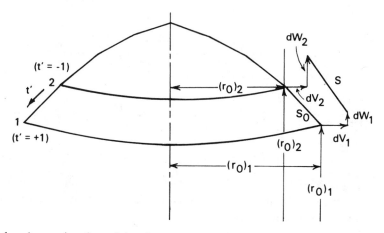

Fig. 1.   Approximation of the sheet geometry into a series of conical frustums.

References pp. 362–363.

global radial position r_0 determined by

$$r_0 = \left(\frac{1 + t'}{2}\right)(r_0)_1 + \left(\frac{1 - t'}{2}\right)(r_0)_2 . \tag{20}$$

The new position r of the same particle is given by

$$r = r_0 + \frac{(1 + t')}{2}v_1 + \frac{(1 - t')}{2}v_2 . \tag{21}$$

The Lagrangian strain increment is approximated, for simplicity, by the logarithmic strain during the small time increment as

$$d\mathbf{E} = \begin{Bmatrix} dE_1 \\ dE_2 \end{Bmatrix} = \begin{Bmatrix} \ln \dfrac{s}{s_0} \\ \ln \dfrac{r}{r_0} \end{Bmatrix} \tag{22}$$

where s_0, s, r_0 and r are expressed by eqs. (18), (19), (20) and (21), respectively. We can express now $\phi^{(m)}$ in terms of nodal values as

$$\phi^{(m)} = \int \bar{\sigma}(\sqrt{\tfrac{2}{3}}t)[d\mathbf{E}^T D \, d\mathbf{E}]^{1/2} \, dA \tag{23}$$
$$+ \tfrac{1}{2}\int H'(\tfrac{2}{3}t)[d\mathbf{E}^T D \, d\mathbf{E}] \, dA - \int \bar{\mathbf{T}}^T N u^{(m)} \, dA$$

for the unit included angle of the element, where $dA = r \, dt'$, t is the sheet thickness, and integration is performed from $t' = -1$ to $t' = +1$. Minimization of the function results in

$$\frac{\partial \phi^{(m)}}{\partial \mathbf{u}^{(m)}} = \int (\sqrt{\tfrac{2}{3}}t)\bar{\sigma}[d\mathbf{E}^T D \, d\mathbf{E}]^{-1/2}\frac{\partial(d\mathbf{E})}{\partial \mathbf{u}^{(m)}} D \, d\mathbf{E} \, dA \tag{24}$$

$$+ \int (\tfrac{2}{3}t)H'D \, d\,\mathbf{E}\frac{\partial(d\mathbf{E})}{\partial \mathbf{u}^{(m)}} - \int N^T\bar{\mathbf{T}} \, dA$$

From eq. (22),

$$\frac{\partial(d\mathbf{E})}{\partial\mathbf{u}^{(m)}} = Q = \begin{bmatrix} \dfrac{\partial(dE_1)}{\partial u_1} & \dfrac{\partial(dE_2)}{\partial u_1} \\[2mm] \dfrac{\partial(dE_1)}{\partial u_2} & \dfrac{\partial(dE_2)}{\partial u_2} \\[2mm] \dfrac{\partial(dE_1)}{\partial u_3} & \dfrac{\partial(dE_2)}{\partial u_3} \\[2mm] \dfrac{\partial(dE_1)}{\partial u_4} & \dfrac{\partial(dE_2)}{\partial u_4} \end{bmatrix} = \begin{bmatrix} \dfrac{r_1 - r_2}{s^2} & \dfrac{1 + t'}{2r} \\[2mm] \dfrac{-(z_2 - z_1)}{s^2} & 0 \\[2mm] \dfrac{-(r_1 - r_2)}{s^2} & \dfrac{1 - t'}{2r} \\[2mm] \dfrac{(z_2 - z_1)}{s^2} & 0 \end{bmatrix} \tag{25}$$

Assembling the eq. (24) for all elements in the finite element scheme, we obtain the non-linear simultaneous equations

$$\frac{\partial \Phi}{\partial \mathbf{u}} = \sum \frac{\partial \phi^{(m)}}{\partial \mathbf{u}^{(m)}} = 0. \tag{26}$$

Equation (26) is solved by the Newton-Raphson method by taking an initial guess of the solution $\mathbf{u}^*$ and neglecting the second and higher terms of $\Delta\mathbf{u} = \mathbf{u} - \mathbf{u}^*$ in the Taylor expansion. Then, we finally obtain the linear stiffness equation by

$$P\,\Delta\mathbf{u} = \mathbf{H} - \mathbf{F} \tag{27}$$

where

$$P = \sum P^{(m)}, \quad \mathbf{H} = \sum \mathbf{H}^{(m)}, \quad \mathbf{F} = \sum \mathbf{F}^{(m)}$$

with

$$P^{(m)} = \frac{2}{3}\int \frac{1}{d\bar{E}}\left\{(\bar{\sigma} + H'\,d\bar{E})\left(K - \frac{2}{3}\frac{\mathbf{b}\mathbf{b}^T}{d\bar{E}}\right) + \frac{2}{3}\frac{H'\,\mathbf{b}\mathbf{b}^T}{d\bar{E}}\right\}t\,dA$$

$$\mathbf{b} = QD\,d\mathbb{E}, \qquad K = QDQ^T$$

$$\mathbf{H}^{(m)} = \frac{2}{3}\int \frac{1}{d\bar{E}}(\bar{\sigma} + H'\,d\bar{E})\mathbf{b}\,t\,dA,$$

$$\mathbf{F}^{(m)} = \int N^T\tilde{\mathbf{T}}\,dA.$$

We evaluate the integrals with the Gaussian quadrature formulae and convergence is checked by the fractional norm.

HYDROSTATIC BULGING

The ductility of sheet metal under biaxial stress is often examined by means of the so-called bulge test. A uniform plane sheet is placed over a die with an aperture and is firmly clamped around the perimeter. An increasing hydrostatic pressure is applied to one side of the sheet, causing it to bulge through the aperture. From the measured profile and thickness of the plastically deformed sheet near the pole, it is possible to calculate the local state of stress in terms of the applied pressure. If, in addition, the state of strain is measured by means of a grid, the stress-strain characteristics of the metal under biaxial tension are obtained. The advantage of this test over any other simple one is that greater range of pre-instability strain can be obtained.

Hydrostatic bulging is not only important as a material property test, but also as a forming operation. Thus, a number of theoretical investigations, dealing with axisymmetric hydrostatic bulging (Fig. 2) have appeared in literature.

The classical analysis of bulging is the one by Hill [17]. His solutions are, however, special ones. Instead of analyzing deformation with a given stress-strain characteristic, Hill first adopted special kinematic assumptions and from them deducted the necessary stress-strain characteristics which satisfy all the governing equations under the prescribed kinematic mode. The kinematic assumptions are first, that any material element describes a circular path which is, moreover, orthogonal to the momentary profile, and second, that circumferential strain is numerically equal to the tangential strain. The required stress-strain characteristic is found to be an exponential type. Hill's other solution for a linear

Fig. 2.   Schematic view of the hydrostatic bulging.

work-hardening solid, uses the method of successive approximation by adopting an approximate yield criterion.

Analyses of work by Woo [18], Yamada [19], and Wang [20] are based upon realistic choices of stress-strain characteristics and the yield criterion. In applying the deformation theory of rigid plasticity, Wang experiences a mathematical difficulty and attributes this to the fact that the differential equations associated with the deformation theory possess a singularity which has the effect of restricting the range of calculation within a certain value of the polar strain. In applying the incremental theory of rigid plasticity, researchers experience a difficulty in satisfying the boundary condition at the fixed edge, i.e., $\dot{\epsilon}_\theta = 0$. To avoid this difficulty, Woo uses the deformation theory, while Yamada reasons that introducing an elastic strain component into the formulation will resolve this "mathematical difficulty" and turns to the elasto-plastic constitutive law. Another theoretical work of interest comes from Wang, using the parametric representation of the stresses.

The only published solution on hydrostatic bulging using the finite-element method is the one by Oseki et al. [21], with the incremental theory of elasto-plasticity.

Computational Procedures—In adopting the rigid-plastic finite-element model to hydrostatic bulging, it is necessary to reconsider the external work increment term, since the pressure is uniform over the entire surface of a closed shell. In this case the increment of external work may be written as

$$\Delta W = p \Delta \bar{V}$$

(Wang [15], Budiansky [22]), where $\Delta \bar{V}$ is the increase of the volume enclosed by the deformed sheet and p is the pressure acting on the deformed configuration.

As an initial condition, Hill's special solution is utilized. In other words, the initial profile of the bulge is assumed to be a part of a sphere whose radius is given by $r = \frac{1}{2}(a^2/h + h)$, where a is the radius of the original blank and h is the polar height at the moment. With this geometry, a pressure p is prescribed. This pressure should be greater, at least, than the pressure which makes the sheet having initial geometry everywhere plastic. The initial guess on the incremental displacement is also obtained from Hill's special solution by assuming normal trajectory of the element particle to the bulge profile.

When a converged solution is obtained for the given pressure, a new bulge profile is determined from the initial bulge profile and incremental displacement field. Then the pressure is assigned a higher value and the converged solution for the previous step is used as the initial guess for the incremental displacement field and the computation continues in this manner.

Results and Discussion—To examine the validity of the present finite-element method for hydrostatic bulging, the solution is compared with those achieved by the elastoplastic FEM.

The following conditions were employed for the comparison with the elasto-plastic FEM:

Workhardening characteristics: $\bar{\sigma} = 105(0.0019 + \bar{\epsilon})^{0.2} \times 9.806 \times 10^6$ N/m²
Thickness: 3.0×10^{-4} m (=0.3 mm)
Radius of the sheet: 2.4×10^{-2} m (= 24 mm)
Anisotropy parameter: 1.0

An identical problem was also solved by Yamada [19], using the finite-difference method with the elasto-plastic theory. Fig. 3 shows the relationship between hydrostatic pressure and the polar thickness strain. The solid line represents the elasto-plastic FEM (and also the finite-difference method) and the points indicate the solution given by the rigid plastic FEM. The deviation of the first circle by the rigid plastic FEM is thought to reflect the approximation involved in the initial condition that the sheet is everywhere plastic and that the initial geometry is a part of a sphere. The solution can be improved numerically by taking a smaller value of h in generating the initial condition. Nevertheless, the solutions after this first step are in extremely good agreement with the elasto-plastic FEM and any disturbance in the initial conditions does not matter after an initial deformation of a small magnitude. The pressure increment is raised by twice

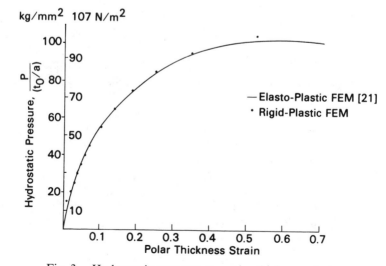

Fig. 3.   Hydrostatic pressure vs. polar thickness strain.

References pp. 362–363.

after some deformation and it is to be noted that the solutions with the larger
pressure increment size are still accurate. This means that the method is computationally economical with a reasonable accuracy. After the last point in the
figure the solution diverges and it is thought that the pressure maximum has been
reached. The convergence is excellent; in every step, five to seven iterations
seem to be sufficient. Figs. 4 and 5 shows the comparisons of strain distributions.
The circumferential strain distributions are in good agreement. The tangential
strain distribution by the rigid plastic FEM deviates somewhat at the edge from
that by the elasto-plastic FEM. The tangential strain is more sensitive to the
method employed than the circumferential strain, but this deviation of tangential
strain is not serious because the solution closely follows that by the finite-difference method and we may conclude that the strain distribution is accurately
predicted. Fig. 6 shows the distributions of stresses when the polar thickness
strain is (-0.4). Again, agreement between the rigid-plastic and elasto-plastic
solutions is very good.

STRETCHING OF A SHEET WITH HEMISPHERICAL PUNCH

Punch stretching is commonly used to assess the pressing quality of sheet
metals. A circular sheet is clamped firmly along the periphery and is stretched
by a rigid punch of hemispherical shape. The depth of the deformed sheet when
it fractures is usually taken as a measure of ductility. See the schematic diagram
in Fig. 7.

From the viewpoint of the deformation analysis, punch stretching is a complicated problem because a moving boundary separates the region in contact with
the punch head from the unsupported one. The friction over the punch head
gives rise to additional complications. One special solution is by Chakrabarty

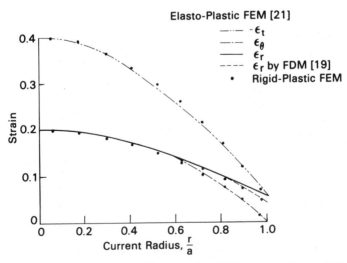

Fig. 4.   Distribution of strains (polar thickness strain $= -0.4$).

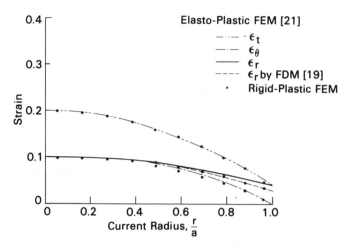

Fig. 5.   Distribution of strains (polar thickness strain $= -0.2$).

[23]. Following the line of Hill's special solution on hydrostatic bulging he obtained an analytical solution for a special material having exponential type stress-strain characteristics. For more general materials the only solutions available are the numerical ones. Numerical solutions of importance are those by Woo [24] and by Wang [25], [26], [27].

Woo's and Wang's solutions were obtained by the finite-difference method. The only solution by the finite-element method on punch stretching is one by Wifi [28]. His elasto-plastic, finite-element model does not neglect the bending moment nor the effect of shear stress and uses two-dimensional triangular ele-

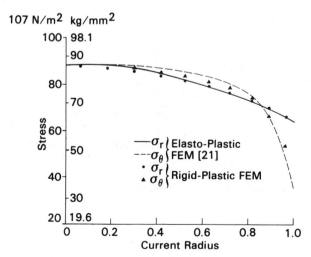

Fig. 6.   Distribution of stresses (polar thickness strain $= -0.4$).

References pp. 362–363.

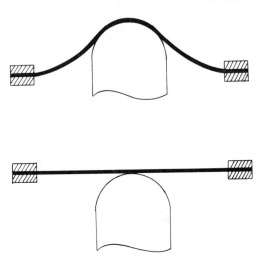

Fig. 7.   Schematic view of the stretching of a sheet with a hemispherical punch.

ments to take the thickness variation into account. Friction, which is of primary significance compared with the secondary effect of bending and thickness, is assumed to be perfect, meaning that once the element touches the punch head, it does not slide over the punch but sticks to it.

Computational Procedures—In applying the finite-element method to punch stretching, a thought should be given to the implementation of boundary conditions. The boundary conditions in punch stretching are stated not only by prescribing tractions and incremental displacements but sometimes by their ratios. In this respect, the problem is similar to the ball indentation problem (Lee et al. [29]).

The radial and axial positions of the material elements in the contact region are not independent but they are related to each other through a mathematical expression for the geometrical requirement that they must be actually on the surface of the punch head. The expression is

$$(r_0 + v)^2 + (c + z_0 + w)^2 = r_p{}^2, \tag{28}$$

where r_0, z_0 are radial and axial positions of the element at the present undeformed configuration; v, w are the increments of radial and axial displacements, and c is a parameter related to the punch height h by the expression (see Fig. 8).

$$c = r_p - h.$$

We linearize the boundary condition (28) to obtain

$$\Delta v = \frac{1}{\alpha} \Delta w + \beta, \tag{29}$$

where

$$\alpha = \frac{(r_0 + v^*)}{(c + z_0 + w^*)}$$

and

$$\beta = \frac{r_p{}^2 - (c + z_0 + w^*)^2 - (r_0 + v^*)^2}{2(r_0 + v^*)}$$

where starred (*) quantities are initial guesses, and Δv, Δw are perturbations. When the finite-element model is implemented, all the tractions are transformed into generalized nodal forces. Therefore, it is convenient to write the boundary condition in terms of the generalized nodal forces $\pi_{(r)}$ and $\pi_{(z)}$, the radial and axial components, respectively. These two nodal forces are related according to the friction condition at the punch head by

$$\pi_{(z)} + \frac{\pi_{(r)}}{\alpha} = \frac{k r_p}{(r_0 + v^*)} . \tag{30}$$

where k is the frictional force at nodes.

If the die has a round profile of the radius r_D, then the requirement for a material element to lie geometrically on the profile is similar to the requirement to be satisfied on the punch head. Therefore, we have

$$\Delta v = \frac{\Delta w}{\gamma} + \Omega, \tag{31}$$

where

$$\gamma = -\frac{(a - r_0 - v^*)}{(r_D - z_0 - w^*)}$$

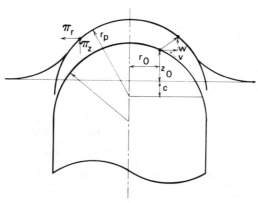

Fig. 8. Geometrical requirement for the node on the contact region.

References pp. 362–363.

and

$$\Omega = \frac{(a - r_0 - v^*)^2 + (r_D - z_0 - w^*)^2 - r_D{}^2}{2(a - r_0 - v^*)}.$$

for the element in contact with the die. The tractional boundary condition over the die profile can be written similarly as

$$\pi_{(z)} + \frac{\pi_{(r)}}{\gamma} = -\frac{k r_D}{(a - r_0 - v^*)}. \tag{32}$$

For the portion of the sheet which is not in contact with the punch head nor with the die profile, the displacement increment in the radial direction and the displacement increment in the axial direction remain as independent variables. Tractions are, however, given the value of zero.

With the advancement of the punch head, the portion of the sheet in contact with the punch or die profile increases and, consequently, the boundary separating this "contact region" from the "unsupported region" changes. The presence of this moving boundary is always a source of complications in the numerical analysis of punch stretching because it requires a basically trial-and-error approach. The treatment of the moving boundary used in the present analysis for punch stretching is given in detail elsewhere [30]. In order to implement Coulomb friction between sheet and punch or die, first we prescribe a tangential friction force S and obtain a converged solution. We then compute the normal component N and the friction coefficient $\mu = \dfrac{S}{N}$ corresponding to the initially prescribed value of S. If the computed friction coefficient is not what is intended, then we

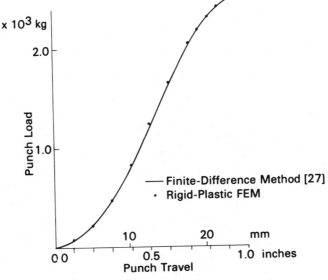

Fig. 9.   Punch load vs. punch travel.

modify the S value and repeat the process. It should be noted here that the correction of frictional force S needs the modification only in the F matrix (Eq. 27), while the stiffness matrix P remains the same.

The deformation step is controlled by the punch head increment, which is designed in the present codes to yield the maximum increment of effective strain roughly equal to a preset value. In the present work the optimum size is shown to be a 0.04 increment of effective strain. The solution generally converged after $10 \sim 15$ iterations for a single step within the fractional norm of 10^{-6}.

Results and Discussion—The present rigid-plastic FEM is compared with the finite-difference methods by Wang [27] and Woo [24].

(1) Comparison with the Finite-Difference Solution by Wang

The parameters used in Wang's example are as follows:

Material: copper
Stress-strain characteristics: $\sigma = 30.5\epsilon^{0.326}$ ton/in.^{2}
$= 4.636\iota^{0.326} \times 10^8$ N/m^2
Anisotropy: $R = 1.0$
Friction: $\mu = 0.04$
Thickness: $t = 0.035$ in. $= 8.89 \times 10^{-2}$ m
Punch radius: $r_p = 1.0$ in. $= 2.54 \times 10^{-2}$ m
Radius of sheet: $a = 1.15$ in. $= 2.921 \times 10^{-2}$ m

The two methods are in excellent agreement in predicting the punch load for a given punch travel. See Fig. 9; the solid line represents Wang's solution and

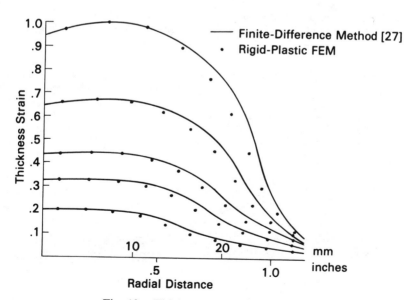

Fig. 10.   Thickness strain distribution.

the points represent the rigid-plastic FEM. Fig. 10 shows the thickness strain distribution. Again, a good agreement between the two solutions is apparent.

(2) Comparison with the Finite-Difference Solution by Woo

The parameters in Woo's example are:

Stress-strain characteristics:
$$\sigma = 5.4 + 27.8\epsilon^{0.504} \text{ ton/in.}^2$$
$$= (0.08208 + 0.422569\epsilon^{0.504}) \times 10^9 \text{ N/m}^2(\epsilon < 0.36)$$
$$= 5.4 + 24.4\epsilon^{0.375} \text{ ton/in.}^2$$
$$= (0.08208 + 0.37089\epsilon^{0.375}) \times 10^9 \text{ N/m}^2 \ (\epsilon > 0.36)$$

Material: copper
Punch radius: 1 in. = 2.54×10^{-2} m
Die profile radius: 0.3 in. = 7.62×10^{-3} m
Radius of sheet: 1.3 in. = 3.302×10^{-2} m
Coefficient of friction: 0.04
Thickness of sheet: 0.035 in. = 8.89×10^{-4} m

Figs. 11 and 12 are the thickness strain distribution and the circumferential strain distribution. Solutions by Woo are represented by solid lines and the solutions by the rigid-plastic FEM are represented by points. Agreement between the two solutions is excellent for most of the deformation. However, at later stages of deformation, a discrepancy is observed around the edges. Reexamining Woo's computational procedure reveals that in order to avoid the difficulty of satisfying boundary conditions exactly along the fixed edge ($\epsilon_\theta = 0$), he allowed a small increment of circumferential string along the edge at each stage. In the present rigid-plastic FEM such difficulty does not arise.

It may be of interest to note that Woo stated that the instability occurs when the resultant tangential stress determined from the strain hardening characteristics

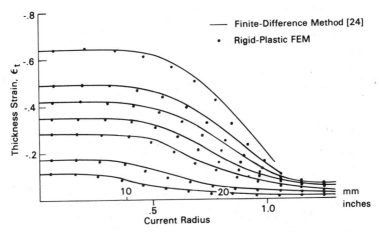

Fig. 11.   Distribution of thickness strain when die profile is considered.

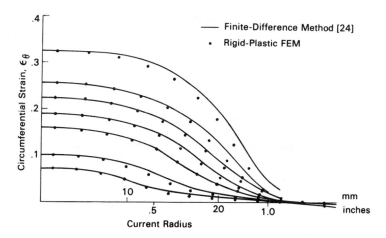

Fig. 12.   Distribution of circumferential strain when die profile is considered.

cannot obtain the value required for the equilibrium. In the present rigid-plastic analysis such an instability is not observed at the point reported by Woo, and the computation continues.

DEEP DRAWING OF A SHEET WITH HEMISPHERICAL PUNCH

In deep drawing a circular sheet of metal is placed between the blank holder and the die and then fully drawn into the shape of a cup by a punch. The formability is then measured by the maximum size of the blank which can be drawn without a failure, or, more often, by its ratio to the punch diameter. This. ratio is called the limiting drawing ratio and this particular kind of test is called the Swift test.

Deep drawing is not only a useful method of material testing, but also one of the basic operations in sheet metal stamping. In practice, various shapes are possible for the bottom of the punch; however, most past investigations are on deep drawing with a flat-bottomed punch.

Among the earlier works on deep drawing are those by Hill [16] and by Chung and Swift [31] using the incremental theory of plasticity. More refined analyses are the finite-difference solutions by Chiang and Kobayashi [32], by Budiansky and Wang [33], and by Chakrabarty and Mellor [34]. Even though such a refinement improves the understanding of the deep drawing process, their works are not complete because they treat the deep drawing problem as an in-plane pure radial drawing and are concerned mostly with the deformation mechanics on the flange. However, it has been observed experimentally (Chung and Swift [31]) that the die profile and the punch profile significantly affect the punch load and the strain distributions and therefore a further refinement is necessary by considering these parameters in the analysis. Woo [35] performs such an analysis.

Contrary to these numerous investigations on deep drawing with a flat-bot-

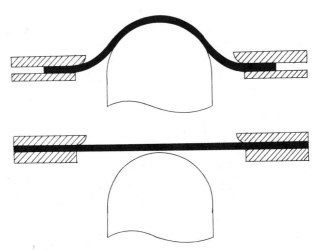

Fig. 13.   Schematic view of the deep drawing of a sheet with the hemispherical head punch.

tomed punch, very few works are reported on the deep drawing of a sheet with a hemispherical head punch (Fig. 13). Woo [36] analyzes this problem by breaking down the deep drawing process into two component processes of the pure radial drawing over the flange and the punch stretching over the hemispherical punch head. He first obtains solutions for pure radial drawing in the flange and then uses this solution at a point initially situated near the die lip as the boundary condition for the stretching problem, and thereby essentially matched the punch stretching component with the pure radial drawing component at a particular point in the die profile region.

Instead of this tedious process of boundary matching, the present finite element method treats the problem in a unified manner. The only finite-element model developed for the deep drawing problem is the elastic-plastic finite element method by Wifi [28] with a limited treatment of friction.

Computational Procedure—The entire sheet in the deep drawing process can be divided into four regions: the contact region with the punch head, the unsupported region, the contact region with the die profile, and the flange over the die. Different kinds of boundary restrictions are imposed depending upon the regions. For example, the flange is constrained to move only horizontally along the die face, while the contact region with the die profile or punch head should satisfy the kind of boundary conditions discussed in the previous section.

The only difference in deep drawing with a hemispherical head punch from the punch stretching with a round die corner is the presence of the flange which is free to slide over the die. The addition of this moving flange is, in effect, equivalent to the addition of the third moving boundary because, even though the boundary separating the flange from the die profile remains stationary in space, it continues to move from the viewpoint of the deforming sheet. To treat this we make an assumption on this third moving boundary and see if it is true by checking the radial positions of the nodes. If the new radial position of any node which is assumed to lie on the flange or the die profile does not fall on the expected region after converged solution is obtained, then the boundary assumption is modified.

Another point to be mentioned is the blank holding condition of which there are two types: clearance holding and force holding. The idealization of the deformation state corresponding to the force blank holding is the plane stress state and the one corresponding to the clearance holding is the plane-strain state. The present rigid-plastic FEM is built to handle the plane stress state deformation. The blank holding force is implemented in the formulation as a tangential friction force acting on the last node located at the rim of the sheet. The increment of deformation is controlled by the punch head movement. The program listing as well as more detailed description of the computation procedure are given in Reference [30].

Results and Discussion—The only available work on the complete analysis of deep drawing with the hemispherical head punch is one by Woo [36]. Along with the numerical solution by the finite-difference method, he also conducted an exper-

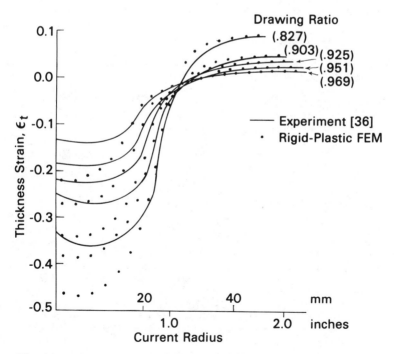

Fig. 14.   Distribution of thickness strains for $\mu_p = 0.04$, $\mu_d = 0.04$.

References pp. 362–363.

iment. The parameters are:

Material: soft copper
Stress-strain characteristics:
   $= 5.4 + 27.8\epsilon^{0.504}$ ton/in.²
   $= (0.08208 + 0.422569\epsilon^{0.504}) \times 10^9$ N/m² ($\epsilon < 0.36$)
   $= 5.4 + 24.4\epsilon^{0.375}$ ton/in.²
   $= (0.08208 + 0.37089\epsilon^{0.375}) \times 10^9$ N/m² ($\epsilon > 0.36$)
Blank radius: 2.2 in. $= 5.588 \times 10^{-2}$ m
Radius of die throat: 2.123 in. $= 5.392 \times 10^{-2}$ m
Radius of die profile: 0.5 in. $= 1.27 \times 10^{-2}$ m
Radius of punch head: 1 in. $= 2.54 \times 10^{-2}$ m
Blank holding force: 0.5 ton $= 500$ kg

As shown in Figs. 14 and 15, the solution by the rigid-plastic FEM is in excellent agreement with the experiment for the flange part; however, over the punch head it predicts more straining than the experiment when the friction coefficient of 0.04 is assigned for the contact region over the punch head and over the die in the numerical analysis. The deviation of the numerical solution from the experimental data gets larger as deformation progresses, which is reflected in the punch load vs. punch depth relationship in Fig. 16.

The lubricant used in the experiment is graphite in tallow and Woo suggested the friction coefficient to be 0.04. In the analysis the practical difficulty always lies in the assignment of a reasonable value of friction coefficient because friction

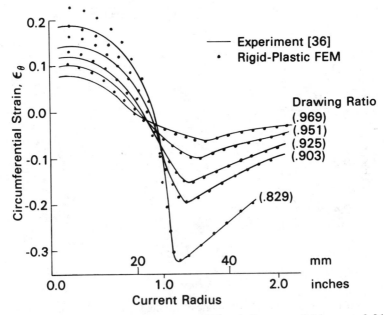

Fig. 15.   Distribution of circumferential strain for $\mu_p = 0.04$, $\mu_d = 0.04$.

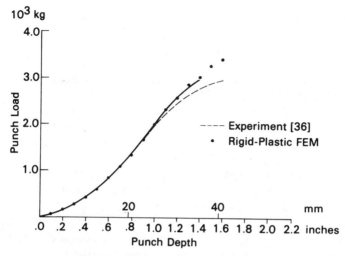

Fig. 16.   Punch load vs. punch depth.

coefficient under a real sheet metal-forming condition is hard to measure and it may even change during deformation.

Comparison of Woo's numerical solution with his experimental data does not yield any better agreement than the present rigid-plastic FEM.

SUMMARY AND CONCLUSIONS

Classical variational formulations for the rigid-plastic solid are not appropriate for solving the sheet metal forming problems. This is due to the nonuniqueness of the deformation mode under certain boundary conditions. This nonuniqueness, however, can be resolved by taking the workhardening rate into consideration. Such an introduction of the workhardening rate into the formulation, on the other hand, necessitates the consideration on the geometry change. An incremental deformation at a generic stage is considered by separating the deformed configuration from the undeformed configuration. The relevant equations are expressed with the deformed configuration at each step as the reference frame and the variational principle is established.

From this variational formulation a finite-element model is developed for the sheet metal forming problems. In many sheet metal forming processes the membrane theory is justifiable and therefore this idealization is introduced in building the model.

Three basic sheet metal forming processes, i.e., the bulging of a sheet subject to the hydrostatic pressure, the stretching of a sheet with a hemispherical head punch, and deep drawing of a sheet with a hemispherical head punch are solved by the method and its solutions are compared with the existing numerical solutions and the experimental data. In hydrostatic bulging the strain distributions and the pressure vs. polar height relationship predicted by the present rigid-

References pp. 362–363.

plastic FEM are in excellent agreement with the available numerical solution by the elasto-plastic FEM.

In punch stretching, to make the problem more tractable, the presence of the die profile is neglected first so that there is only one moving boundary. Taking the die profile into consideration is equivalent to introducing another moving boundary, and while handling two moving boundaries simultaneously could be time consuming, the present rigid-plastic FEM again proves to be efficient and reliable. The strain distributions and the punch load vs. punch depth relationship predicted by the present rigid-plastic FEM are in excellent agreement with the numerical solutions by the finite-difference model.

Then the method is extended to the deep drawing problem. The strain distribution predicted by the present rigid-plastic FEM is in excellent agreement with the experimental data over the flange of the sheet; however, over the punch head, agreement is not as good. An improvement in the prediction seems possible by giving the friction coefficient a proper value. Although confirmation of the validity of the present rigid-plastic FEM for deep drawing analysis needs further investigation, it is concluded that the present rigid-plastic FEM can treat the sheet metal forming problems with efficiency and reasonable accuracy.

ACKNOWLEDGEMENTS

The authors wish to thank the Mathematics Department, General Motors Corp. for its grant and the Air Force Materials Laboratory for its contract F33615-77-C-5111, under which the present investigation was possible. They also wish to thank Mrs. Carmen Marshall for preparing the manuscript.

REFERENCES

[1] C. H. Lee and S. Kobayashi, J. Eng. Ind. (Trans. ASME, B), 95(1973), 865.
[2] C. H. Lee andS. Kobayashi, Proc. North Amer. Metalworking Res. Conf., Hamilton, Canada, 1 (1973), 185.
[3] S. N. Shah, C. H. Lee and S. Kobayashi, Proc. Int. Conf. Prod. Engrs., Tokyo (1974), 295.
[4] S. N. Shah and S. Kobayashi, Proc. 15th Int. Mach. Tool Design and Res. Conf. (1975), 603.
[5] S. H. Lee and S. Kobayashi, Proc. 15th Int. Mach. Tool Design & Res. Conf. (1975), 561.
[6] S. N. Shah and S. Kobayashi, J. Engrg. Prod., 1 (1977), 73.
[7] S. I. Oh, C. C. Chen and S. Kobayashi, to be published in J. Engrg. Prod. (Trans. ASME, B) (1978).
[8] R. Hill, J. Mech. Phys. Solids, 4 (1956), 247.
[9] R. Hill, J. Mech. Phys. Solids, 5 (1956), 1.
[10] R. Hill, J. Mech. Phys. Solids, 5 (1957), 133.
[11] R. Hill, J. Mech. Phys. Solids, 5 (1957), 302.
[12] P. Larsen, Ph.D. dissertation, University of California, Berkeley (1971).
[13] L. Malvern, Introduction to the Mechanics of a Continuous Medium, Prentice-Hall (1969).
[14] A. E. Green and P. M. Naghdi, Arch. Rat. Mech. Anal., 18 (1965) 251.
[15] N. M. Wang, Research Publication GMR-1038, General Motors Research Laboratories, Warren, Mich. (1970).
[16] R. Hill, The Mathematical Theory of Plasticity, Oxford Press (1950).
[17] R. Hill, Philos. Mag., 41 (1950), 1133.

[18] D. M. Woo, Int. J. Mech. Sci., 6 (1964), 303.

[19] Y. Yamada and Y. Yokouchi, Manuf. Res., 21 (1969), 26 (in Japanese).

[20] N. M. Wang and M. R. Shammamy, J. Mech. Phys. Solids, 17 (1969), 43.

[21] H. Iseki, T. Jimma, and T. Murota, Bull. JSME, 17 (1974).

[22] B. Budiansky, J. Appl. Mech. (Trans. ASME, E), 35 (1968), 393.

[23] J. Chakrabarty, Int. J. Mech. Sci., 12 (1970), 315.

[24] D. M. Woo, The Engineer, Lond., 220 (1965), 876.

[25] N. M. Wang and W. J. Gordon, Research Publication GMR-744, General Motors Research Laboratories, Warren, Mich. (1968).

[26] N. M. Wang, Research Publication GMR-862, General Motors Research Laboratories, Warren, Mich. (1969).

[27] N. M. Wang, J. Appl. Mech. (Trans. ASME, E), 37 (1970), 431.

[28] A. S. Wifi, Int. J. Mech. Sci., 18 (1976), 23.

[29] C. H. Lee, S. Masaki and S. Kobayashi, Int. J. Mech. Sci., 14 (1972), 417.

[30] J. H. Kim, Ph.D. Dissertation, University of California, Berkeley (1977).

[31] S. Y. Chung and H. W. Swift, Inst. of Mech. Engrs. (Lond.), 165 (1951), 199.

[32] D. C. Chiang and S. Kobayashi, J. Eng. Ind. (Trans. ASME, B), 88 (1966), 443.

[33] B. Budiansky and N. M. Wang, J. Mech. Phys. Solids, 14 (1966), 357.

[34] J. Chakrabarty and P. B. Mellor, Proc. 5th Biennial Congress, IDDRG; Metall. Ital. (1968), 791.

[35] D. M. Woo, J. Mech. Engrs., 6 (1964), 116.

[36] D. M. Woo, Int. J. Mech. Sci., 10 (1968), 83.

DISCUSSION

S. C. Tang *(Ford Motor Company)*

In your formulation you have incremental plastic strains expressed in terms of stresses by the flow rule. But, you have to invert that matrix, namely, you have to express the incremental stresses in terms of the plastic strains. That matrix is singular, and I wonder how did you really invert that matrix?

Kobayashi

We don't have to deal with stress increment in the rigid-plastic formulation. As far as stress is concerned, we first get the strain distribution from the variational principle, and then calculate the stresses at various points. In fact, in rigid-plastic formulation, we could start from zero stress state because of that.

H. A. Armen *(Grumman Aerospace Corporation)*

Is that the main advantage of the rigid-plastic analysis that it allows you to take larger step size? In the usual elastic-plastic constitutive relations, it requires small steps.

Kobayashi

Yes.

Tang

I have another comment about the maximum load in hydrostatic bulging. I don't think to decrease the pressure after maximum load will get the result because when you decrease the pressure, it rarely converges on the other side (post maximum load side). What people usually do is to add another vector to avoid the singularity and get convergent results on the other side.

M. A. Bragard *(Centre de Recherches Metallurgiques, Belgium)*

If you increase the coefficient of friction, you can surely obtain a better physical correlation. The question is: what value of μ will you need to have a good correlation?

Kobayashi

Yes! We can do anything if I can get the formulation, say, for example μ as a function of coordinates, etc. I will be happy to do the calculation.

N.-M Wang *(General Motors Research Laboratories)*

I have a couple of comments. First, the rigid-plastic analysis allows you to use larger steps. However, for each step, it is a non-linear problem which has to be solved iteratively. Therefore, there is a trade-off in numerical efficiency. The second comment is that the logarithmic strain definition used in your paper is exact rather than what you termed approximate. The Lagrangian strain increment being expressed in principal coordinates and integrated from one step to the next will give the logarithmic form.

Kobayashi

Yes. Thank you.

B. Budiansky *(Harvard University)*

How do you handle unloading in your rigid-plastic formulation?

Kobayashi

I don't know right now. As long as unloading occurs, I don't think we can really rely on the solution with confidence. But, in a very engineeringly approximate way, we can deal with rigid body as such: When the strain rate in an element is very, very small compared to the rest of the sheet, and if you keep iterating, it blows up; so what we can do is that whenever a local strain rate reaches say 10^{-5} compared to 10^{-1}, we keep that number in the matrix which means even for rigid body, we allow a small plastic strain to take place. So, the stress calculation may not be correct, but the singularity associated with the rigid body is avoided. If the boundary condition is a force specified type and contains the rigid body, then we will have trouble. That is to say to some extent we may have to make approximations.

ELASTIC-VISCOPLASTIC ANALYSES OF
SIMPLE STRETCH FORMING PROBLEMS

N.-M. WANG and M. L. WENNER

General Motors Research Laboratories, Warren, Michigan

ABSTRACT

Plane strain and axisymmetric punch stretching problems incorporating Coulomb friction are considered. Based on a general rate-sensitive flow theory and the nonlinear theory of membrane shells, the formulation includes work hardening and normal anisotropy. Two numerical schemes are presented. One is an extension of a previously developed finite element method (restricted to axisymmetric or plane strain problems) while the other is a nonlinear iterative method (restricted to plane strain). Experimental data on aluminum-killed steel are used to devise an explicit form for the rate dependence, and these schemes are applied to plane strain punch stretching (by a flat bottomed punch) and to hemispherical punch stretching. It is shown that a theoretical instability associated with the elastic-plastic analysis in plane strain stretching is removed by inclusion of the strain rate effect, while a better fit to experimental data on strain distribution in hemispherical punch stretching is obtained.

INTRODUCTION

It has long been known that the plastic response of some metals is strongly affected by strain rate, and constitutive equations designed to account for such effects have been studied by such authors as Hohenemser and Prager [1], Malvern [2], Perzyna [3] and Cristescu [4]. These studies have been applied in the past primarily to problems involving the propagation of plastic stress waves. It is now clear, however, that rate sensitivity can be an important factor in sheet metal deformation when varying strain rates exist in the material. Such strain rate variations can very easily arise in simple problems from inhomogeneities in the initial sheet thickness or in material properties. This device has been used by Marciniak, Kuczynski and Pokora [5], Hutchinson and Neale [6] and Ghosh [7] to examine limiting strain behavior for various nominal biaxial strain states. One aim of this type of analysis is to understand better the physical phenomena which govern the Forming Limit Diagram (FLD).

References pp. 390–391.

In general sheet metal forming problems, strain rate variations are caused by friction, geometry and boundary conditions. Solution of such problems via rate insensitive theory (see, e.g., Wang and Budiansky [8]) suggests that strain rates may differ by several orders of magnitude in different regions, and hence, strain rate sensitivity may be an important factor in the strain distribution. Ghosh and Hecker [9] have already asserted that rate sensitivity decreases strain localization, and aids in producing more uniform strain distributions.

We have two major goals in the present paper. First, we wish to develop computational schemes (based on the phenomenological theory of plasticity) for the analysis of punch-stretching problems. Second, these computational methods are used to evaluate the effect of strain rate sensitivity in two simple forming problems, i.e., plane strain and hemispherical punch stretching. These methods and their applications to sheet metal processing problems would be useful to both producers and consumers of sheet materials. For producers, such as the steel and aluminum industries, computational methods could be used in conjunction with experimental forming studies to evaluate the effects of different material parameters in designing more highly formable metals. For consumers, such as the automobile industry, computational methods can assist part designers and processing engineers in evaluating the severity of the forming problem. They would also aid in choosing the most economical material and in designing the sequence of operations required to fabricate the part. Thus, we feel that analytical methods can augment experimental techniques both in designing new materials and in using these materials effectively.

We now describe the content of this paper. We begin by listing the strain-displacement and equilibrium equations for plane strain and axisymmetric problems. The punch-sheet interface conditions, which incorporate Coulomb friction, are written down, and we also discuss a set of constitutive equations for normally anisotropic, rate-sensitive materials. The strain-rate effect is initially left quite general, but later we describe two rather simple forms for describing this effect. By examining experimental data for aluminum-killed (AK) steel, we choose one of these forms and establish the value of certain material parameters.

The two numerical methods used are very briefly compared, and it is shown that they yield nearly identical results for plane strain punch stretching. Detailed descriptions of the methods are relegated to Appendices. Numerical results for plane strain punch stretching and hemispherical punch stretching are presented next. In these discussions, we examine the effects of various material properties on strain distributions, with particular emphasis on strain rate sensitivity.

BASIC EQUATIONS

We list here the basic equations and assumptions used in the present paper. Two problems are considered, namely, plane strain deformation (labeled as Case I), and axisymmetric deformation (Case II). The equations presented below are those appropriate for the membrane theory of shells. Thus, we employ the usual plane stress assumption that the stress component normal to the sheet is negligible with respect to the other components. It is also assumed that the thickness of the sheet metal is much smaller than the smallest radius of curvature of the punch or die. Finally, since the coordinates used here are the principal directions, only two stress and strain components are used in the basic equations.

Strain-Displacement Relations—The sheet is assumed to be initially flat and lies in the $z = 0$ plane of the (ρ, z) coordinate system (Fig. 1). The material points are identified by their initial distance ξ from the z axis. The current horizontal and vertical displacements are u and w, respectively, and they are regarded here as functions of ξ. Fig. 1 serves for both plane strain (in which there is no variation in any quantity in the direction normal to the paper) and for axisymmetric deformation (in which case the z-axis is the axis of symmetry). The principal logarithmic strain components are

$$\epsilon_1 = \ln \left[\left(1 + \frac{du}{d\xi} \right)^2 + \left(\frac{dw}{d\xi} \right)^2 \right]^{1/2}, \quad \text{(Cases I, II)}, \tag{1}$$

and

$$\epsilon_2 = 0, \quad \text{(Case I)},$$

$$\epsilon_2 = \ln \left(1 + \frac{u}{\xi} \right). \quad \text{(Case II)}, \tag{2}$$

The direction indicated by subscript 1 is tangential to the sheet in Fig. 1, while the second direction is normal to the paper. A thickness strain can also be

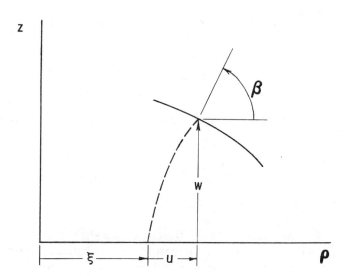

Fig. 1.   Notation and geometry.

defined,

$$\epsilon_h = \ln\left(\frac{h}{h_0}\right), \tag{3}$$

where h, h_0 are the current and initial sheet thicknesses, respectively.

Equilibrium Equations—Let σ_1 and σ_2 be the principal Cauchy stresses defined in the same directions as above. The membrane theory of shells yields the following equilibrium equations: in the tangential direction

$$\frac{d}{d\rho}(h\sigma_1) = -F/\sin\beta, \qquad \text{(Case I)},$$
$$\tag{4}$$
$$\frac{d}{d\rho}(h\rho\sigma_1) = h\sigma_2 - \rho F/\sin\beta \qquad \text{(Case II)},$$

where F is the applied tangential load per unit of deformed area of the sheet. Angle β is defined in Fig. 1. Equilibrium in the normal direction requires

$$h\kappa_1\sigma_1 = P, \qquad \text{(Case I)},$$
$$\tag{5}$$
$$h(\kappa_1\sigma_1 + \kappa_2\sigma_2) = P, \qquad \text{(Case II)},$$

where κ_1 and κ_2 are the principal curvatures of the sheet and P is the normal pressure.

Since Kirchhoff stress components will be used in the constitutive equations, it is convenient to introduce this stress into (4), (5). The Kirchhoff stress $\tau_i (i = 1, 2)$ are related to σ_i by

$$h_0 \exp(-\epsilon_1)\tau_i = h\sigma_i, \qquad \text{(Case I)},$$
$$\tag{6}$$
$$h_0 \exp(-\epsilon_1 - \epsilon_2)\tau_i = h\sigma_i. \qquad \text{(Case II)},$$

Using (6) in (4) and (5), we have

$$\frac{d}{d\rho}(h_0 e^{-\epsilon_1}\tau_1) = -F/\sin\beta \qquad \text{(Case I)},$$
$$\tag{7}$$
$$\frac{d}{d\rho}(h_0 e^{-\epsilon_1 - \epsilon_2}\rho\tau_1) = h_0 e^{-\epsilon_1 - \epsilon_2}\tau_2 - \rho F/\sin\beta, \qquad \text{(Case II)},$$

and

$$h_0\kappa_1 e^{-\epsilon_1}\tau_1 = P, \qquad \text{(Case I)},$$
$$\tag{8}$$
$$h_0 e^{-\epsilon_1 - \epsilon_2}(\kappa_1\tau_1 + \kappa_2\tau_2) = P. \qquad \text{(Case II)},$$

Interface Conditions—During punch stretching, the sheet may be divided into a punch-contact region, a die-contact region and a non-contact region. In the non-contact region, F and P both vanish in (7) and (8). In the punch-contact region the displacements u and w must satisfy the constraint equation

$$w(t) = S(\xi + u(t), t), \tag{9}$$

where t is time and S denotes the surface equation of the punch. Since the punch is assumed to be rigid and moving along the z-axis with a velocity $V(t)$, S may be represented by

$$S(\rho, t) = S_0(\rho) + \int_0^t V(t') \, dt', \tag{10}$$

where S_0 is the equation of the punch at $t = 0$.

In any contact region, the material points can be further classified into "slip" and "non-slip" categories. The slip condition holds when the sheet has a relative velocity tangent to the punch, while in the non-slip case the sheet moves with the punch as a rigid body. It is entirely possible that at the same time, part of the contact region is slipping over the punch while the remainder is locked to it. Since Coulomb friction is assumed, we have

$$\begin{aligned} |F| &= \mu |P|, \quad \text{(Slip)}, \\ |F| &< \mu |P|, \quad \text{(Non-Slip)}, \end{aligned} \tag{11}$$

where μ is the coefficient of friction between sheet and punch. In the present formulation the sign of F may be written as

$$\text{sgn}(F) = -\text{sgn}(\dot{u}), \tag{12}$$

where a dot denotes the time derivative at a constant value of ξ. The die-contact region is handled precisely the same way, except that the die surface is stationary.

Constitutive Equations—We assume that the total strain rates $\dot{\epsilon}_i$ are obtained as a sum of the elastic and viscoplastic parts,

$$\dot{\epsilon}_i = \dot{\epsilon}_i^{(e)} + \dot{\epsilon}_i^{(vp)}, \tag{13}$$

where the elastic strain rates are given by

$$\begin{bmatrix} \dot{\epsilon}_1^{(e)} \\ \dot{\epsilon}_2^{(e)} \end{bmatrix} = \frac{1}{E} \begin{pmatrix} 1 & -\nu \\ -\nu & 1 \end{pmatrix} \begin{bmatrix} \dot{\tau}_1 \\ \dot{\tau}_2 \end{bmatrix}, \tag{14}$$

and the viscoplastic strain rates are

$$\begin{bmatrix} \dot{\epsilon}_1^{(vp)} \\ \dot{\epsilon}_2^{(vp)} \end{bmatrix} = \frac{\dot{\bar{\epsilon}}}{\tau_e} \begin{pmatrix} 1 & -R/(1+R) \\ -R/(1+R) & 1 \end{pmatrix} \begin{bmatrix} \tau_1 \\ \tau_2 \end{bmatrix}, \tag{15}$$

In (14) E and ν are Young's modulus and Poisson's ratio respectively. The functions τ_e and $\dot{\bar{\epsilon}}$ in (15) are the effective stress and effective strain rates defined by

$$\tau_e = \left(\tau_1^2 + \tau_2^2 - \frac{2R}{1+R} \tau_1 \tau_2 \right)^{1/2}, \tag{16}$$

$$\dot{\bar{\epsilon}} = \frac{1+R}{\sqrt{1+2R}} \left[\dot{\epsilon}_1^{(vp)^2} + \dot{\epsilon}_2^{(vp)^2} + \frac{2R}{1+R} \dot{\epsilon}_1^{(vp)} \dot{\epsilon}_2^{(vp)} \right]^{1/2}. \tag{17}$$

The constant R is the normal anisotropy parameter as obtained from Hill's [10]

References pp. 390–391.

theory of anisotropic plasticity, and equations (15) are precisely the classical form of flow theory (except for the anisotropy described by R). In rate-independent theory, τ_e is expressed as a function of the current effective strain, $\bar{\epsilon}$, where

$$\bar{\epsilon}(t) = \int_0^t \dot{\bar{\epsilon}}(t') \, dt'. \tag{18}$$

In contrast, we assume here that the effective stress depends upon both $\bar{\epsilon}$ and $\dot{\bar{\epsilon}}$,

$$\tau_e = \tau_e(\bar{\epsilon}, \dot{\bar{\epsilon}}). \tag{19}$$

Although the assumption (19) is a simple extension of inviscid flow theory, it is also quite general, and with suitable interpretations it can incorporate the "overstress" theory of viscoplasticity as well as other representations of strain rate sensitivity.

CONSTITUTIVE EQUATIONS

In order to specify fully our constitutive equations, there remains only the task of choosing the function $\tau_e(\bar{\epsilon}, \dot{\bar{\epsilon}})$ in Equation (19). Obviously, experiments are needed in order to write down this relationship for any particular material. Before presenting some experimental results, however, let us consider several alternatives for the form of the function $\tau_e(\bar{\epsilon}, \dot{\bar{\epsilon}})$. Malvern [2] and Sokolovsky [11], following Hohenemser and Prager [1], introduced the notion of "overstress" into viscoplasticity. Perzyna [3, 12] later generalized these results and showed that some phenomena of rate-dependent materials could be adequately described by such a model. Within the context of our constitutive equations presented previously, and under the assumption of isotropic work hardening, we now briefly describe the overstress model. Let $\kappa(\bar{\epsilon})$ be the hardening parameter in isotropic work hardening. In rate insensitive materials, we would have $\tau_e = \kappa$ at any stage of plastic deformation. The overstress postulate is that a rate sensitive material can attain stress states for which the effective stress exceeds $\tau(\bar{\epsilon})$ by an amount which depends upon the effective strain rate, and, perhaps, the effective strain. Symbolically,

$$\tau_e = \kappa(\bar{\epsilon}) + g(\bar{\epsilon}, \dot{\bar{\epsilon}}), \tag{20}$$

where $g(\bar{\epsilon}, \dot{\bar{\epsilon}})$ is a material dependent function which is non-negative for all strain states. At any state for which $\tau_e < \kappa(\bar{\epsilon})$, the material response is required to be elastic. It is easily seen from (20) that the constitutive equations of inviscid flow theory are recovered as the function $g(\bar{\epsilon}, \dot{\bar{\epsilon}})$ reduces to zero. Two particularly simple forms of (20) are often used in practice. They are:

$$\tau_e = \kappa(\bar{\epsilon}) + \psi(\dot{\bar{\epsilon}}/\gamma), \tag{21}$$

$$\tau_e = \kappa(\bar{\epsilon})[1 + \phi(\dot{\bar{\epsilon}}/\gamma)], \tag{22}$$

where ψ, ϕ are material functions and γ is a reference effective strain rate. The form (22) represents a material whose work-hardening is effectively increased with strain rate, i.e., $d\tau_e/d\bar{\epsilon}$ increases with $\dot{\bar{\epsilon}}$. Equation (21), on the other hand, exhibits no increase in work hardening with strain rate. In a uniaxial tensile test, for example, (21) would produce stress-strain curves which are higher than, but parallel to, the static curve.

It is important to note that use of an overstress model to describe a particular material requires the determination of the static "yield function", $\kappa(\bar{\epsilon})$. This function is easily found in a uniaxial relaxation test. Suppose, for example, that the material has been deformed in a testing program to an effective plastic strain $\bar{\epsilon}_0$ at time t_0. Subsequently, the strain component ϵ_1 is held fixed, and the material relaxes to a stress τ_1. During the test, the effective plastic strain rate decreases to zero, and (21) and (22) reduce to

$$\tau_e(t \to \infty) = \tau_1(t \to \infty) = \kappa\left(\bar{\epsilon}_0 + \frac{\tau_1(t_0) - \tau_1(t \to \infty)}{E}\right) \tag{23}$$

Neglecting the change in the effective plastic strain due to the elastic deformation, we write

$$\tau_1(t \to \infty) = \kappa(\bar{\epsilon}_0). \tag{24}$$

Equation (24) effectively determines the function $\kappa(\bar{\epsilon})$, and uniaxial tensile tests carried out at constant strain rates can now be used to find the function ψ or ϕ.

Another commonly used form for the function τ_e is

$$\tau_e = K\bar{\epsilon}^N(\dot{\bar{\epsilon}}/\gamma)^m. \tag{25}$$

In this formulation, no reference is made to a static yield curve; hence, it is not an "overstress" model, and no restriction is placed on the magnitude of τ_e. Equation (25) has several advantages over (21) and (22). It is simpler in form, it uses a measure of strain rate sensitivity (the value of m in (25)) which is often used by experimentalists, and it does not require relaxation data. Of course, it is not as flexible as (21) in matching experimental data.

With this background, let us consider some experimental data for aluminum-killed (AK) steel, a material which is known to be strain rate sensitive. Dr. Robert Ayers of GMR's Physics Department performed relaxation tests and uniaxial tensile tests on samples of this material. The tensile tests were run at constant engineering strain rates ranging from 0.0002/s to 0.2/s. All tests were performed on an MTS hydraulic servo-feedback machine which has the capability of monitoring and adjusting the strain rate on the test section of the material specimen. The monitoring device is an extensometer which measures strain over a 50.8 mm (two inch) gage length. All tests were carried out with the stress parallel to the rolling direction. Due to machine limitations, the maximum strain rate available was 0.2/s.

It was found that the stress strain curves were approximately parallel, and hence, Equation (21) was better able to reproduce them than was (22). Using the relaxation data and the fact that the static curve should be parallel to the other curves for equation (21)*, the function $\kappa(\bar{\epsilon})$ was determined to be

$$\kappa(\bar{\epsilon}) = K(\bar{\epsilon} + \epsilon_0)^n$$

$$K = 482.7\,\text{MPa}, \quad \epsilon_0 = 0.007707, \tag{26}$$

$$n = 0.22.$$

The parameter n in this equation may be regarded as the static work-hardening exponent. With $\kappa(\bar{\epsilon})$ given by this equation, the uniaxial stress-strain curves were fairly well matched by taking for ψ the function

$$\psi = K\eta \ln(1 + \dot{\bar{\epsilon}}/\gamma),$$

$$\eta = 0.018, \tag{27}$$

$$\gamma = 0.000137/\text{s}.$$

The experimental curves together with the corresponding results from (26) and (27) are given in Fig. 2. It is seen that the lower strain rates are matched quite well, while for $\dot{e}_1 = 0.02/\text{s}$ the experimental curve lies under the predicted curve. At the highest rate, the experimental curve is no longer parallel to the others,

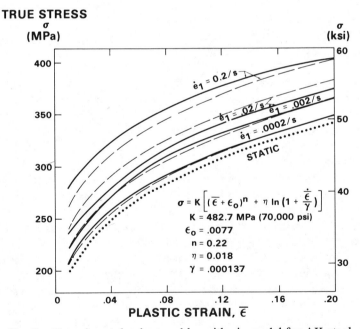

Fig. 2. Experimental values and logarithmic model for AK steel.

* Strictly speaking, $\kappa(\bar{\epsilon})$ should be determined only from relaxation data. However, these data were sketchy and somewhat scattered, and since our real aim was to match stress-strain curves at finite strain rates, we used results at 0.0002/s to help in the construction of $\kappa(\bar{\epsilon})$.

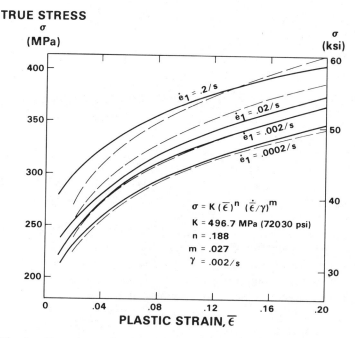

Fig. 3.   Experimental values and power law model for AK steel.

and the model matches it only at higher strains. Altogether, the agreement can only be called fair, but it is sufficiently accurate for our purposes. We feel that the final characterization of the strain rate sensitivity of AK steel (or any other material) will require an extensive testing program.

We also attempted to match the experimental data with Equation (25). However, it proved to be impossible to maintain accuracy at both high and low strains. Hence, we chose K and N to approximate the $\dot{e}_1 = 0.002/s$ curve*, and then found m such that the stresses would be as accurate as possible at $\bar{\epsilon} = 0.1$. We found the values $K = 496.7$ MPa, $N = 0.188$ and $m = 0.027$. A comparison with the experimental data is shown in Fig. 3 where it is seen that the lower curves are matched quite well, but deviations are present for higher strain rates. However, it seems clear that the overstress model is more accurate than (25). This is not surprising, of course, since more parameters were used.

In the following, our computations are carried out using the model

$$\tau_e = K[(\bar{\epsilon} + \epsilon_0)^n + \eta \ln(1 + \dot{\bar{\epsilon}}/\gamma)] \tag{28}$$

and sensitivity to strain rates are varied by means of the parameter η. However, neither of the numerical procedures are dependent on this representation, and, in principle, any well-behaved function could be used for $\tau_e(\bar{\epsilon}, \dot{\bar{\epsilon}})$.

---

* In equation (25), γ was taken to be 0.002/s.

References pp. 390–391.

NUMERICAL METHODS—A COMPARISON

The basic equations presented above were used to develop two numerical methods. One is a finite element method (described in Appendix A), while the other is a nonlinear iteration method (Appendix B), which is used here only for plane strain punch stretching. The finite element method is far more flexible, in the sense that it can easily be adapted to a wide range of problems. The iterative method, on the other hand, is much more economical in computer time (by a factor of at least 10) for the plane strain problem.

As a check on both of the programs, a test problem in plane strain stretching was chosen and analyzed by each method. The basic geometry is shown in Fig. 4. The sheet is initially flat and it is stretched over a die by a flat bottomed punch with radii ground at its edges. The sheet is clamped at the outside edges, and the punch moves vertically with a constant velocity. Friction is assumed to exist at the radii on both punch and die. The friction coefficients between punch and sheet and between die and sheet are μ_P, μ_D, respectively. Since this is a plane strain problem, no quantity varies in the direction perpendicular to the paper in Fig. 4, and the strain component in this direction is zero everywhere.

The parameters shown in Fig. 4, together with a value of 0.018 for η (typical of AK steel), were used in each program. The punch velocity was taken to be 1 mm/s. Strain distributions at three values of punch depth are plotted in Fig. 5. For moderate punch depths, the results are virtually identical for each method. As peak strain increases above 30%, however, the iterative method predicts slightly higher strains in the unsupported region between the punch and die. The maximum difference is about 0.7% strain at a peak strain of 50%. In Fig. 6 the peak strain and the floor strain (i.e., the strain under the punch) are plotted as

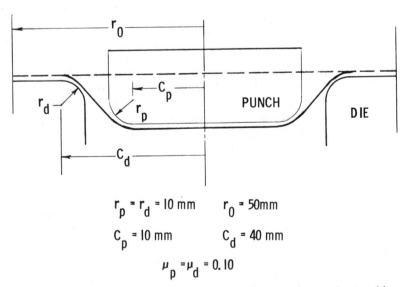

$$r_p = r_d = 10 \text{ mm} \qquad r_0 = 50 \text{mm}$$

$$C_p = 10 \text{ mm} \qquad C_d = 40 \text{ mm}$$

$$\mu_p = \mu_d = 0.10$$

Fig. 4.   Configuration and dimensions for plane strain punch stretching.

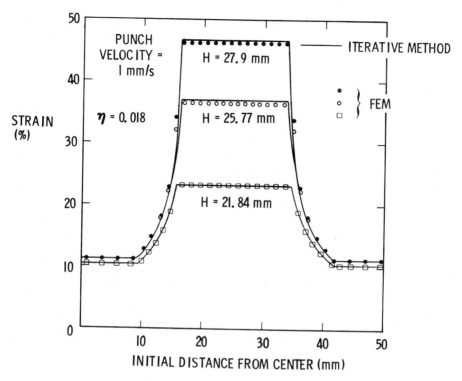

Fig. 5. Comparison of strain distributions in plane strain punch stretching for finite element and iterative methods.

functions of punch depth. Note that the predictions for floor strain are virtually identical for each method. While there are slight differences in the peak strain, these are not very important. For example, an important practical problem is to predict punch depth at failure. If failure occurs at a certain value of peak strain (as predicted by the use of forming limit diagrams, e.g.), then the difference in failure punch depths from the two methods will be less than 1%.

Figs. 5 and 6 indicate that the two methods give similar results, and hence, that both programs are functioning properly. In the following, the nonlinear iterative method (Appendix B) is used to study the plane strain problem, while the finite element method of Appendix A is employed for hemispherical punch stretching.

PLANE STRAIN PUNCH STRETCHING

The basic problem of plane strain punch stretching, with its associated geometry, has already been described. This problem is both simple and technically interesting. For example, the process of stamping stiffening channels in an otherwise flat panel can be represented by a plane strain analysis. We now present

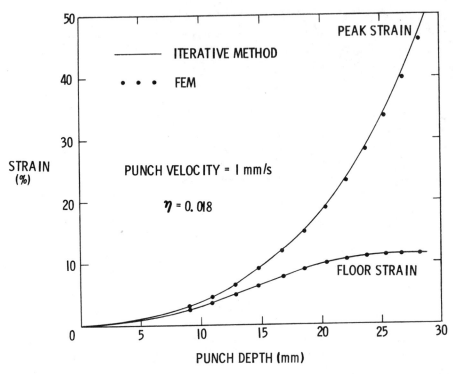

Fig. 6. Comparison of peak strain and floor strain versus punch depth in plane strain punch stretching for finite element and iterative methods.

some further results on this problem. In particular we study the effects of material properties, punch velocities and initial inhomogeneities.

In the following we always use the geometrical parameters and friction coefficients shown in Fig. 4. Also, as previously indicated, we use (28) to describe the material. To describe some basic results, we now return to Figs. 5 and 6. Note from Fig. 5 that as punch depth increases above 20 mm, most of the additional strain is concentrated in the unsupported area between the punch and die. This, of course, is due to friction, and such concentrations of incremental strain are typical of punch stretching problems with friction. This effect is shown graphically in Fig. 6, where it is seen that after a punch depth of 27 mm, there is no further straining under the punch. This material sticks to the punch and moves as a rigid body, while strain in the unsupported region increases very rapidly.

The effect of changing the strain rate sensitivity is described in Fig. 7. Here, we show strain distributions at a punch height of 26.7 mm for three different values of η. As strain-rate sensitivity increases, the peak strain decreases. At the same time, of course, the floor strain becomes larger. That is, a material which is highly strain rate sensitive tends to produce a strain distribution which is more uniform than one which has small rate sensitivity. If the punch depth is limited by the maximum strain in the part, then strain rate sensitive materials will obviously produce deeper parts without failure.

Fig. 8 shows peak strain and floor strain versus punch depth for the same parameters. In addition, we include the case $\eta = 0$, which corresponds to inviscid elastic-plastic flow theory. In this latter case, the computations stop at a strain $\epsilon_1 \approx n = 0.22$. At this point the material sustains its greatest load, and further increases in punch depth cannot be supported. Hence, there is no longer a unique solution to the problem. However, if the material has even a very small strain rate sensitivity, the increased loads are matched by increased strain rates, and computation can continue. Since it is found in practice that AK steel can sustain strains far exceeding the n value in plane strain, Figs. 7 and 8 show that we might tentatively ascribe this fact to strain-rate sensitivity. At the very least, incorporating strain-rate sensitivity allows us to extend our computations to strain states not available to inviscid elastic-plastic theory.

In a sense, however, these viscoplastic results are too successful. That is, strains greatly exceeding those found in practice are easily reached in the numerical calculations. Hence, an initial inhomogeneity was introduced into the sheet material in the peak strain region, as shown in Fig. 9. In the numerical computations, the inhomogeneity was incorporated by assigning to the initial sheet thickness at one node a value equal to $1 - f$ times the thickness of the remainder of the sheet. The strain inside the inhomogeneity e_i is plotted versus the outside strain e_0 in Fig. 10 for several values of f. The material constants are those of AK steel and the punch velocity is 1 mm/s. As expected, the strain inside the inhomogeneity rises almost vertically as punch depth increases. The

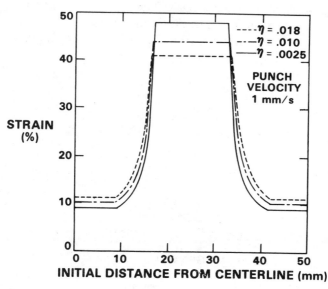

Fig. 7. Effect of strain rate sensitivity on strain distributions for plane strain punch stretching.

References pp. 390–391.

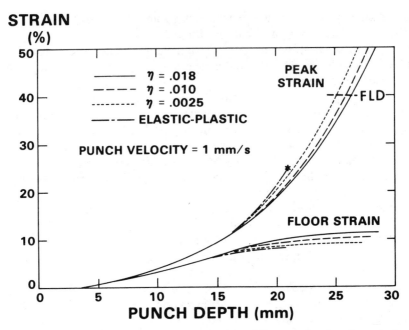

Fig. 8.   Effect of strain rate sensitivity on peak and floor strains for plane strain punch stretching.

asymptotic values of e_0 are plotted for several punch velocities in Fig. 11. At punch velocities of 1, 0.1 and 0.01 mm/s the curves are quite close. For smaller velocities, however, e_0 decreases substantially. For reference purposes, we also include results from inviscid elastic-plastic flow theory, using the same values of K, n, ϵ_0 and R. However, in this case, we plot the value of e_0 when e_i reaches a value corresponding to a load maximum. It is clear that an inhomogeneity has

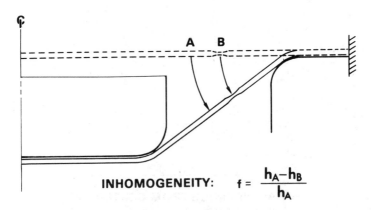

Fig. 9.   Thickness inhomogeneity in sheet.

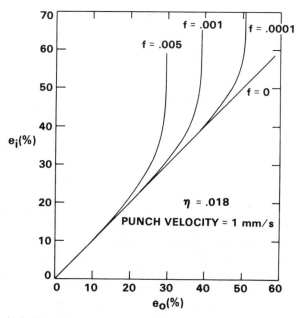

Fig. 10.   Strain inside inhomogeneity versus strain in uniform neighboring region.

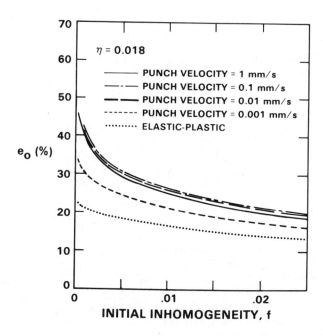

Fig. 11.   Limiting strains versus initial inhomogeneity for several punch velocities.

References pp. 390–391.

a much stronger effect in viscoplasticity than in plasticity, and that the limit strains are greatly increased by incorporating strain-rate sensitivity. We point out, however, that the curves in Fig. 11 do not flatten out very rapidly as f increases (compared to results in shell buckling with imperfections, for example), and it is not immediately clear from this diagram what the proper limiting value of e_0 should be.

We remark here that other authors [5, 6, 7] have studied the problem of inhomogeneity growth in rate-sensitive materials*, and have produced results very similar in form to Fig. 11. However, these studies used a rate sensitivity law of the form

$$\tau_e = K(\bar{\epsilon} + \epsilon_0)^n \dot{\bar{\epsilon}}^m, \tag{29}$$

and did not solve for the limiting strain in the context of a punch stretching problem. With these assumptions, their results on the limiting strains are independent of strain rate, in contrast to the findings here. Although our results are not greatly affected by punch velocities in the range 0.01 to 1 mm/s, it is clear that they must reduce to the rate-insensitive curve in Fig. 11 as punch velocity (i.e., strain rate) becomes very small. This statement is not true for (29), which reproduces the inviscid results only as m approaches zero, regardless of strain rate.

We have also examined the effects of the parameter n and R on strain distribution. It is seen from Fig. 12 that increasing n (which corresponds to increasing

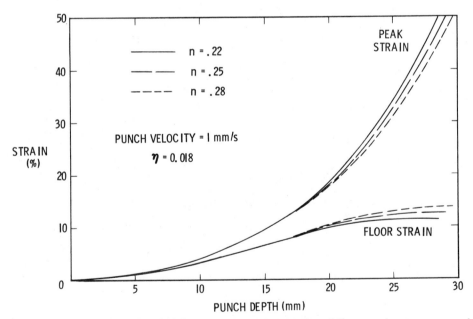

Fig. 12.   Effect of work hardening parameter n on peak and floor strains versus punch depth in plane strain punch stretching.

* These authors also treated stress states different from plane strain.

the rate of work-hardening) also tends to produce a more uniform strain distribution. The anisotropy parameter R, however, has virtually no effect at all on the strain distribution for plane strain problems.

As discussed in [8], a considerable simplification occurs in the constitutive equations if Poisson's ratio, ν, is put equal to $R/(1 + R)$. To test the accuracy of this approximation, several test cases in plane strain punch stretching were run with $\nu = .3$ (the appropriate value for steel) and with $\nu = R/(1 + R)$. It turned out that the strain distributions were virtually indistinguishable, as plotted on the scale of Figs. 5 or 6. The stresses, on the other hand, were different at small punch depths (as one might easily guess), but they quickly converged. This stress behavior is shown in Fig. 13, where we plot the maximum stress in the

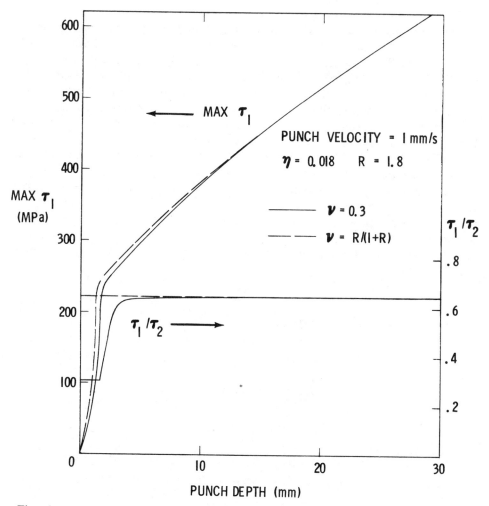

Fig. 13.   Maximum stress components versus punch depth for realistic and approximate values of Poisson's Ratio.

sheet, and the ratio of the maximum τ_1 to the maximum τ_2 versus punch depth. It is easily seen that in the elastic region the stress response is different for the two values of ν. However, as the punch depth increases, these differences gradually become trivial. We conclude that putting ν equal to $R/(1 + R)$ has no discernible effect on the computations at larger punch depths (or strains), and hence, for our purposes, it is a perfectly reasonable approximation to make.

HEMISPHERICAL PUNCH STRETCHING

The configuration used for hemispherical punch stretching is shown in Fig. 14, which also includes the values of the geometrical parameters used throughout this section. The basic material law is again given by (28) with the parameters listed in (26) and (27), and a value of 1.8 for R. In the following, however, we shall examine the consequences of changing some of these parameters.

Fig. 15 shows distributions of engineering strains in the radial and circumferential directions for four different punch depths. The coefficient of friction is 0.2, and the material parameters are those of AK steel. The punch velocity is taken to be 0.423 mm/s (1 in/min), a standard laboratory punch speed. Note that friction causes a distinct peak in the radial strain. In fact, at larger punch depths most of the incremental radial strain occurs in a relatively small area of the sheet.

Strain distributions at a constant punch depth are shown in Figs. 16(a) and (b) for two values of the strain rate sensitivity parameter η and for two values of the coefficient of friction. We also include results for the corresponding rate-insensitive calculations. Although the circumferential strain is not greatly affected by the rate sensitivity, it is clear that the peak in the radial strain is reduced with increasing η. Note that even a value of 0.010 for η is effective in reducing the magnitude of the peak strain and in producing a generally more uniform strain distribution.

In Fig. 17 are plotted the peak strains together with the associated circumfer-

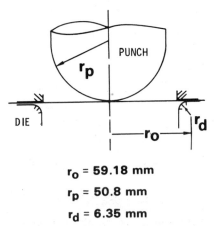

$r_o = 59.18$ mm

$r_p = 50.8$ mm

$r_d = 6.35$ mm

Fig. 14. Configuration and dimensions for hemispherical punch stretching.

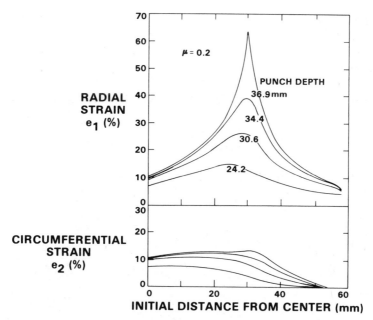

Fig. 15. Radial and circumferential strain distributions for hemispherical punch stretching—AK steel.

ential strains at the same material point as a function of punch depth. Again, we include two values of friction, two values for η, and elastic-plastic results. The effect of rate sensitivity in reducing the peak strain at a given punch depth is again shown. To illustrate the practical significance of this result, we have indicated on Fig. 17 the points at which the various peak strain curves cross the

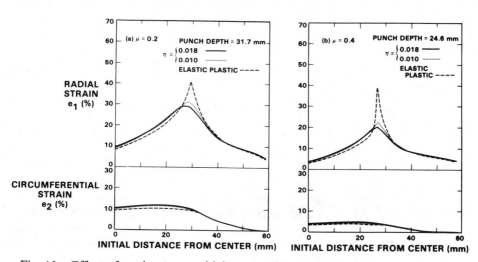

Fig. 16. Effect of strain rate sensitivity on radial and circumferential strain distributions in hemispherical punch stretching.

References pp. 390–391.

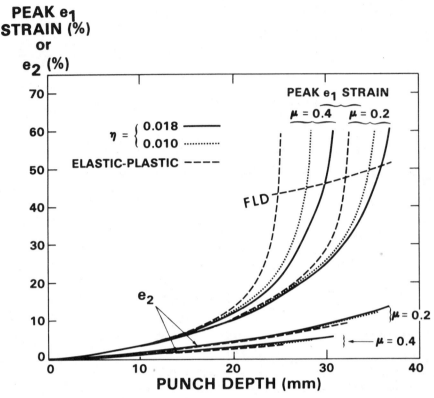

Fig. 17.   Effect of strain rate sensitivity on peak strain versus punch depth in hemi-spherical punch stretching.

forming limit diagram for AK steel. Note that the predicted punch depths at failure are about 3–5 mm greater for $\eta = 0.018$ than for inviscid elastic-plastic theory. This corresponds to approximately a 10% increase in depth, a result of some importance.

It is also clear from Fig. 17 that increasing the frictional coefficient μ causes the peak strain to increase more rapidly with respect to punch depth.

The effects of n and R on the strain distributions are given in Fig. 18. An increase in the static work-hardening exponent n also causes a more uniform strain distribution. Note, however, that strain near $\xi = 0$ is not changed. The normal anisotropy parameter R has a much different effect. As R increases, the peak strain also increases, and the pole strain decreases. Thus, a material with a large value of R yields a more non-uniform distribution than a material with small R. This situation stands in contrast to the case of plane strain stretching, where R has little effect, and to deep-drawing, where a large R is desirable.

In the remainder of this section we shall discuss a comparison between theory and some experimental results obtained by Dr. Amit Ghosh while he was with General Motors Research Laboratories. Before this comparison is begun, how-

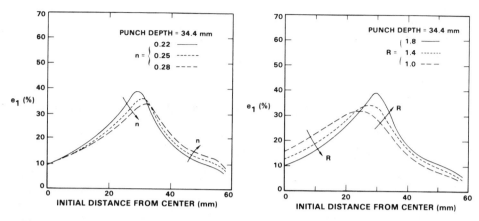

Fig. 18.   Effect of n and R on strain distributions in hemispherical punch stretching.

ever, we must point out that at the present time a completely unambiguous comparison is not possible. The reason for this situation is that we do not know what value of coefficient of friction to use in representing a particular test condition. We handle this ambiguity by choosing μ so that a plot of *peak* strain versus punch depth for a particular test matches the experimental results in a reasonable way. Using this frictional coefficient, we then compare strain *distributions* in the sheet at several punch depths.

Before comparing with the rate-sensitive theory, we show in Fig. 19 peak strain versus punch depth curves at three values for μ, using the rate-insensitive elastic-plastic theory. Experimental values for a test in which no lubrication was used between punch and sheet are also given. Note that a line which interpolates the experimental data does not lie on any of the analytical curves. In other words, it is not possible to choose a single value for μ which reproduces the experimental points at all punch depths, although we obviously can choose μ so that the peak strain is correctly predicted at a single value of punch depth.

In Fig. 20 we show the corresponding results for the viscoplastic theory using the values of the parameters given in (26) and (27). In addition, experimental results for a test in which the punch-sheet interface is well lubricated* are also given. Note that the experimental points for the unlubricated case now follow an analytical curve much more closely. In fact, we choose $\mu = 0.4$ to match the dry test data. The lubricated condition is modeled by taking $\mu = 0.25$. We point out here that the lubricated test could also be fairly well matched by the inviscid theory. However, the strain rate variations are much larger in the dry test, and it is encouraging that incorporating strain rate sensitivity into the analysis gives an improvement in a test for which improvement is needed, and in which one would expect strain rate sensitivity to be more important.

---

* *The lubrication consisted of a teflon coating on the sheet and a layer of polyethelene between the punch and sheet.*

References pp. 390–391.

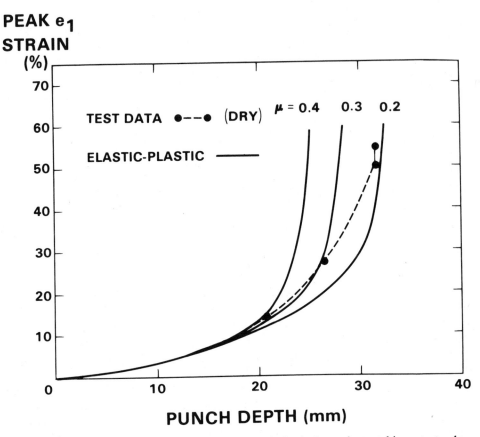

Fig. 19.   Peak strain versus punch depth in hemispherical punch stretching—test values and elastic-plastic results.

Strain distributions for the test data and the viscoplastic theory are shown in Fig. 21. These results are fairly good, but at large punch depths, the calculated location of the peak strain in both the lubricated and dry cases differ from the measured locations by about 3 mm. Also, the calculated strains generally underestimate the experimental values near the edge. Referring to Fig. 18, it is seen that the effect of increasing n would be to move the peak strain to the right, and to increase the strain near the edge. But these are precisely the changes needed to produce results in better agreement with test data in Fig. 21. Moreover, there is evidence to indicate that AK steel has a higher value of n in balanced biaxial tension than in uniaxial tension [13]. Since the stress state in most of the sheet during hemispherical punch stretching is much closer to balanced biaxial than to uniaxial stress, it would appear to be reasonable to increase n. Hence, we have calculated strain distributions for $n = 0.25$ (the other parameters remaining constant) and plotted the results in Fig. 22. The calculated results are now much closer to the experimental values. In particular, the location of the peak strain

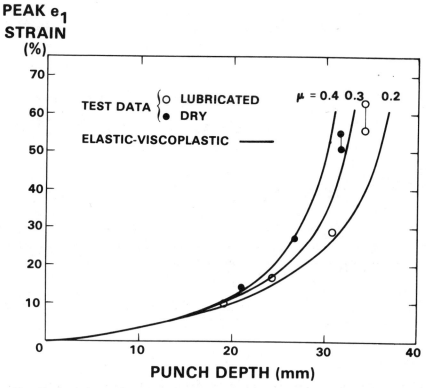

Fig. 20.   Peak strain versus punch depth in hemispherical punch stretching—test values and elastic-viscoplastic results.

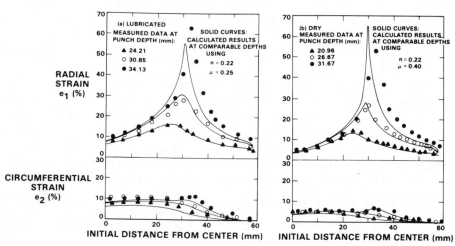

Fig. 21.   Strain distributions in hemispherical punch stretching-test values and elastic-viscoplastic results.

References pp. 390–391.

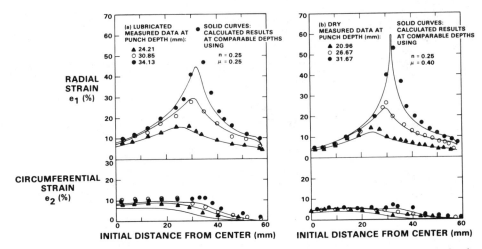

Fig. 22. Strain distributions in hemispherical punch stretching-test values and elastic-viscoplastic results using an n-value appropriate for biaxial deformation.

now matches the experimental location, and the discrepancy near the edge has been greatly reduced.

CONCLUSIONS

We have shown in this paper that a relatively simple extension of elastic-plastic flow theory can be used to describe strain rate sensitivity. This extension includes some commonly used representations of strain rate sensitivity, and, in particular, the overstress model. Experimental data for uniaxial deformation of AK steel is fairly well modeled by a simple viscoplastic law, but more accurate (and more elaborate) representations are needed. There are, in fact, two main difficulties in using uniaxial data in a biaxial problem. First, the data are restricted to strain levels which are roughly half the magnitude of those found in realistic biaxial problems, and second, the plastic response under biaxial conditions may differ from that in uniaxial stress.

It is clear that rate sensitivity yields some benefits in punch-stretching problems. In plane strain, the instability at a strain near the value of n is removed. In hemispherical punch-stretching, a better fit with experimental results is given. Moreover, a generally more uniform strain distribution is predicted when rate sensitivity is included. This has been shown to have an important bearing on the prediction of maximum punch depth. This latter point especially allows us to conclude that when rate sensitivity exists in a material, it should be included in punch-stretching calculations.

REFERENCES

[1] K. Hohenemser and W. Prager, Z. Angew. Math. Mech., *12* (1932), 216.
[2] L. E. Malvern, J. Appl. Mech. (Trans. ASME, E), *18* (1951), 203.

[3] P. Perzyna, Q. Appl. Math., *20* (1963), 321.
[4] N. Cristescu, Dynamic Plasticity, North-Holland (1967).
[5] Z. Marciniak, K. Kuczynski and T. Pokura, Int. J. Mech. Sci., *15* (1973), 789.
[6] J. W. Hutchinson and K. W. Neale, "Sheet Necking—III. Strain rate effects," this volume.
[7] A. K. Ghosh, "Plastic flow-properties in relation to localized necking in sheets," this volume.
[8] N. M. Wang and B. Budiansky, J. Appl. Mech. (Trans. ASME, E), *45* (1978), 73.
[9] A. K. Ghosh and S. S. Hecker, Metall. Trans., *6A* (1975), 1065.
[10] R. Hill, The Mathematical Theory of Plasticity, Oxford Press (1950).
[11] V. V. Sokolovsky, Doklady Akad. Nauk. SSSR, *60* (1948), 775 (in Russian).
[12] P. Perzyna, Adv. Appl. Mech., *9* (1966).
[13] A. K. Ghosh and J. V. Laukonis, Proc. 9th Biennial Congress IDDRG, Ann Arbor (1976), 167.
[14] B. Budiansky, J. Appl. Mech. (Trans. ASME, E), *35* (1968), 393.
[15] O. C. Zienkiewicz and I. C. Cormeau, Arch. Mech., *24* (1972), 873.
[16] A. Ralston, A First Course in Numerical Analysis, McGraw-Hill, New York (1965).

APPENDIX A—A FINITE ELEMENT PROCEDURE

In order to develop a finite element procedure for the problems considered in this paper, we write the virtual work theorem:

$$h_0 \int_{A_0} (\tau_1 \delta \dot\epsilon_1 + \tau_1 \delta \dot\epsilon_2)\, dA_0 = \int_{A_0} (f_1 \delta \dot u + f_2 \delta \dot w)\, dA_0, \tag{30}$$

where f_1, f_2 are components in the horizontal and vertical directions, respectively, of the prescribed surface loads per unit undeformed area. The differential area in (30) is given by

$$\begin{aligned} dA_0 &= d\xi, & \text{Case I} \\ dA_0 &= 2\pi\xi\, d\xi. & \text{Case II} \end{aligned} \tag{31}$$

It may be noted that the term $\tau_2 \delta \dot\epsilon_2$ vanishes for Case I since $\epsilon_2 = 0$ by definition. Following the standard procedure [14], we obtain a variational equation which involves stress rates and loading rates as follows:

$$h_0 \int_{A_0} (\dot\tau_1 \delta \dot\epsilon_1 + \dot\tau_2 \delta \dot\epsilon_2)\, dA_0 - 2h_0 \int_{A_0} (\tau_1 \dot\epsilon_1 \delta \dot\epsilon_1 + \tau_2 \dot\epsilon_2 \delta \dot\epsilon_2)\, dA_0$$

$$+ h_0 \int_{A_0} \left\{ \tau_1 \frac{\dot u_\xi \delta \dot u_\xi + \dot w_\xi \delta \dot w_\xi}{[(1 + u_\xi)^2 + w_\xi^2]} + \tau_2 \frac{\dot u \delta \dot u}{(\xi + u)^2} \right\} dA_0 \tag{32}$$

$$= \int_{A_0} (\dot f_1 \delta \dot u + \dot f_2 \delta \dot w)\, dA_0$$

This equation together with the strain-displacement relations (1)–(2), the interface conditions (10)–(12) and the constitutive equations (13)–(19) are the basic equations for the derivation of the present finite element procedure.

Referring to Fig. 1, we let the initial sheet domain be decomposed into an assemblage of flat membrane elements for Case I and truncated cone elements for Case II. Let (ρ_1, z_1) and (ρ_2, z_2) be the current nodal coordinates of a typical element whose initial nodal coordinates are ξ_1, ξ_2. We assume that within the element the displacements u and w are linear in ξ, i.e.,

$$\begin{pmatrix} u \\ w \end{pmatrix} = \frac{\xi_2 - \xi}{\Delta\xi} \begin{pmatrix} U^{(1)} \\ W^{(1)} \end{pmatrix} + \frac{\xi - \xi_1}{\Delta\xi} \begin{pmatrix} U^{(2)} \\ W^{(2)} \end{pmatrix}, \tag{33}$$

where $\Delta\xi = \xi_2 - \xi_1$ and $U^{(i)}$, $W^{(i)}(i = 1, 2)$ refer to the horizontal and vertical nodal displacements, respectively.

The elemental strain-displacement relations may be obtained by substituting (33) into (1)–(2) and differentiating the resulting expressions with respect to t:

$$\dot{\epsilon} = H\dot{u}, \tag{34}$$

where

$$\dot{\epsilon} = \begin{pmatrix} \dot{\epsilon}_1 \\ \dot{\epsilon}_2 \end{pmatrix}, \qquad H = \begin{pmatrix} H_1 \\ H_2 \end{pmatrix},$$

$$H_1 = \left(-\frac{(1 + u_\xi)}{\Delta\xi\, G}, \frac{-w_\xi}{\Delta\xi\, G}, \frac{1 + u_\xi}{\Delta\xi\, G}, \frac{w_\xi}{\Delta\xi\, G} \right),$$

$$H_2 = \begin{cases} (0, 0, 0, 0) & \text{, Case I} \\ \left(\dfrac{\xi_2 - \xi}{\Delta\xi\,(\xi + u)}, 0, \dfrac{\xi - \xi_1}{\Delta\xi\,(\xi + u)}, 0 \right), & \text{Case II} \end{cases}$$

$$\dot{u}^T = (\dot{U}^{(1)}, \dot{W}^{(1)}, \dot{U}^{(2)}, \dot{W}^{(2)}), \qquad G = (1 + u_\xi)^2 + w_\xi^2.$$

Inverting the constitutive equations (13)–(15) and assuming as in [8] that $\nu = \dfrac{R}{1 + R}$, we obtain

$$\dot{\tau} = D\dot{\epsilon} - \dot{\epsilon}C\tau, \tag{35}$$

where

$$D = \frac{E(1 + R)}{(1 + 2R)} \begin{pmatrix} 1 & R/(1 + R) \\ R/(1 + R) & 1 \end{pmatrix},$$

and

$$C = E/\tau_e.$$

Substituting (34) and (35) into (32) gives the elemental equilibrium equations

$$K_e\dot{u} = \dot{f}^{(vp)} + \dot{f} \tag{36}$$

where

$$K_e = h_0 \int (H^T D H)\, dA_0 - 2h_0 \int (H_1{}^T \tau_1 H_1)\, dA_0$$
$$+ h_0 \int (B_1{}^T \tau_1 B_1 - H_2{}^T \tau_2 H_2)\, dA_0,$$

$$B_1 = \frac{1}{\Delta\xi\sqrt{G}} \begin{pmatrix} -1 & 0 & 1 & 0 \\ 0 & -1 & 0 & 1 \end{pmatrix}$$

$$\dot{f}^{(vp)} = h_0 \int \dot{\epsilon}C H^T \tau\, dA_0,$$

and

$$\mathbf{f}^T = (\dot{R}^{(1)}, \dot{Z}^{(1)}, \dot{R}^{(2)}, \dot{Z}^{(2)}),$$

with $\dot{R}^{(i)}$, $\dot{Z}^{(i)}(i = 1, 2)$ denoting the nodal force rates in the horizontal and vertical directions, respectively. In (36), $\mathbf{f}^{(vp)}$ may be termed a viscoplastic force-rate vector [15] which is calculable from τ_e, $\bar{\epsilon}$ and (19). Note that the contributions due to the surface loads f_1, f_2 are absorbed in nodal force rates.

Assembling Equation (36) for all elements gives the global stiffness equations

$$K\dot{U} = \dot{\mathbf{F}}^{(vp)} + \dot{\mathbf{F}}, \tag{37}$$

where K is the global stiffness matrix; $\dot{U}$ is the global nodal velocity vector; and $\dot{\mathbf{F}}^{(vp)}$, $\dot{\mathbf{F}}$ are the global viscoplastic force rate and nodal force-rate vectors, respectively.

The interface conditions (10)–(12) must be introduced into (37). For each node in contact with the punch, the constraining equation (10) requires

$$\dot{W} - \frac{dS_0}{d\rho} \dot{U} = V, \tag{38}$$

where the surface derivative is evaluated at the current horizontal distance of the node and V is the instantaneous punch velocity. If the node is in the slipping category, the first of equations (11) requires that that horizontal and vertical components of the nodal force satisfy the following relations:

$$\begin{pmatrix} R \\ Z \end{pmatrix} = \begin{pmatrix} -\dfrac{dS_0}{d\rho} - (\mathrm{sgn}\ \dot{u})\mu \\ 1 - (\mathrm{sgn}\ \dot{u})\mu \dfrac{dS_0}{d\rho} \end{pmatrix} \dfrac{P}{N}, \tag{39}$$

where

$$N = \sqrt{1 + (dS_0/d\rho)^2},$$

and P is the normal component of the contact force. Differentiating (39) with respect to t gives the corresponding equation in rates:

$$\begin{aligned}
\begin{pmatrix} \dot{R} \\ \dot{Z} \end{pmatrix} &= \begin{pmatrix} -dS_0/d\rho - (\mathrm{sgn}\ \dot{u})\mu \\ 1 - \mathrm{sgn}(\dot{u})\mu\ dS_0/d\rho \end{pmatrix} \frac{\dot{P}}{N} \\
&+ \frac{P}{N^3} \begin{pmatrix} -\dfrac{d^2S_0}{d\rho^2} + (\mathrm{sgn}\ \dot{u})\mu \dfrac{dS_0}{d\rho} \dfrac{d^2S_0}{d\rho^2} \\ -\dfrac{dS_0}{d\rho} \dfrac{d^2S_0}{d\rho^2} - (\mathrm{sgn}\ \dot{u})\mu \dfrac{d^2S_0}{d\rho^2} \end{pmatrix} \dot{U}.
\end{aligned} \tag{40}$$

Equations (40) can now be used to eliminate the term $\dot{\mathbf{F}}$ in (37). The introduction of the additional unknown $\dot{P}$ is compensated for by the availability of the additional equation (38). If the node is in the non-slip category ($\dot{U} = 0$), the constraining equation (38) becomes simply $\dot{W} = V$.

Similar treatment applies to die-contact nodes. For nodes not in contact with either the punch or the die, $\dot{R} = \dot{Z} = 0$.

Initial and Boundary Conditions—To initialize the calculation, we use the same method as in [8], i.e., we set $(\tau_1, \tau_2) > 0$ at $t = 0$. For the two numerical examples discussed in this paper, we let

$$\begin{cases} \tau_1 = \theta K \\ \tau_2 = [R/(1 + R)]\tau_1 \end{cases} \quad \text{Case I}$$

$$\tau_1 = \tau_2 = \theta K \quad \text{Case II}$$

where θ is a small positive number. Numerical experiments show that the initial stress for $0.01 \leq \theta \leq 0.1$ has no effect on the calculated strain distributions. Moreover, the differences in initial stress totally disappear when strain reaches 1%.

The boundary conditions for the numerical examples are trivial. For the boundary node at $\xi = 0$, we require $\dot{U} = 0$ and $\dot{W} = V$ since the node follows the punch all the time. For the boundary node clamped in the die-contact region, $\dot{U} = \dot{W} = 0$.

Integration Scheme—Solution of (37) at time t gives instantaneous nodal velocities $\dot{U}$. Strain rates and stress rates are next calculated from (34) and (35). To calculate the displacements, strains and stresses at $t + \Delta t$, we have used a half-step forward integration scheme which involves the following steps:

1. Extrapolate to $t + \Delta t/2$, e.g.,

$$\mathbf{U}\left(t + \frac{\Delta t}{2}\right) = U(t) + \dot{U}(t)\frac{\Delta t}{2},$$

2. Reassemble $K\left(t + \dfrac{\Delta t}{2}\right)$, $\dot{\mathbf{F}}^{(vp)}\left(t + \dfrac{\Delta t}{2}\right)$ and $\dot{\mathbf{F}}(t + \Delta t)$

3. Compute

$$\dot{\mathbf{U}}\left(t + \frac{\Delta t}{2}\right) = \left[K\left(t + \frac{\Delta t}{2}\right)\right]^{-1}\left[\dot{\mathbf{F}}^{(vp)}\left(t + \frac{\Delta t}{2}\right) + \dot{\mathbf{F}}\left(t + \frac{\Delta t}{2}\right)\right],$$

4. Extrapolate to $t + \Delta t$, e.g.,

$$\mathbf{U}(t + \Delta t) = \mathbf{U}(t) + \dot{\mathbf{U}}\left(t + \frac{\Delta t}{2}\right)\Delta t.$$

To assure solution accuracy, the step size Δt must be controlled. In obtaining the numerical results presented here, we have used in addition to the several criteria discussed in [8] for rate-insensitive calculations, a new criterion which places a limit on the size of incremental effective stress by $\text{Max}(\dot{\tau}_e)\Delta t \leq \delta_\sigma K$, with $\delta_\sigma = 0.003$.

APPENDIX B—NONLINEAR ITERATIVE SOLUTION-PLANE STRAIN

The numerical method described in this Appendix requires repeated integration of the constitutive equations for a prescribed strain path. To avoid later interruption in our description of the main features of the method, we begin by discussing this integration. Solving the constitutive equations for the stress rates, we find

$$\dot{\tau}_1 = D\left[\dot{\epsilon}_1 + \nu\dot{\epsilon}_2 - \frac{k_1\dot{\bar{\epsilon}}}{\tau_e}(\tau_1 - k_2\tau_2)\right],$$

$$\dot{\tau}_2 = D\left[\dot{\epsilon}_2 + \nu\dot{\epsilon}_1 - \frac{k_1\dot{\bar{\epsilon}}}{\tau_e}(\tau_2 - k_2\tau_1)\right],$$

(41)

where

$$k_1 = 1 - \frac{\nu R}{1 + R}, \qquad k_2 = \frac{1}{k_1}\left(\frac{R}{1 + R} - \nu\right),$$

$$D = \frac{E}{1 - \nu^2}.$$

(42)

We also solve Equation (19) for $\dot{\bar{\epsilon}}$, and write the result as*

$$\dot{\bar{\epsilon}} = g(\tau_e, \bar{\epsilon}),$$

(43)

where g is a known function. Now if $\dot{\epsilon}_1$, $\dot{\epsilon}_2$ are given as functions of t over the interval $t_1 \leq t \leq t_2$, and if the values of τ_1, τ_2 and $\bar{\epsilon}$ are available at t_1, then (41) and (43) can be integrated numerically to give $\tau_1(t)$, $\tau_2(t)$ and $\bar{\epsilon}(t)$. During elastic response, k_1 is put equal to zero in (41) and $\dot{\bar{\epsilon}}$ remains constant.

It was found that Runge-Kutta or predictor-corrector methods were not suitable for this integration because exceedingly small step sizes were required to maintain numerical stability**. Hence, Simpson's rule, a highly stable method, was used. That is, for an equation of the form

$$\dot{Y} = F(Y, t)$$

the solution at t_{i+1} is found approximately by solving

$$Y_{i+1} = Y_i + \frac{t_{i+1} - t_i}{2}[F(Y_i, t_i) + F(Y_{i+1}, t_{i+1})],$$

(44)

for Y_{i+1}, where we use the notation

$$Y_i = Y(t = t_i).$$

Equation (44) is solved iteratively via Newton's method, the initial estimate being found by a quadratic extrapolation using the known values Y_{i-1}, Y_i and F_i. An

* It is not necessary that we be able to solve (19) analytically for $\dot{\bar{\epsilon}}$, only that a unique $\dot{\bar{\epsilon}}$ exists for given τ_e and $\bar{\epsilon}$, and that we can find it (numerically, if necessary).
** In many respects (41)–(43) acts like a "stiff" system of differential equations.

automatic step size generator is incorporated by using error estimates for (44) and for the initial estimate (see, e.g., Sec 5.5-3 of [16]). This feature is really a necessity in the present problem because step sizes need to be quite small at small values of $\bar{\epsilon}$, but can be greatly increased as deformation proceeds. One further point should be made concerning the solution of (41) and (43). Suppose that the given functions $\epsilon_1(t)$, $\epsilon_2(t)$ depend upon a parameter, α. Then by differentiating the appropriate versions of (44) with respect to α at each time step, we may compute the values

$$\frac{d\tau_1(t_2)}{d\alpha}, \frac{d\tau_2(t_2)}{d\alpha}, \frac{d\bar{\epsilon}(t_2)}{d\alpha}. \tag{45}$$

These quantities will be needed in the following.

We now consider the basic equations for plane strain punch stretching. Let us denote by s the arc length along the sheet at the current time. Then it is seen that ϵ_1 in Equation (1) can be written as

$$\epsilon_1 = \ln\left(\frac{ds}{d\xi}\right). \tag{46}$$

From incompressibility of the viscoplastic strains,

$$\frac{h}{h_0} = \exp\left[\frac{1-2\nu}{E}(\tau_1 + \tau_2) - \epsilon_1\right], \tag{47}$$

where h is the current sheet thickness and h_0 is the initial thickness. Finally, for $|\dot{u}| > 0$, the equilibrium equations can be combined to yield

$$\frac{d(h_0 e^{-\epsilon_1}\tau_1)}{ds} = \mu\kappa_1(h_0 e^{-\epsilon_1}\tau_1).$$

But the product $\mu\kappa_1$ is piecewise constant in s, and this equation can be integrated exactly:

$$(h_0 e^{-\epsilon_1}\tau_1) = (h_0 e^{-\epsilon_1}\tau_1)_{s_0} \exp[\mu\kappa_1(s - s_0)], \tag{48}$$

where $s_0 < s$ is some fixed value of s, and $\mu\kappa_1$ is constant from s_0 to s^*.

A mesh is imposed on the initial configuration with N nodal points, ξ_i, where

$$0 = \xi_1 < \xi_2 \cdots \xi_{N-1} < \xi_N = r_0.$$

A subscript subsequently denotes the value of a quantity at the material point ξ_i. The solution is generated at finite time intervals. Suppose that all variables are known at time t_1, and for $t_2 = t_1 + \Delta t$, we have computed the solution from ξ_1 to $\xi_i(i < N)$. We describe now an iterative method for computing the values at ξ_{i+1}. The iterated variable is taken to be the strain at ξ_{i+1}, which we temporarily denote by x. The strain path at ξ_{i+1} from t_1 to t_2 is obtained by a quadratic interpolation which matches the strain and strain rate at t_1, as well as the assumed

---

* (48) is easily generalized to the case where $\mu\kappa_1$ has a discontinuity between s_0 and s.

value at t_2. Hence, after differentiating,

$$\dot{\epsilon}_1(\xi_{i+1}, t) = g_1 + g_2 t, \tag{49}$$

where

$$g_2 = \frac{2}{(\Delta t)^2} [x - \epsilon_1(\xi_{i+1}, t_1) - \Delta t \dot{\epsilon}_1(\xi_{i+1}, t_1)]$$

$$g_1 = \dot{\epsilon}_1(\xi_{i+1}, t_1) - g_2 t_1. \tag{50}$$

We can now integrate the constitutive equations from t_1 to t_2 at ξ_{i+1} using (49) and $\dot{\epsilon} = 0$. This yields τ_1 and τ_2 and (47) can be used to compute the thickness, $h(\xi_{i+1}, t_2)$. The current arc length from ξ_i to ξ_{i+1} is obtained by solving (46) for $ds/d\xi$ and applying Simpsons rule:

$$s_{i+1} - s_i = \frac{(\xi_{i+1} - \xi_i)}{2} [\exp(x) + \exp \epsilon_1(\xi_i, \tau_1)]. \tag{51}$$

The function $F(x)$ is now defined to be

$$F(x) = (h_0 e^{-\epsilon} \tau_1)_{i+1} - (h_0 e^{-\epsilon} \tau_1)_i [\exp \mu \kappa_1 (s_{i+1} - s_i)]. \tag{52}$$

If $F(x)$ equals zero, the equilibrium equation is satisfied and we proceed to the next point, ξ_{i+2}. If $F(x)$ is not zero, the estimate for x (or $\epsilon_1(\xi_{i+1}, t_2)$) was incorrect, and a new estimate is obtained by applying Newton's method to (52). It is for this purpose that the values in (45) were computed. This procedure is followed for each material point.

In order to begin the above procedure, it is necessary to estimate the value of $\epsilon_1(\xi_1 = 0, t_2)$. Then equations such as (49), (50) are written for this point and the constitutive equations are integrated. The scheme in the above paragraph is used to compute results at points ξ_2 through ξ_N, and the boundary condition at ξ_N, namely that $\dot{u} = 0$ is checked. If this condition is not satisfied, the estimate for $\epsilon_1(\xi_1, t_2)$ is modified, and all the results at t_2 are recomputed. This iterative loop continues until the boundary condition is satisfied. Using results (45), Newton's method may also be used in this "global" iteration to aid in the determination of the new estimate for $\epsilon_1(\xi_1, t_2)$.

The initial conditions imposed are that at $t = 0$, the stresses, strains and strain rates are zero, and $h = h_0$ everywhere. Any initial inhomogeneity is easily handled by stipulating the appropriate values of $h_0(\xi_i)$.

It is found that as the frictional forces become very large, parts of the sheet may stick to the punch or die, or both. If material point ξ_J, for example, sticks to the punch, we assume that all ξ_i, $i < J$ also stick. Similarly, if point ξ_K contacts the die without moving relative to it, then all ξ_i for $i > K$ are also fixed. The numerical scheme outlined above is carried through precisely as indicated, except that the global iteration is carried out on the strain at ξ_J, and the boundary condition $\dot{u} = 0$ is imposed at ξ_K. This approach to the non-slip condition is somewhat crude, since it requires the boundary between the slip and non-slip regions to lie on a mesh point. This fact, together with the finite-difference approximation (51) and the strain representation (49) can lead to anomalous situations in which a material point has positive $\dot{u}$ under the punch, but a negative strain rate.

Numerical experiments indicate that taking about 20 mesh points under the punch and die radii is easily sufficient for maintaining spatial accuracy. The time steps are chosen so that the peak strain advances by about one percent each step. Under these conditions, accurate solutions are obtained for a wide range of geometrical factors, material properties and friction. They seem to be conservative enough that the simplified non-slip conditions mentioned above do not affect the results for any reasonable punch depth.

DISCUSSION

H. A. Armen *(Grumman Aerospace Corporation)*

In the finite element formulation, you treat variable contact and possible separation of material and die, and you said you reformulated the stiffness matrix in every step. If you didn't have new points coming into contact or separation, would you still reformulate the entire matrix? If the changes are small, could you take that into account as a structural modification and solve accordingly?

Wang

Yes, we formulate the entire stiffness matrix in every step. Even though no new points come into punch or die contact, the incremental rotations and stresses are quite large, and we can't make much use of the previous calculations.

B. Budiansky *(Harvard University)*

In other words, the matrix does get reformulated every step whether contact points have changed or not.

R. A. Ayres *(General Motors Research Laboratories)*

Throughout the two-day Symposium, people have talked about strain rate sensitivity and always assumed that the material has positive strain-rate sensitivity. The fact is that most aluminum alloys which are of interest to automobile applications, have negative strain-rate sensitivity. Could you comment how you think negative strain-rate sensitivity will effect your calculations?

Wang

This procedure, so far, works only for positive strain-rate sensitivity.

S. Kobayashi *(University of California, Berkeley)*

I just want to ask one or two simple questions. Why did you compare partic-ularly the radial strain distribution rather than the thickness strain distribution? Is there any particular reason?

Wang

The radial strain distributions are more critical. In the written report we do also compare the circumferential strain distributions.

Kobayashi

Could you give me some idea about what is the step size compared to the dimension you dealt with in the viscoplastic calculation?

Wang

The elastic-viscoplastic calculation right now is quite expensive relatively speaking. Let me just give you a relative order of magnitude as far as computa-tions are concerned. Suppose an elastic-plastic calculation takes 1 minute of machine time; the elastic-viscoplastic calculation would take 10 to 15 times of that. The average step size in the viscoplastic calculation for the hemispherical punch case is about 10^{-4} of the punch radius.

E. H. Lee *(Stanford University)*

You mentioned the advantage of elastic-plastic theory being able to do the unloading problems. In this connection, I was wondering if the membrane theory is adequate for calculating the springback problem; i.e. when you unload from the pressing. It seems likely that the bending stresses (which you ignored) become significant in the springback calculation.

Wang

The unloading which I discussed occurs in the forming stage where part of the material is trapped somewhere and will not deform (plastically) any further. Really I am speaking in that context. At the completion of the forming process as the punch retreats from the press, the material will respond elastically, and I agree with you that the bending stress will be more important at that stage.

B. S. Levy *(Inland Steel Company)*

Did I read your matrix properly when you were talking about incorporating the strain-rate sensitivity and you were doing separate strain-rate sensitivity for each of the two principal directions?

Wang

The strain-rate used is referred to the effective strain which takes account of the strains in both principal directions.

R. P. Nimmer *(General Electric Company)*

We heard quite a bit about the relative merits of applying deformation theory and flow theory to different set of problems. I was wondering if you had any experiences with deformation theory at all, and could you comment on the relative merits of using deformation theory in finite element applications?

Wang

We started applying deformation theory for sheet metal forming analysis many years ago. We found, however, some difficulties in using that theory. For example, if material unloading occurs during the forming process, the description of the deformation theory becomes physically inadequate, and that perhaps is one of the major difficulties of deformation theory.

M. L. Devenpeck *(U. S. Steel Corporation)*

Have you considered running problems which do not involve friction, thus permitting you to evaluate the constitutive equations including strain-rate effect?

Wang

We investigated the hydrostatic bulge test of steel some time ago, using the rate-insensitive incremental rigid-plastic theory. One thing we found out is that the calculated strain distributions did not match the experimental data very well. However, the comparison was unfair in the sense that the calculation was based on the rate-insensitive assumption, while the material is rate sensitive. Your comment is a good one. We are thinking of recalculating the hydrostatic bulge test including rate sensitivity and comparing that with experiment.

K. W. Neale *(Université de Sherbrooke, Canada)*

I was wondering if you made any calculations by varying the punch velocity and, if you did, has this any effect at all on your results?

Wang

Yes. We have made some calculations based on the logarithmic overstress model, and the results show some velocity dependence in the strain distribution. But, the effect is rather small. For example, for three decades of punch velocity, we found only a few percent of differences in maximum strain.

Unidentified Questioner

What is the range of strain-rate encountered in typical problems here?

Wang

For the particular examples discussed here, the maximum strain-rate is about 0.05/s. In actual stamping press, the strain-rate could be as high as 10/s. In other words, the data for the four decades of strain-rates we used to construct the model would not be enough for actual stamping analyses.

Lee

In connection with your comments on deformation theory, you mentioned that one would run into trouble on unloading. It seems to me that the unloading could be elastic and the real difficulty arises if you load up again and go through yield, if you are using deformation theory the strain jumps suddenly. It seems to me that unloading *only* would simply be elastic deformation and I don't see why that would cause trouble.

Wang

Oh, the unloading is in the process. Take, for example, Professor Kobayashi's deep drawing of a circular blank by a hemispherical punch. If the initial blank is much larger than the so-called limiting drawing ratio, then as the punch comes down, the flange area will become unloaded while the stretching process still goes on over the punch head. I don't know how to handle this situation by using deformation theory.

Budiansky

Although you agreed that the rigid-plastic deformation theory has problem in handling unloading, but the elastic-plastic deformation theory could handle the situation just as well.

Lee

The elastic-plastic theory deformation is all right as long as you load along one direction. If you unload, and reload and reach the yield surface at a different stress state, then by deformation theory you must supplement a jump in plastic strain appropriate to the yield stress. So in general you have discontinuity whenever you reload again. The fundamental problem which is brought up is that deformation theory is physically unreal.

Budiansky

We can get around that by tricks. Just pretending you are starting from scratch with new increments instead of insisting that the total strain suddenly jumps, in the sense of incrementally on the basis of deformation theory—incremental deformation theory.

SYMPOSIUM SUMMARY*

B. BUDIANSKY

Harvard University

Now comes the moment you've all been waiting for, when, as the presenter of the Summary, I will wrap everything up into a neat little package—or several neat little packages for the various kinds of members of the audience. You will all go home understanding everything, with everything reconciled. In fact, if I heard our electrifying Keynote speaker correctly, I think he threw down a challenge to me to present a consensus viewpoint together with instructions on precisely what should be done from now on. Well! I have neither the temerity nor the effrontery to offer such instructions, especially since it would be coming from somebody who is, perhaps, just as parochial as many others here. I am on one of those peaks; I guess I'm hanging-on by my fingernails to the slopes of the Mathahorn these days.

But, what I think I can do is this: I might tell you about the point-of-view that brought me to this Symposium. It's a point-of-view or, if you like, a dream that I had about fifteen years ago when I first made contact with General Motors. I imagined a black box—a black computation box that was incredibly powerful and into which we could feed a mathematical description of what the stylists envisioned for a certain sheet metal shape; feed in the thickness of a sheet from which the part will be made; put in the catalog number of the material—and then push a button. The computer then spits out the die shape, the blank configuration needed, the drawbeads and their orientation and configuration. Oh! Let's see, we put the lubricant catalog number in too! If it's possible to make the part—the computer program tells us this. If it's not possible, it tells us that too! Even if it is possible, perhaps that's not enough information, and so the computer program will give us the probability of success, if we have fed in variabilities in thicknesses, moduli, shapes of stress-strain relations, and so on.

Although Stu Keeler said in his address that we have to all work together, I think he cut the stylists out of the loop. We have to do what they tell us! But, in *my* dream I had the black box occasionally throwing out an answer that said, "Why don't you try this shape instead? It might be easier." Or "Why not use last year's fender?"

Well, that's my fantasy—"a modest proposal," to quote an earlier, non-metallic

---

* As transcribed from the oral presentation.

Swift, and I think it doesn't hurt to dream, especially, if you accept that it doesn't have to be an all-or-nothing proposal. In addition, we can have faith that if we bend our efforts toward finding out what we need to know to achieve our goal, good things will result. It was from this point-of-view that I listened to the talks.

We heard a lot of talks, and as the meeting went on, I found that the duty of presenting a Summary would be a pleasant one, because I could say, without lying, that I really felt it was a great symposium. Stimulating! and I learned a lot!

We can go back to the first talk following the electric Keynote. We heard Dr. Miyauchi's presentation of the paper by Yoshida and Miyauchi on how they were systematically correlating material properties with experimental facts concerning forming behavior. And it struck me there is a new kind of individual of which he is the prototype: namely the scientific artisan as opposed to the man of the press shop. He is becoming the kind of artisan who can—not via our black box, *not* via the computer program at this stage, but via the intuition, or, if you like, the accumulated experience that he has gained—actually be able to influence, in a rational way, the development of the manufacturing technology that's required.

We heard Professor Mellor give us an example of how one has to confront the results of continuum analysis with the facts of experiment and establish the limits of our skill to do modeling—analytical modeling—and to focus on the areas of ignorance that lie before us. And at the same time—it emerged in his talk as in Dr. Miyauchi's—we must focus on new material parameters that suddenly pop out that we hadn't considered previously. What I have in mind, of course, is the X-parameter, the N-parameter of Hutchinson. Indeed, it was rather remarkable, I felt, that the first few talks all converged from completely different directions to have this commonality of focus with respect to new ways of describing anisotropic materials—new ways that are essential.

Professor Toxawa's work—actually *pioneering* work—on determining constitutive relations in the very large strain domain represents the kind of work for which there is a crying need, and which, unfortunately, is being done less and less. When excitement about the theory of plasticity was really in its heyday—shortly after World War II—there were many people doing experiments in order to determine basic constitutive relations. Interest in doing so seems to have reached the vanishing point among academics. And in any event, most of the earlier work was in strain regions well below those we are interested in. I can only say that the work of Prof. Tozawa really carries with it a challenge to other experimenters to do similar things. It contains within it a challenge to theoreticians to discover and formulate mathematical laws—easily used mathematical laws—that will incorporate the phenomena which he has discovered. It embodies a challenge to engineers either to discover or invent rapid ways to do tests to evaluate those special parameters needed to characterize the material so that they can be used in analysis that requires the right stress-strain laws.

Professor Hutchinson gave us a nice guided tour to new analytical pathways. There was an indication of how, perhaps, a theoretician could discover the main features of laws of deformation of anisotropic materials by once again tackling the age-old problem of deducing polycrystalline behavior on the basis of single-crystal properties. It shows the kind of thing one has to do to be free from the constraint of the too-simple theories that are generally used: the simple J_2 flow theory without time effects and—what shall I say?—the biased insistence that we have to work with materials that have smooth yield surfaces.

Well, when we moved on to the afternoon and I heard the masterful and scholarly presentation by Prof. Wilson on lubrication, my dream almost turned into a nightmare. How are we going to handle friction when it seems to be such a complicated phenomenon? It did occur to me, especially in conversation later, that maybe there is a hope that the apparent complexities are involved only in the very high pressures that occur in three-dimensional fabrication, and that perhaps the pressures are sufficiently lower in two-dimensional sheet forming to make the problem much easier. I see no assent [looking at Prof. Wilson] so perhaps that isn't true. Well, we'll do what is necessary. If we can't get away with an average coefficient of friction, we'll do whatever is necessary to improve on this. And, in fact, as long as we are building fantasies, as long as we think that we can have a great, general computer program that can predict what the sheet material does when the die hits it and follow it through the complexities of the kind that Norman Wang discussed, why shouldn't we solve the equations of lubrication at the same time, and let the flow of the lubricating material interact with the deformations of the sheet.

Harmon Nine showed us an example of how very clever and careful testing can be focused on a special problem with results that will be useful and will have good fallout. They will lead to a better understanding of how friction works, they will promote analyses—theoretical analyses—that might not otherwise have been undertaken on special problems like: "What are the restraint effects of beads on the drawing of sheet material?" And they may well lead to standardized testing techniques to discover gross frictional properties of die and sheet contact.

Moving on to this morning, we had the Gang of Four discussing material instability. It is no accident that these four talks involving the leaders of the world in the problem of material instability were focused on this central question. These presentations were wide ranging, provocative, controversial, incisive—and here I have to say that maybe I'll temper my optimism a little bit. Perhaps, in the short run, when we contemplate computing what happens in sheet-metal processing, we'll just have to analyze the gross deformation and the development of instabilities in series, rather than in parallel. Why do I say this? Because it looks like it is going to be easier to get constitutive relations that are reasonably reliable in the preinstability range than within the instability range. We mustn't, however, give up the quest for constitutive relations that will work all the way through. It is really extraordinary (as Prof. Hutchinson pointed out) how this question of constitutive relations—flow versus deformation theory—keeps rearing its ugly head. Maybe finally we are going to meet this question head-on and solve it. You know the same thing, as Hutchinson said, arose in buckling thirty years ago. Great controversy! and the question didn't get resolved. The people who do buckling have gotten around it, but they really have not met it head-on. Here maybe we are in a situation where we should not ask what plasticity can do for metal forming but rather what metal forming can do for plasticity. It may be that finally there will be enough interest provoked to solve this problem.

Well then we go to this afternoon and the extraordinary performance by Dr. Ishigaki who I think showed that the kind of scientific artisanship, to which I alluded earlier, can work—I'll never be able to look at my car again without seeing the strain pattern on it.

Finally, we concluded with the two papers that were closest to the black box idea. The approaches that we heard by Kobayashi and by Wang were, of course, quite different but both will help us achieve our ultimate goal. They are competitive but that's all right because, as you may know, it is written that, "The jealousy of scholars brings out wisdom."

So, to conclude, I think we had a very highly stimulating meeting. I hope you keep up the momentum of communication. I think that some of us have learned to talk to our opposite members on these other peaks that we heard so much about and I hope we keep it up, not only in formal conferences but in informal discussions. I know there are continuing informal meetings among various groups concerned with metal forming, but perhaps they should broaden their attendance lists so that wider communication can continue among metallurgists, mathematicians, numerical analysts, engineers and, God help us, administrators too. Because there is a very big potential payoff and it's a terrifically good bet to spend a little money on brains. Okay, that's my Summary—go forth and do good!

PARTICIPANTS

Agnew, W. G.
General Motors Research Laboratories
Warren, Michigan

Altan, T.
Battelle Laboratories
Columbus, Ohio

Amann, C. A.
General Motors Research Laboratories
Warren, Michigan

Armen, H. A.
Grumman Aerospace Corporation
Bethpage, New York

Astill, C. J.
National Science Foundation
Washington, DC

Ayres, R. A.
General Motors Research Laboratories
Warren, Michigan

Babcock, S. G.
General Motors Manufacturing Staff
Warren, Michigan

Bartell, B. E.
Chevrolet Motor Division, GMC
Parma, Ohio

Bathe, K.-J.
Massachusetts Institute of Technology
Cambridge, Massachusetts

Baxter, W. J.
General Motors Research Laboratories
Warren, Michigan

Beaman, R. T.
General Motors Research Laboratories
Warren, Michigan

Bowden, R. L.
Fisher Body Division, GMC
Warren, Michigan

Bidwell, J. B.
General Motors Research Laboratories
Warren, Michigan

Bird, J. E.
Aluminum Company of America
New Kensington, Pennsylvania

Bragard, M. A.
Centre de Recherches Metallurgiques
Liège, Belgium

Brazier, W. G.
Fisher Body Division, GMC
Warren, Michigan

Bucher, J. H.
Jones & Laughlin Steel Corporation
Pittsburgh, Pennsylvania

Budiansky, B.
Harvard University
Cambridge, Massachusetts

Butterworth, A. V.
General Motors Research Laboratories
Warren, Michigan

Buzan, L. R.
General Motors Research Laboratories
Warren, Michigan

Caddell, R. M.
  University of Michigan
  Ann Arbor, Michigan

Chang, D. C.
  General Motors Research Laboratories
  Warren, Michigan

Chatfield, D. A.
  National Steel Corporation
  Weirton, West Virginia

Chen, K.-K.
  General Motors Research Laboratories
  Warren, Michigan

Chenea, P. F.
  General Motors Research Laboratories
  Warren, Michigan

Conrad, H.
  University of Kentucky
  Lexington, Kentucky

Devenpeck, M. L.
  U.S. Steel Corporation
  Monroeville, Pennsylvania

Dodd, G. G.
  General Motors Research Laboratories
  Warren, Michigan

Dubey, R. N.
  University of Waterloo
  Waterloo, Canada

Duncan, J. L.
  McMaster University
  Hamilton, Canada

Eary, D.
  General Motors Institute
  Flint, Michigan

Ebert, L. J.
  Case Western Reserve University
  Cleveland, Ohio

Elliott, W. A.
  General Motors Engineering Staff
  Warren, Michigan

Fredericks, D. R.
  AC Spark Plug Division, GMC
  Flint, Michigan

Frey, W. H.
  General Motors Research Laboratories
  Warren, Michigan

Furubayashi, T.
  Nissan Motor Company Ltd.
  Yokohama, Japan

Gardels, K. D.
  General Motors Research Laboratories
  Warren, Michigan

Gegel, H. L.
  Air Force Materials Laboratory
  Dayton, Ohio

Ghosh, A. K.
  Rockwell International
  Thousand Oaks, California

Gibala, R.
  Case Western Reserve University
  Cleveland, Ohio

Goodwin, G. M.
  Chrysler Corporation
  Detroit, Michigan

Grube, W. L.
  General Motors Research Laboratories
  Warren, Michigan

Hall, C.
  University of Pittsburgh
  Pittsburgh, Pennsylvania

Hall, D. A.
  Chevrolet Motor Division, GMC
  Warren, Michigan

Hart, D. E.
General Motors Research Laboratories
Warren, Michigan

Havner, K.
North Carolina State University
Raleigh, North Carolina

Hays, D. F.
General Motors Research Laboratories
Warren, Michigan

Hecker, S. S.
Los Alamos Scientific Laboratory
Los Alamos, New Mexico

Heimbuch, R. A.
General Motors Manufacturing Staff
Warren, Michigan

Henry, A. K.
Fisher Body Division, GMC
Warren, Michigan

Hiam, J. R.
Dominion Foundries and Steel Ltd.
Hamilton, Canada

Hilsen, R. R.
Inland Steel Company
East Chicago, Indiana

Hockett, J. E.
Los Alamos Scientific Laboratory
Los Alamos, New Mexico

Hodge, P. G., Jr.
University of Minnesota
Minneapolis, Minnesota

Hollyer, R. N.
General Motors Research Laboratories
Warren, Michigan

Holzwarth, J. C.
General Motors Research Laboratories
Warren, Michigan

Hook, R. E.
Armco Steel Corporation
Middletown, Ohio

Hosford, W. F.
University of Michigan
Ann Arbor, Michigan

Hunter, J. E.
General Motors Research Laboratories
Warren, Michigan

Hutchinson, J. W.
Harvard University
Cambridge, Massachusetts

Ishigaki, H.
Toyota Motor Company
Toyota, Japan

Jalinier, J. M.
Université de Metz
Metz, France

Jamerson, F. E.
General Motors Research Laboratories
Warren, Michigan

James, K. F.
General Motors Manufacturing Staff
Warren, Michigan

Johnson, W.
University of Cambridge
Cambridge, England

Jonas, J. J.
McGill University
Montreal, Canada

Justusson, J. W.
General Motors Research Laboratories
Warren, Michigan

Kaftanoglu, B.
The Middle East Technical University
Ankara, Turkey

Kalpakjian, S.
Illinois Institute of Technology
Chicago, Illinois

Kamal, M. M.
General Motors Research Laboratories
Warren, Michigan

Kasper, A. S.
Chrysler Corporation
Detroit, Michigan

Keeler, S. P.
National Steel Company
Ecorse, Michigan

Key, S. W.
Sandia Laboratories
Albuquerque, New Mexico

Kobayashi, S.
University of California
Berkeley, California

Kocks, U. F.
Argonne National Laboratory
Argonne, Illinois

Koistinen, D. P.
General Motors Research Laboratories
Warren, Michigan

Laukonis, J. V.
General Motors Research Laboratories
Warren, Michigan

Lee, D.
General Electric Company
Schenectady, New York

Lee, E. H.
Stanford University
Stanford, California

LeRoy, G.
McMaster University
Hamilton, Canada

Levy, B. S.
Inland Steel Company
East Chicago, Indiana

Lindholm, U. S.
Southwest Research Institute
San Antonio, Texas

Litzke, H.
Fried. Krupp Huttenwerke AG
Bochum, West Germany

Magee, C. L.
Ford Motor Company
Dearborn, Michigan

Mancewicz, T. A.
General Motors Manufacturing Staff
Warren, Michigan

Marciniak, Z.
Technical University of Warsaw
Warsaw, Poland

Mattavi, J. N.
General Motors Research Laboratories
Warren, Michigan

McClintock, R.
General Motors Research Laboratories
Warren, Michigan

McCullough, D. G.
Pontiac Motor Division, GMC
Pontiac, Michigan

McDonald, G. C.
General Motors Research Laboratories
Warren, Michigan

McDonald, R. J.
General Motors Research Laboratories
Warren, Michigan

McLaughlin, B. D.
Alcan International Ltd.
Kingston, Canada

McMillan, M. L.
General Motors Research Laboratories
Warren, Michigan

Mellor, P. B.
University of Bradford
Bradford, England

Miller, E. J.
General Motors Research Laboratories
Warren, Michigan

Miyauchi, K.
The Institute of Physical and Chemical
Research, Tokyo, Japan

Morris, L.
Alcan International Ltd.
Kingston, Canada

Muench, N. L.
General Motors Research Laboratories
Warren, Michigan

Neale, K. W.
Université de Sherbrooke
Sherbrooke, Canada

Needleman, A.
Brown University
Providence, Rhode Island

Neimeier, B. A.
Reynolds Metals Company
Richmond, Virginia

Nemat-Nasser, S.
Northwestern University
Evanston, Illinois

Ni, C.-M.
General Motors Research Laboratories
Warren, Michigan

Nimmer, R. P.
General Electric Company
Schenectady, New York

Nine, H. D.
General Motors Research Laboratories
Warren, Michigan

Oh, H. L.
General Motors Research Laboratories
Warren, Michigan

Rashid, M. S.
General Motors Research Laboratories
Warren, Michigan

Rasmussen, G. K.
AC Spark Plug Division, GMC
Flint, Michigan

Rhodes, C. J.
GMC Truck & Coach Division
Pontiac, Michigan

Rice, J. R.
Brown University
Providence, Rhode Island

Robinson, G. H.
General Motors Research Laboratories
Warren, Michigan

Rogers, H. C.
Drexel University
Philadelphia, Pennsylvania

Sajewski, V. F.
Fisher Body Division, GMC
Warren, Michigan

Sang, H.
Alcan International Ltd.
Kingston, Canada

Shabaik, A.
University of California
Los Angeles, California

Smith, E. J.
National Steel Corporation
Weirton, West Virginia

Smith, G. W.
General Motors Research Laboratories
Warren, Michigan

Sorensen, E. P.
General Motors Research Laboratories
Warren, Michigan

Stevenson, R.
General Motors Research Laboratories
Warren, Michigan

Stine, P. A.
General Electric Company
Louisville, Kentucky

Tanaka, T.
Nippon Kokan K.K.
New York City, New York

Tang, S. C.
Ford Motor Company
Dearborn, Michigan

Taylor, B.
General Motors Manufacturing Staff
Warren, Michigan

Thomas, J. F., Jr.
Wright State University
Dayton, Ohio

Tozawa, Y.
Nagoya University
Nagoya, Japan

Tracy, J. C.
General Motors Research Laboratories
Warren, Michigan

Vail, C. F.
General Motors Engineering Staff
Warren, Michigan

VanderVeen, P.
Bethlehem Steel Corporation
Bethlehem, Pennsylvania

Vigor, C. W.
General Motors Research Laboratories
Warren, Michigan

Wang, N.-M.
General Motors Research Laboratories
Warren, Michigan

Webbere, F. J.
General Motors Research Laboratories
Warren, Michigan

Weber, B. C.
Fisher Body Division, GMC
Warren, Michigan

Wenner, M. L.
General Motors Research Laboratories
Warren, Michigan

Wilson, W. R. D.
University of Massachusetts
Amherst, Massachusetts

Woo, D. M.
University of Sheffield
Sheffield, England

Zimmerer, R.
Cadillac Motor Car Division, GMC
Detroit, Michigan

SUBJECT INDEX

Yield surface
- aluminum killed (or stabilized) steel, 99.
- annealed aluminum, 99.
- annealed steel, 94.
- brass, 94.
- excessive hardening phenomenon, 104.
- Hill's criterion for anisotropic sheets,
  56, 67, 86, 88, 302, 344, 371.
- prestrain direction, 100.
- prestrained sheets, 99, 101, 102, 107.
- Tresca criterion, 92.
- vertex model, 238.
- von Mises criterion, 92.
Yield vertex
- bifurcation, 291.
- effect on critical hardening rate, 251.

Yoshida's X-value, 29, 67.